Synthesis Lectures on Digital Circuits & Systems

Series Editor

Mitchell A. Thornton, Southern Methodist University, Dallas, USA

This series includes titles of interest to students, professionals, and researchers in the area of design and analysis of digital circuits and systems. Each Lecture is self-contained and focuses on the background information required to understand the subject matter and practical case studies that illustrate applications. The format of a Lecture is structured such that each will be devoted to a specific topic in digital circuits and systems rather than a larger overview of several topics such as that found in a comprehensive handbook. The Lectures cover both well-established areas as well as newly developed or emerging material in digital circuits and systems design and analysis.

Steven F. Barrett

Arduino VIII

Portenta Machine Control

Steven F. Barrett
University of Wyoming
Laramie, WY, USA

ISSN 1932-3166 ISSN 1932-3174 (electronic)
Synthesis Lectures on Digital Circuits & Systems
ISBN 978-3-031-85943-4 ISBN 978-3-031-85944-1 (eBook)
https://doi.org/10.1007/978-3-031-85944-1

© The Editor(s) (if applicable) and The Author(s), under exclusive license to Springer Nature Switzerland AG 2026

This work is subject to copyright. All rights are solely and exclusively licensed by the Publisher, whether the whole or part of the material is concerned, specifically the rights of translation, reprinting, reuse of illustrations, recitation, broadcasting, reproduction on microfilms or in any other physical way, and transmission or information storage and retrieval, electronic adaptation, computer software, or by similar or dissimilar methodology now known or hereafter developed.
The use of general descriptive names, registered names, trademarks, service marks, etc. in this publication does not imply, even in the absence of a specific statement, that such names are exempt from the relevant protective laws and regulations and therefore free for general use.
The publisher, the authors and the editors are safe to assume that the advice and information in this book are believed to be true and accurate at the date of publication. Neither the publisher nor the authors or the editors give a warranty, expressed or implied, with respect to the material contained herein or for any errors or omissions that may have been made. The publisher remains neutral with regard to jurisdictional claims in published maps and institutional affiliations.

This Springer imprint is published by the registered company Springer Nature Switzerland AG
The registered company address is: Gewerbestrasse 11, 6330 Cham, Switzerland

If disposing of this product, please recycle the paper.

Preface

This book is about the Arduino microcontroller and the Arduino concept. The visionary Arduino team of Massimo Banzi, David Cuartielles, Tom Igoe, Gianluca Martino, and David Mellis launched a new innovation in microcontroller hardware in 2005, the concept of open source hardware. Their approach was to openly share details of microcontroller-based hardware design platforms to stimulate the sharing of ideas and promote innovation. This concept has been popular in the software world for many years. Their efforts resulted in a global phenomenon of making computing accessible for all.

I was quite excited when I heard Arduino was extending the concept of accessible computing to the industrial and Internet of Things (IoT) sectors. Originally I planned a book including both of the Opta series of programmable logic controllers (PLCs) and the Portenta Machine Controller (PMC). As the book evolved, it became quite clear there was too much information for a single text. Instead, a complementary set of books was planned: *Arduino VII: Industrial Control* and *Arduino VIII: Portenta Machine Control*. Although the books are a complementary set, each is independent in the information contained.

This book, *Arduino VIII: Portenta Machine Control*, is an accessible primer on industrial control and programmable logic controller concepts for those without a deep instrumentation background. An understanding of basic circuit theory is an appropriate prerequisite for the book. The three main goals for the book are: explore accessible Arduino Portenta Machine Control industrial control products; learn the fundamentals of programming using ladder logic; and explore related sensors and interface concepts. We use multiple examples throughout the book and conclude with an instrumented greenhouse project. Throughout the book we concentrate on remote, direct current (DC) powered systems. We develop systems that operate on positive polarity (e.g. supplied by solar panels and batteries).

Approach of the Book

The book has been divided into a series of six chapters and several appendices to accomplish the book's goals. The book follows these chapters:

- Chapter 1. Operational Technology and the Arduino Portenta Machine Control
- Chapter 2. Arduino Portenta Machine Control features
- Chapter 3. Connectivity and Communications
- Chapter 4. Input Sensors, Output Actuators, and Interfacing
- Chapter 5. Programming with the Arduino PLC IDE
- Chapter 6. Application: Instrumented Greenhouse
- Appendix A: Safety
- Appendix B: Embedded Systems Design
- Appendix C: Getting Started–Arduino UNO R3

Throughout the book we provide numerous hardware and software examples. A tutorial on safety concepts is readily available in Appendix A and referenced throughout the book. We recommend reading this appendix first (now) and regularly as you progress through the book. Appendix B provides a tutorial on system design concepts and tools. In Appendix C we provide information on how to use the Arduino UNO R3 processor. It is used in several examples.

For completeness and independence, this volume contains tutorial information contained in some of the other volumes in the Arduino series and related works completed for Morgan and Claypool and Springer Nature. Chapter footnotes identify the source of this information contained elsewhere in the series. The book series thus far includes:

- *Arduino I: Getting Started*
- *Arduino II: Systems*
- *Arduino III: Internet of Things*
- *Arduino IV: DIY Robots–3D Printing, Instrumentation, Control*
- *Arduino V: AI and Machine Learning*
- *Arduino VI: Bioinstrumentation*
- *Arduino VII: Industrial Control*
- *Arduino VIII: Portenta Machine Control*

In the rapidly evolving Arduino world, I anticipate other books in the series.

Acknowledgments

A number of people have made this book series possible. I would like to thank Massimo Banzi of the Arduino design team for his support and encouragement in writing the first edition of this book: *Arduino Microcontroller: Processing for Everyone!*

I would also like to acknowledge Joel Claypool for his publishing expertise and support on a number of writing projects. His vision and expertise in the publishing world has made this book possible. Joel "retired" in September 2022 after 40 plus years of service to the U.S. Navy and the publishing world. On behalf of the multitude of writers you have provided a chance to become published authors, we thank you!

I would also like to thank Charles (Chuck) Glaser, Editorial Director at Springer Nature, for his encouragement and support on this project. If you have a good idea for a book, I highly recommend contacting Chuck. He will assist you in converting your idea into a finished, professional book product.

I would also like to thank Ramasubramaniyan Velu and Vishnu Muthuswamy of Straive for their expertise in converting the final draft into a finished product. You provide outstanding service.

Finally, and most importantly, I would like to thank my wife and best friend of many (almost 50) years, Cindy.

Laramie, WY, USA Steven F. Barrett
June 2025

Contents

1 **Operational Technology and the Arduino PMC** 1
   1.1 Overview 1
   1.2 Internet of Things–IoT 2
   1.3 Information Technology Versus Operational Technology 2
   1.4 Operational Technology 4
   1.5 IoT Architecture 5
   1.6 IoT Technology 7
      1.6.1 Smart Home Technology 7
      1.6.2 Industrial Internet of Things (IIoT) 7
   1.7 Cybersecurity 8
   1.8 IoT and IIoT Security 10
   1.9 Arduino Portenta Machine Control Features 11
   1.10 Getting Started with the Arduino IDE 11
   1.11 PMC Blink Program 12
   1.12 Application: Portable Lab Configuration 19
   1.13 Summary 19
   1.14 Chapter Problems 22

2 **Arduino PMC Features** 23
   2.1 Overview 23
   2.2 PMC and the Portenta H7 Processor 23
   2.3 Digital Output 24
   2.4 Digital Input 24
      2.4.1 Switch Debounce 27
   2.5 Programmable Digital I/O 31
   2.6 Analog Input 33
      2.6.1 Analog–To–Digital Conversion (ADC) 34
      2.6.2 Arduino PMC Examples 37
   2.7 Analog Output 43
   2.8 Real Time Clock 50
   2.9 Interrupts 51

		2.9.1	Foreground and Background Processing	54
	2.10	System and Data Integrity		58
		2.10.1	Cyclic Redundancy Check (CRC)	58
	2.11	Advanced Encryption Standard (AES)		61
	2.12	Application I: Programming Practice		65
	2.13	Application II: RGB Color		67
	2.14	Summary		68
	2.15	Chapter Problems		69
3	**Connectivity**			**71**
	3.1	Overview		71
	3.2	Serial Communications		73
	3.3	Serial Communication Terminology		73
	3.4	Inter–Integrated Circuit (I2C)		74
		3.4.1	Programming the I2C Subsystem with the Arduino IDE	76
		3.4.2	I2C LCD	77
	3.5	Universal Serial Bus (USB)		81
	3.6	Bluetooth Low Energy (BLE)		85
		3.6.1	Arduino BLE Library	88
	3.7	UART/RS–232 Communication Protocol		95
	3.8	Internet Concepts		100
		3.8.1	A Big Picture of the Internet	100
		3.8.2	Internet Cloud	104
		3.8.3	Internet Protocol Models	105
		3.8.4	Internet Addressing Techniques	105
	3.9	PMC Ethernet 10/100BASE T Port		112
	3.10	Wi–Fi 802.11 B/G/N		121
	3.11	Controller Area Network		126
	3.12	RS–485 Communication		131
	3.13	Application I: Encryption/Decryption		143
	3.14	Application II: USB Data Logger		143
	3.15	Summary		147
	3.16	Chapter Problems		147
4	**PMC Peripheral Interfacing**			**149**
	4.1	Overview		149
	4.2	Power Requirements		150
		4.2.1	AC Operation	151
		4.2.2	DC Operation	151
	4.3	Input Sensors		152
		4.3.1	Digital Input Sensors	153
		4.3.2	Switches	154

	4.3.3	Optical Encoder	155
4.4	Analog Input Sensors		159
	4.4.1	Flex Sensor	161
	4.4.2	Ultrasound Sensor	161
	4.4.3	Temperature Sensors	164
	4.4.4	Fluid Depth Sensors	175
	4.4.5	Light Sensor	176
	4.4.6	Tilt Sensor	183
	4.4.7	Environmental Sensors	185
	4.4.8	Joystick	190
	4.4.9	Greenhouse Sensors	193
4.5	Output Devices and Actuators		194
4.6	Light Emitting Diodes (LEDs)		194
4.7	Annunciators–Sonalerts, Beepers, Buzzers		194
4.8	Electromechanical Devices		195
4.9	DC Motors		195
4.10	DC Motor Speed and Direction Control		196
	4.10.1	Pulse Width Modulation	197
	4.10.2	H Bridge Direction Control	200
4.11	Linear Actuator		200
4.12	Servo Motor Control		203
4.13	Stepper Motor Control		207
4.14	DC Solenoid Control		213
4.15	Transducer Interface Design (TID)		217
4.16	Operational Amplifier Overview		219
	4.16.1	Operational Amplifier Origins	220
	4.16.2	Ideal Characteristics	221
	4.16.3	Nonideal Characteristics	222
	4.16.4	Configurations	224
4.17	Application I: DC Motor Speed Control		226
	4.17.1	Motor Control Software Configuration	227
4.18	Application II: PCA9685 16-Channel PWM and Servo Driver		229
	4.18.1	Application IIa: PCA9685 16-Channel Servo Driver	230
	4.18.2	Application IIb: PCA9685 16-Channel PWM	233
4.19	Application III: PMC Weather Station		235
	4.19.1	Structure Chart	235
	4.19.2	Circuit Diagram	235
	4.19.3	UML Activity Diagram	244

		4.19.4 Bottom-Up Implementation	244
4.20	Summary		246
4.21	Chapter Problems		246

5 Arduino PLC IDE and Ladder Logic ... 249
- 5.1 Overview ... 249
- 5.2 Arduino PMC Programming Tools ... 250
- 5.3 Getting Started–Arduino PLC IDE ... 250
- 5.4 Structure of Arduino PLC IDE Program ... 253
  - 5.4.1 Contacts, Coils, Branches, and Blocks ... 253
  - 5.4.2 LD Editor ... 257
- 5.5 LD Program Examples ... 258
- 5.6 Ladder Logic Examples ... 263
  - 5.6.1 LM34 Temperature Sensor ... 263
  - 5.6.2 Smoke Detector ... 265
- 5.7 Application I: Sequencer Control Logic ... 267
  - 5.7.1 Stepper Motor Control–Ladder Logic Sequencer ... 268
- 5.8 Application II: Test Fixture ... 271
- 5.9 Application III: Greenhouse Temperature Sensing System ... 272
- 5.10 Summary ... 275
- 5.11 Chapter Problems ... 275

6 Application: IoT Greenhouse ... 277
- 6.1 Objective ... 277
- 6.2 Greenhouse Theory ... 278
- 6.3 Water Harvesting ... 280
- 6.4 Greenhouse Control System Requirements ... 280
- 6.5 Solar Power Subsystem ... 283
- 6.6 Greenhouse Control Subsystem ... 283
  - 6.6.1 Milone E–Tape Fluid Sensor ... 286
  - 6.6.2 Humidity Sensor ... 286
  - 6.6.3 Soil Moisture Sensor ... 288
  - 6.6.4 Thermocouple/RTD Interior/exterior Greenhouse Temperature Sensors ... 288

	6.6.5	Misting System and LED	289
	6.6.6	Vent Fan and LED	289
	6.6.7	GCS System Code	290
	6.6.8	GCS Printed Circuit Board	298
	6.6.9	Enclosure	298
6.7	Testing		298
6.8	Application: Greenhouse Control System–Ladder Logic		302
6.9	Summary		304
6.10	Chapter Problems		305
References			309

Appendix A 311

Appendix B 317

Appendix C 325

Index 329

About the Author

Steven F. Barrett Ph.D., P.E. received the BS Electronic Engineering Technology from the University of Nebraska at Omaha in 1979, the M.E.E.E. from the University of Idaho at Moscow in 1986, and the Ph.D. from The University of Texas at Austin in 1993. He was formally an Active Duty Faculty Member at the United States Air Force Academy, Colorado. He now serves as Associate Dean for Undergraduate Programs in the College of Engineering and Physical Sciences at the University of Wyoming and Professor of Electrical and Computer Engineering. He is Member of IEEE (Life Senior) and Tau Beta Pi (Chief Faculty Advisor). His research interests include digital and analog image processing, computer–assisted laser surgery, and embedded controller systems. He is a registered Professional Engineer in Wyoming and Colorado. In 2004, Barrett was named "Wyoming Professor of the Year" by the Carnegie Foundation for the Advancement of Teaching and in 2008 was Recipient of the National Society of Professional Engineers (NSPE) Professional Engineers in Higher Education, Engineering Education Excellence Award. In 2023 Barrett received the National Council of Examiners for Engineering and Surveying (NCEES) Distinguished Examination Service Award.

Operational Technology and the Arduino PMC

Objectives: After reading this chapter, the reader should be able to do the following:

- Define Information Technology (IT) and Operational Technology (OT);
- Describe the features of a programmable logic controller (PLC) based OT system;
- Provide a working definition of the Industrial Internet of Things (IIoT);
- Execute a "blink" program on the Arduino Portenta Machine Control; and
- Construct a lab environment for the Arduino Portenta Machine Control.

1.1 Overview

This book, "Arduino VIII: Portenta Machine Control," is an accessible primer on industrial control and programmable logic controller concepts for those without a deep instrumentation background in the operational technology world. The Portenta Machine Control (PMC) is designed as an industrial control for equipment and machines. It couples a powerful dual processor microcontroller with industrial grade inputs and outputs. It also is equipped with a rich complement of onboard systems.

In this chapter we begin our exploration of the Operational Technology (OT) world. We start with a basic introduction to the Internet of Things (IoT). Within IoT there is a close relationship between Information Technology (IT) and Operational Technology (OT). We explore this relationship in some detail. The reader is assumed to have a solid grounding in basic IT concepts.[1] The pervasiveness of IoT is then examined in industry or the Industrial Internet of Things (IIoT). We then shift our focus to OT and basic PLC concepts. We conclude the chapter with an introduction to the Arduino PMC.

1.2 Internet of Things–IoT

The term Internet of Things was first used by Kevin Ashton in a 1999 Proctor and Gamble presentation. Mr. Ashton's presentation discussed concepts on using the existing internet infrastructure to support P&G's supply chain (Greer, Hanes). From this early start, applications within business and industry have become quite pervasive. A review of the literature provides a feature list describing the Internet of Things systems concept (Rajkumar, Hanes, Greer):

- An IoT system connects things to the internet;
- Each thing or device has its own unique identifier or address;
- Communication between things is provided via the internet;
- An IoT system provides for interrelated and integrated computing devices and physical processes;
- An IoT system provides the ability to measure, process, and transfer to and from remote locations; and
- IoT processes are monitored, coordinated, and controlled.

Interestingly the concept of Cyber–physical Systems (CPS) share many of the same features. The National Institute of Standards and Technology (NIST) performed a study to examine the relationship between IoT and CPS and noted although the concepts originated in different industries, they are substantially equivalent concepts. A unified perspective of the two concepts was provided: "Internet of Things and Cyber–Physical Systems comprise interacting logical, physical, transducer, and human components engineered for function through integrated logic and physics (Greer)."

1.3 Information Technology Versus Operational Technology

A key concept within IoT is the close relationship between IT and OT. The relationship between IT and OT is shown in Fig. 1.1.

IT communications usually consist of short, frequent communications that are broken into packets and communicated globally. IT provides a wide variety of message traffic including

---

[1] A basic introduction to IT concepts is provided in "Arduino III: Internet of Things," S.F. Barrett, Springer Nature, 2021. Portions of this chapter have been adapted with permission for use with the Arduino Portenta Machine Control.

1.3 Information Technology Versus Operational Technology

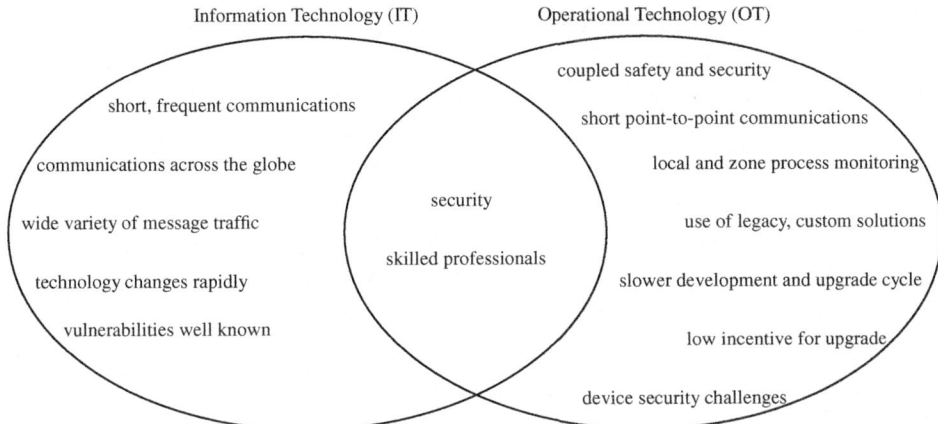

Fig. 1.1 Information technology (IT) versus operational technology (OT) (Hanes)

e-mails, requests and response for information from websites, and multiple other types. IT technology developments continue to rapidly evolve with vulnerabilities well known, documented, combatted, and corrected (Hanes).

Operational Technology (OT) provides for process control within many areas of industry. As shown in Fig. 1.1, industrial safety and security are intertwined. OT communications are typically short, point–to–point communications on a factory floor or within an industrial process. Monitoring via a Supervisory Control and a Data Acquisition (SCADA) system is typically performed within a local and confined zone. Although OT developments are actively taking place, adoption timelines are slower than with IT. Since OT governs a proprietary and custom solution for a given industrial process, typically there is a low incentive for technology upgrade. Although many IT and OT concepts are related but different, IT and OT both enjoy the dedication of skilled professional practitioners (Hanes).

IT and OT share the requirement for robust security protection and countermeasures. Many of the security concepts discussed for IT also apply for OT. In the industrial world IT and OT systems are often linked to share information among related processes. For example, a remote oil drilling platform may be controlled via OT processes. If an oil company has multiple remote platforms, they may be linked via IT processes to share production data. Some form of isolation, an "air gap," is typically provided between IT and OT related processes for security purposes. This helps prevent a nefarious actor from accessing a critical industrial process via the internet (Hanes).

1.4 Operational Technology

Operational Technology is used to control industrial processes. The fundamental OT building block is the programmable logic controller (PLC). A PLC diagram is provided in Fig. 1.2a.

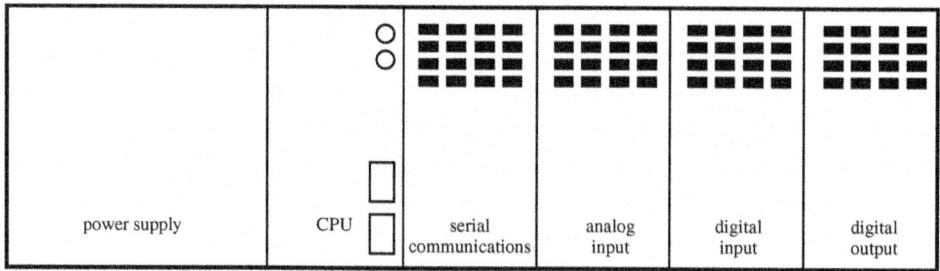

a) modular programmable logic controller system

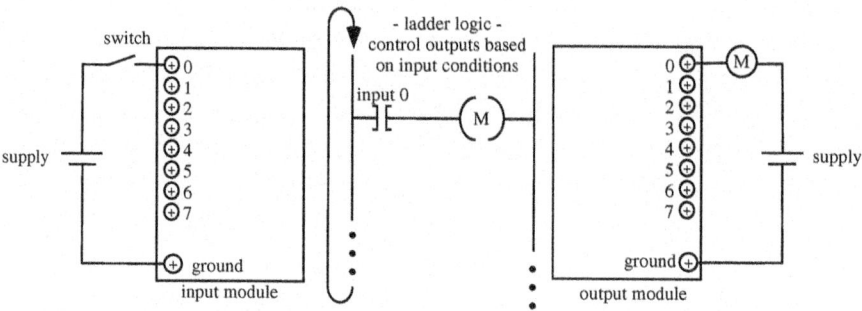

b) ladder logic links control outputs based on the status of input conditions and the linking PLC instructions (Stenerson).

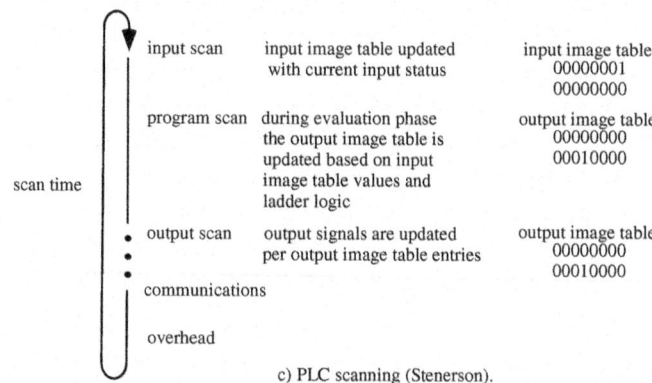

c) PLC scanning (Stenerson).

Fig. 1.2 PLC overview (Stenerson)

1.4 Operational Technology

A PLC is an industrial hardened microcontroller. As shown in Fig. 1.2a, a PLC is typically a rack mounted collection of modules. Each module provides a critical subsystem for the PLC. The PLC subsystems share many of the same functions typically found in most microcontrollers. For example, a typical PLC system consists of power supply, central processing unit (CPU), serial communications, analog input, digital input and output, and timer modules. A custom system is assembled by choosing modules to meet system requirements.

PLC systems are typically programmed using ladder logic techniques. A ladder logic program resembles a ladder with two vertical side rails linked by a number of rungs. As shown in Fig. 1.2b, the real world inputs (switches, sensors, etc.) are interfaced to the PLC via input modules. Output real world devices such as indicators, audible alarms, motors, actuators, etc. are interfaced to the PLC via output modules. The PLC ladder logic rungs represent steps of a program that link input device conditions to output control signals.

As shown in Fig. 1.2c, a ladder logic program goes through a scan consisting of multiple stages. The scan begins with an input scan. During the input scan, the status of inputs is checked and an input image table is updated in the PLC CPU memory. The input status is fixed in the input image table for the remainder of the scan time. With input status updated, the program scan commences. This is called the evaluation phase where the output image table is updated based on the input image table values and the ladder logic rungs connecting input values to output control signals. Each rung in the ladder logic program is evaluated sequentially starting with the top rung and progressing down the ladder. With the completion of the evaluation stage, output signals are generated per the output image table. The final two steps of the scan include related serial communications and any required PLC housekeeping. Upon completion of the scan, the scan is repeated beginning again with the input scan (Stenerson).

1.5 IoT Architecture

The Internet of Things (IoT), as first described by Mr. Ashton in 1999, initiated the movement to provide a link between the IT and OT worlds. There are multiple models available to describe this vital link. Hanes et al. provides the model shown in Fig. 1.3a.

The model provides three layers linking IoT "things" to applications via internet–based communication channels. The "things" are the sensors and actuators interfacing to a physical world process. The sensors and actuators provide for the monitoring and control signals for the process. The application, which may be physically distant from the process, takes in as input the sensor information and provides output control signals based on the control algorithm (Hanes).

Example. I have always wanted to build a greenhouse. I find it quite fascinating that the sun's energy may be captured, stored, and employed at a later time to extend and stabilize the growing season for vegetables. Part of the fascination may be related to spending much of my life in northern climes (Newfoundland, Nebraska, North Dakota, Montana, and Wyoming).

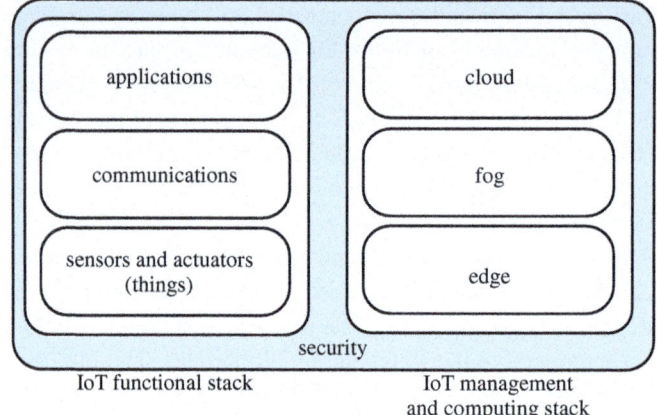

a) Simplified IoT model (Hanes).

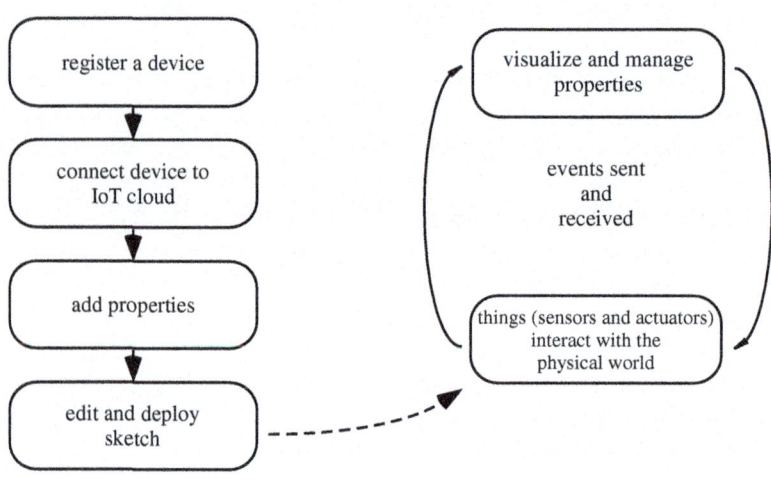

b) Arduino IoT deployment model (arduino.cc).

Fig. 1.3 IoT models (Hanes, www.arduino.cc)

Applying the IoT model described, the "things" of the greenhouse would be the sensors used to measure the vital signs of the greenhouse. For example, we might measure the following parameters: indoor temperature, outdoor temperature, humidity, soil moisture, stored water level, backup battery voltage level, etc. The actuator "things" of the greenhouse would be those devices used to change the greenhouse configuration: a vent fan when the indoor greenhouse temperature becomes too high, a water pump to mist the vegetables when appropriate conditions are met (e.g. plant soil too dry, etc.). An Arduino–based sketch may be developed to visualize and manage greenhouse properties. For example, the greenhouse indoor and outdoor temperatures may be logged and displayed over a long period of time (e.g. the winter months). The internet infrastructure with Wi–Fi access may be used to allow the sending and receiving of greenhouse events. Also, a Bluetooth link to a cell phone might be helpful. In Chap. 6 we explore this IoT application in detail.

1.6 IoT Technology

To support IoT deployment, a number of technologies have been developed to support project level IoT applications, smart home concepts, and industrial level Industrial Internet of Things (IIoT) applications. The dividing lines between these applications are blurry. It is more of a continuum of applications rather than categories of applications.

1.6.1 Smart Home Technology

A smart home uses technology to efficiently monitor and control home parameters such as temperature, humidity, lighting, security, lawn health, etc. In 2005 the Z–Wave Alliance was established to provide a standard configuration and control protocol for smart home applications. The Z–wave protocol provides for the wireless mesh networking of smart objects within a home. The protocol provides for a data rate between configured devices of 100 kbps. Devices communicate securely at frequencies of 908.4 or 916 MHz using Advanced Encryption Standard (AES) 128 encryption.[2] Network activities are coordinated by a smart hub that is connected to the internet. The smart hub can control up to 232 devices within a home or small business environment at a range up to 328 ft. Each smart home network has a unique network identification and each device within the home has node identification. The node identification is provided using the IPv6 address space. This provides for non–interference between smart configured homes within a neighborhood (www.z-wavealliance.org).

1.6.2 Industrial Internet of Things (IIoT)

IoT technology has found its way into a number of industries as shown in Fig. 1.4. This merger of IoT concepts and processes applied to industry has resulted in the Industrial Internet of Things or IIoT. As an end of chapter assignment, we ask you to investigate one of these areas.

---

[2] The AES 128 data encryption standard is discussed in Chap. 2.

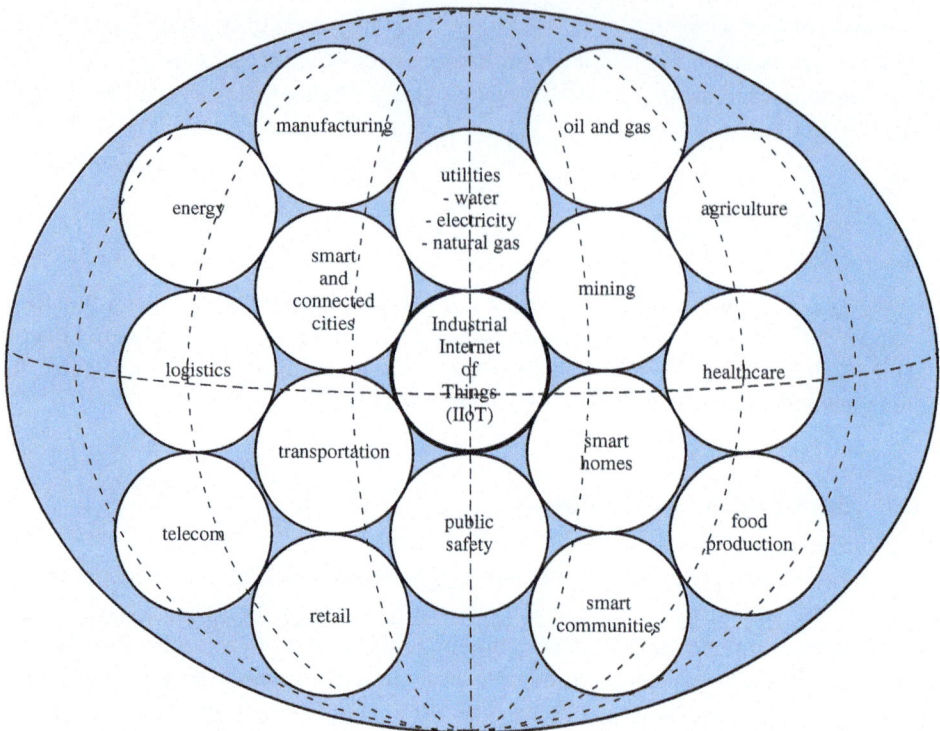

Fig. 1.4 IIoT applications

1.7 Cybersecurity

One of my favorite books is the *The Once and Future King* by T. H. White. It is the tale of the young boy, "the Wart," becoming King Arthur and the many adventures along the way. I have read this book every several years since I was young. Early in the book, White provides a description of the Wart's guardian's castle. He describes how the castle is protected from marauders by a moat (deep ditch) filled with water. To get access to the castle, a drawbridge is lowered across the moat and then raised again to secure the castle.[3]

There are many dangers surrounding a network or a computer on the internet or within an operational environment. As shown in Fig. 1.5 the dangers are in the form of malicious software (malware) or the nefarious efforts of computer hackers. These dangers and challenges include (Kurose, Levine, Lowe):

- botnet–network of infected computers controlled from an external source to perform coordinated nefarious activities on target computers;

---

[3] T. H. White, "The Once and Future King".

1.7 Cybersecurity

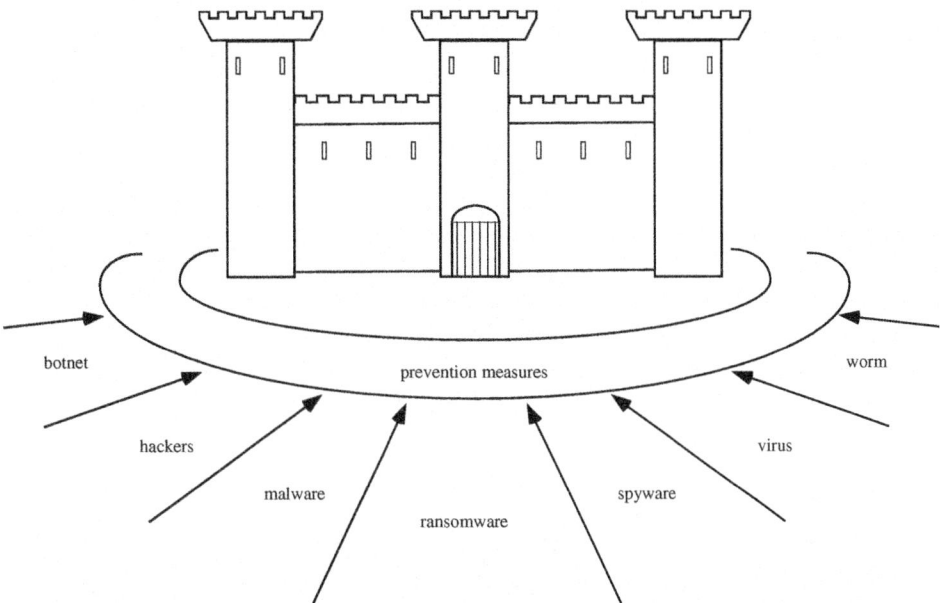

Fig. 1.5 Network threats

- hackers–individuals who try to overcome computer protection measures and procedures to gain personal data;
- ransomware–a computer attack where files are encrypted and held for ransom. If the ransom is paid, the files are returned to normal service;
- spyware–software that is accidently downloaded while browsing the internet. The software spies on your computer activities and reports back to its source;
- virus–a nefarious software program spread as an e–mail attachment. When the e–mail attachment is executed it goes to your computer's address bank and sends out e–mails with the virus program as an attachment masquerading as you. Using this technique, the virus may be spread to a number of computers. The nefarious intent of the virus may be activated by a specific event such as reaching a particular date and time; and
- worm–a worm creeps into a computer by means of flaws within network programs. Once onboard your computer, the worm looks for password and credit card information.

As in the analogy, the castle is protected from marauders by the surrounding moat and securing the drawbridge. The moat and drawbridge for a network and its computer assets include preventive countermeasures including (Kurose, Levine, Lowe):

- Firewall–A firewall protects network resources from external dangers. It applies policies to determine message traffic that may enter a protected network.

- Antivirus programs (AVP)–Each computer on the network should have an AVP installed. The AVP should be current with all software updates applied.
- Operating system updates–Regular operating system updates are sent to computer users. These updates should be made when received. They may contain updates to correct a security flaw.
- Passwords–You should employ a strong password to protect your computer assets. IoT hardware devices are sometimes configured with a default password. You should replace the default password with a strong password.
- File backups–Computer files should be backed up on a regular basis.
- User awareness–Users should be skeptical of e–mails that appear questionable. An e–mail with an executable attachment should not be opened.

1.8 IoT and IIoT Security

IoT and IIoT security borrows many of the same measures from the IT world. In addition, the International Society of Automation (ISA) and the International Electrotechnical Commission (IEC) have jointly developed a suite of security processes, procedures, and standards for control systems as shown in Fig. 1.6 (www.isa.org). The literature contains documentation of nefarious actors penetrating a secure system. Typically, these attacks have breached the "air gap" and vulnerabilities between the IT and OT components of the system.

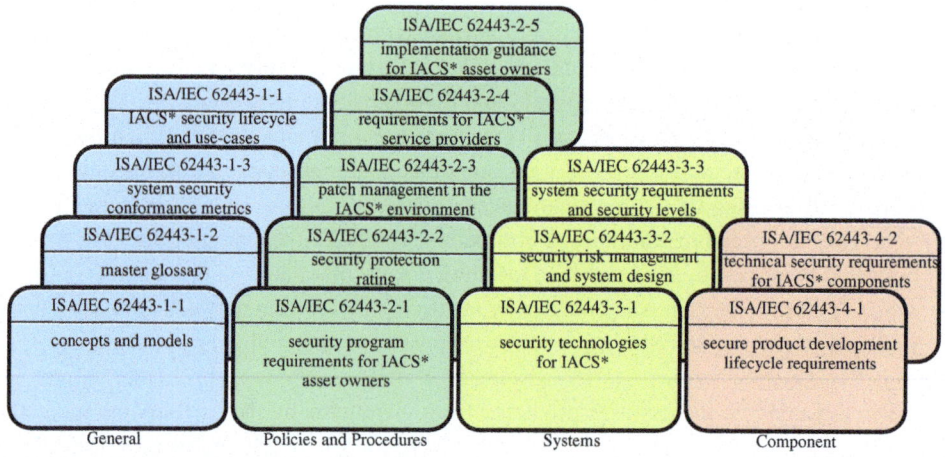

Fig. 1.6 ISA/IEC 62443 control system security (www.isa.org)

1.9 Arduino Portenta Machine Control Features

The Arduino Portenta Machine Control (PMC) is shown in Fig. 1.7. The PMC is equipped with a rich complement of subsystems for use in machine control, industrial control, Internet of Things (IoT), and Industrial Internet of Things (IIoT) applications. For communications, the PMC is equipped with RS–485, Controller Area Network (CAN), Inter–Integrated Circuit (I2C), Wi–Fi, and Bluetooth Low Energy (BLE) subsystems. We discuss these communication protocols in Chap. 3.

The PMC is also equipped to measure a wide variety of input signals including temperature, analog input, digital input, and encoder signals. The PMC is also equipped to generate digital and analog output signals.

1.10 Getting Started with the Arduino IDE

Most microcontrollers are programmed with some variant of the C programming language. The C programming language provides a nice balance between the programmer's control of the controller hardware and time efficiency in program writing. As an alternative, the Arduino Integrated Development Environment (IDE) provides a user–friendly interface to

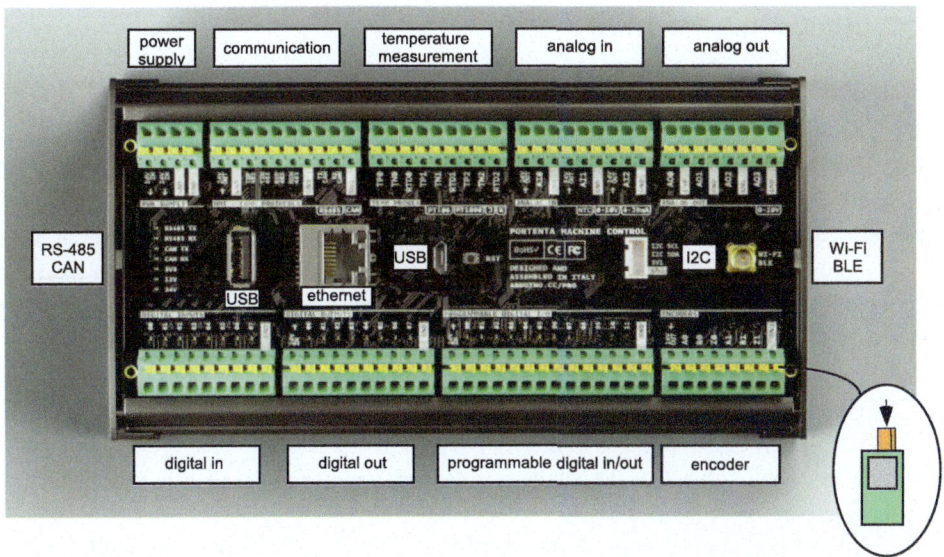

Fig. 1.7 Arduino Portenta Machine Control. Images used courtesy of the Arduino Team (CC BY–NC–SA) (www.arduino.cc)

quickly develop a program or sketch, transform the sketch to machine code, and then load the machine code into the Arduino processor in several simple steps.

The first version of the Arduino IDE was released in August 2005. It was developed at the Interaction Design Institute in Ivrea, Italy to allow the ability to quickly put processing power to use in a wide variety of projects. Since that time, updated versions incorporating new features, have been released on a regular basis (www.arduino.cc).

At its most fundamental level, the IDE is a user–friendly interface to allow one to quickly write, load, and execute code on an Arduino microcontroller or PMC. A barebones program need only consist of a setup() and loop() function. The Arduino IDE adds the other required pieces such as header files and the main program construct. The IDE is written in Java and has its origins in the Processor programming language and the Wiring Project (www.arduino.cc).

The Arduino IDE may be downloaded from the Arduino website's front page at www.arduino.cc. Versions are available for Windows, Mac OS X, and Linux. When the IDE is successfully installed, install the following libraries:

- Arduino Machine Control Library
- Arduino mbed–enabled Boards Library.

using the Library Manager within the Arduino IDE.

1.11 PMC Blink Program

In this section we configure the Arduino PMC for basic operation and blink an onboard LED. The PMC is connected to the support PC or laptop via a USB micro cable as shown in Fig. 1.8. The PMC is powered from a laboratory power supply set for 24 VDC. The PMC may also be powered from a DIN rail mounted PLC power supply such as the Meanwell MDR 60–24. Provide a jumper wire from the Digital OUTPUTS 24V IN to the 24 VDC power supply. The PMC is equipped with special connectors to allow for rapid prototyping. To connect a wire (e.g. 22 solid AWG), press down on the orange pushbutton with a small screw driver, insert the wire with approximately 5 mm of conductor exposed, and release the orange pushbutton as shown in the inset.

Using the Arduino IDE, compile and upload the following sketch.

- Under the Tools tab select the evaluation **Board** you are using and the **Port** that it is connected to.
- Upload and execute the program by asserting the "Upload" (right arrow) button.
- When uploaded the small DIGITAL OUTPUTS red LED will blink.

1.11 PMC Blink Program

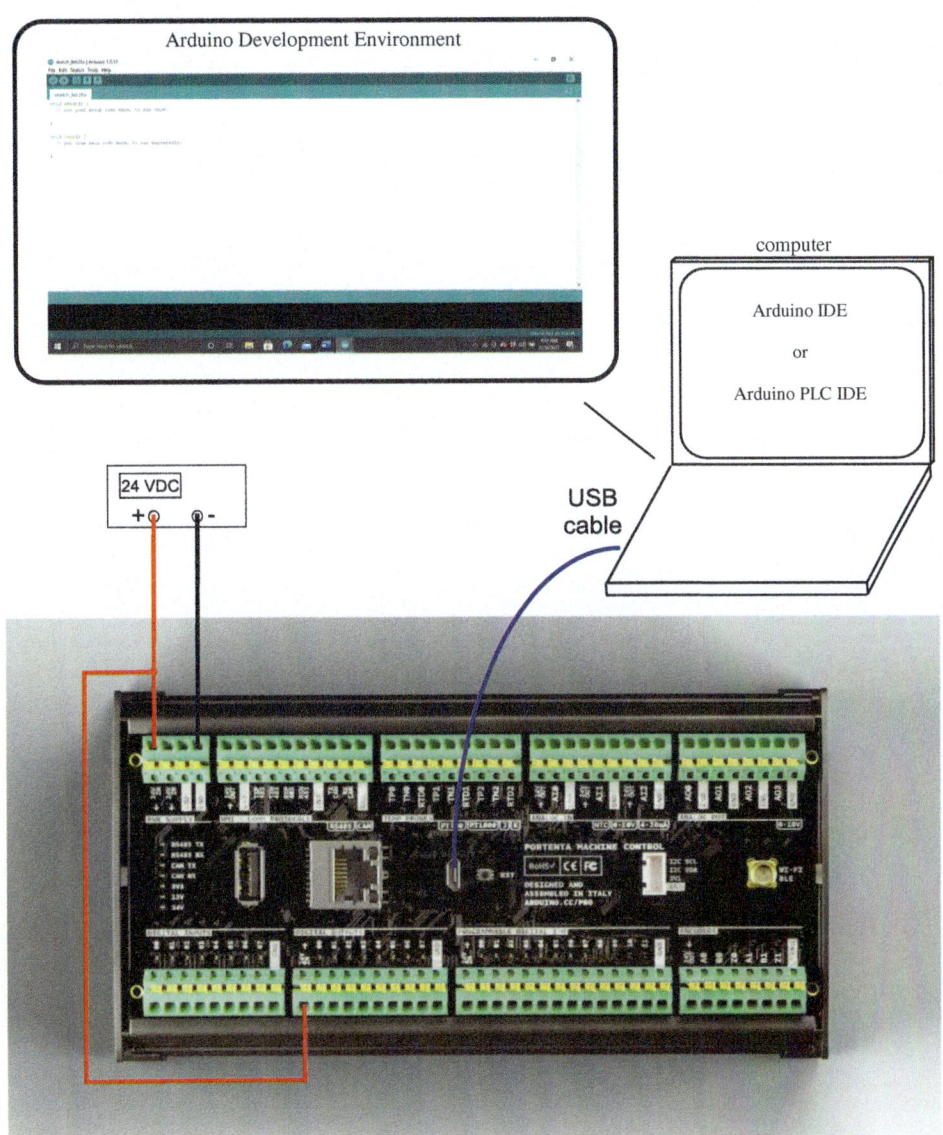

Fig. 1.8 PMC barebones quickstart. Images used courtesy of the Arduino Team (CC BY–NC–SA) (www.arduino.cc)

```
//*******************************************************************
//PMC_blink
//
//Configuration:
//   - PMC connected to 24 VDC power supply
//   - Digital OUTPUTS 24V IN connected to 24 VDC power supply
//
//This example code is in the public domain.
//*******************************************************************

#include <Arduino_PortentaMachineControl.h>

void setup()
{
MachineControl_DigitalOutputs.begin();        //init PMC digital outputs
}

void loop()
{
MachineControl_DigitalOutputs.write(0, HIGH); //digital out 00 on
delay(1000);                                  //delay in ms
MachineControl_DigitalOutputs.write(0, LOW);  //digital out 00 off
delay(1000);                                  //delay in ms
}

//*******************************************************************
```

We modify the sketch so the LEDs sequentially cycle left to right. A for loop is used to change the LED being flashed for each loop iteration.

```
//*******************************************************************
//PMC_blink_sequence
//
//Configuration:
//   - PMC connected to 24 VDC power supply
//   - Digital OUTPUTS 24V IN connected to 24 VDC power supply
//
//This example code is in the public domain.
//*******************************************************************

#include <Arduino_PortentaMachineControl.h>

void setup()
{
MachineControl_DigitalOutputs.begin();        //init PMC digital outputs
}

void loop()
```

1.11 PMC Blink Program

```
{
unsigned int LED_number;

for(LED_number = 0; LED_number < 8; LED_number++)
  {
  MachineControl_DigitalOutputs.write(LED_number, HIGH); //digital out
on delay(100);                                           //delay in ms
  MachineControl_DigitalOutputs.write(LED_number, LOW);  //digital out
off delay(100);                                          //delay in ms
  }
}

//*****************************************************************
```

In the next example we interface an external 10 mm LED to digital output 00 as shown in Fig. 1.9. We provide details of the interface circuit in Chap. 4. The PMC_blink sketch is used for this example.

We next modify the PMC_blink sketch to generate a 1 kHz square wave with a 50% duty cycle. The Arduino IDE function delayMicroseconds is used to control the on time and off time of the signal as shown in Fig. 1.10. Note the rise time and fall time of the resulting signal.

```
//*****************************************************************
//PMC_square_wave - generates 1 kHz square wave
//
//Configuration:
//   - PMC connected to 24 VDC power supply
//   - Digital OUTPUTS 24V IN connected to 24 VDC power supply
//
//This example code is in the public domain.
//*****************************************************************

#include <Arduino_PortentaMachineControl.h>

void setup()
{
MachineControl_DigitalOutputs.begin();     //init PMC digital outputs
}

void loop()
{
MachineControl_DigitalOutputs.write(0, HIGH); //digital out 00 on
delayMicroseconds(500);                       //delay in us
MachineControl_DigitalOutputs.write(0, LOW);  //digital out 00 off
delayMicroseconds(500);                       //delay in us
}
```

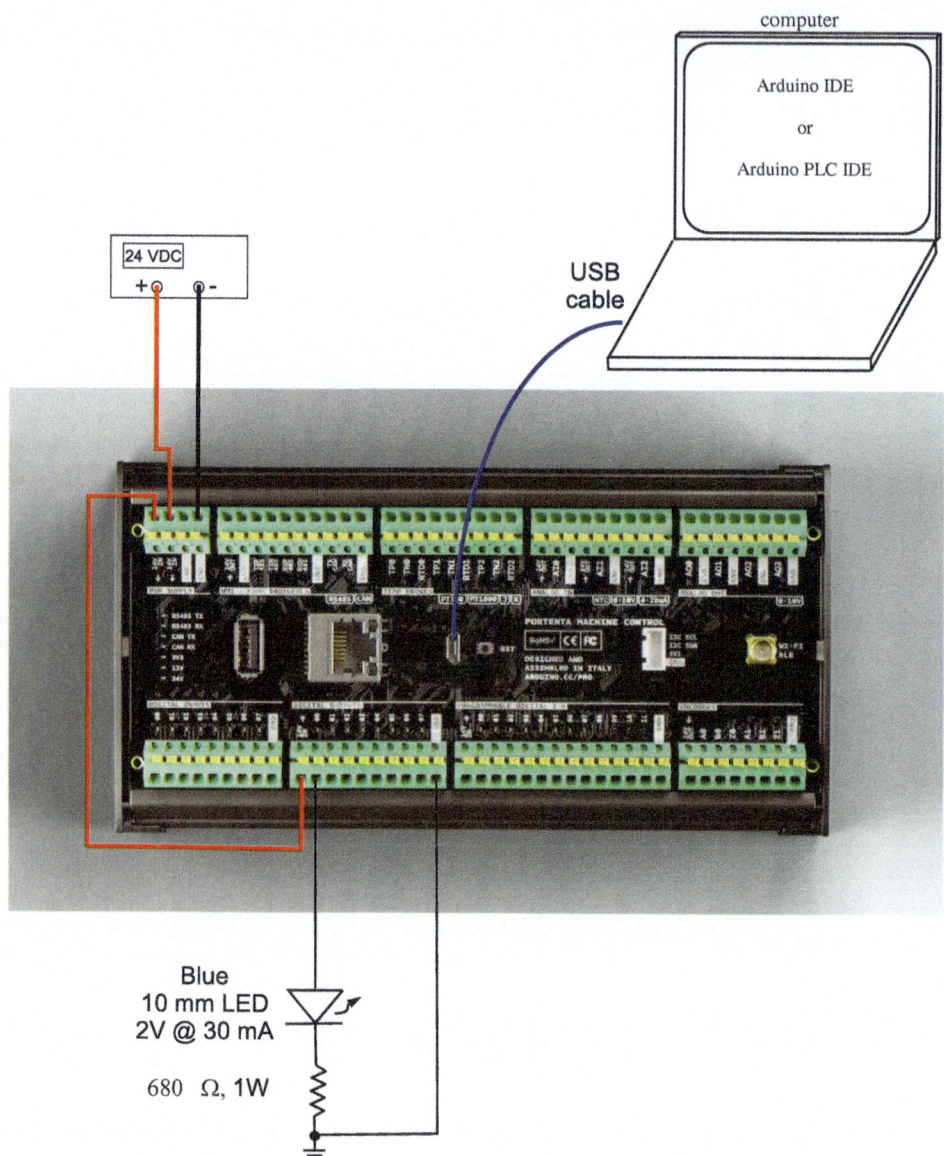

Fig. 1.9 PMC controlling external LED. Images used courtesy of the Arduino Team (CC BY–NC–SA) (www.arduino.cc)

1.11 PMC Blink Program

//***

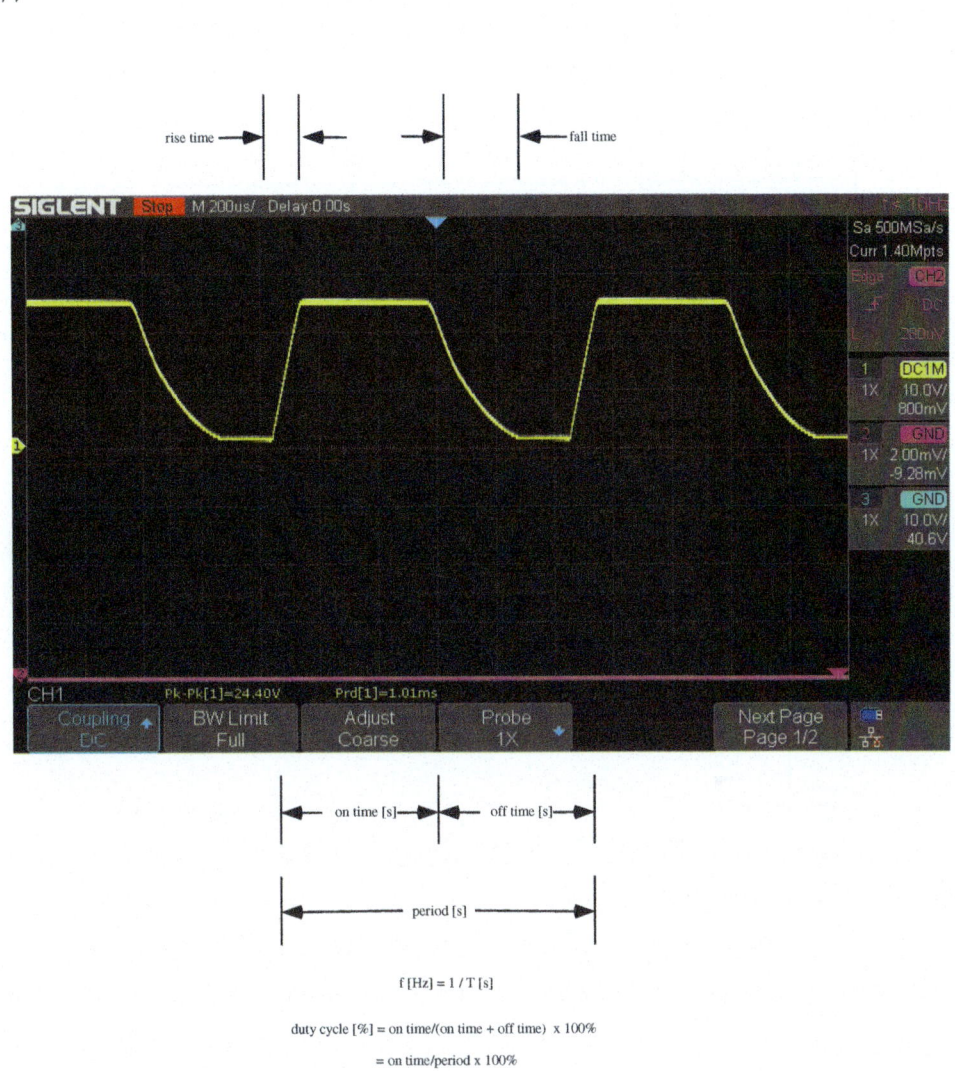

Fig. 1.10 Square wave

Pulse width modulation (PWM) is a technique to vary a signal's duty cycle over time as shown in Fig. 1.11. At zero percent duty cycle the effective voltage delivered to a device is 0 V. As the signal's duty cycle is increased the effective signal voltage is also increased as shown in a). The signal's duty cycle may be adjusted over time to adjust the effective voltage delivered to a device as shown in b). This technique may be used to control the intensity of an LED or the speed of a motor.

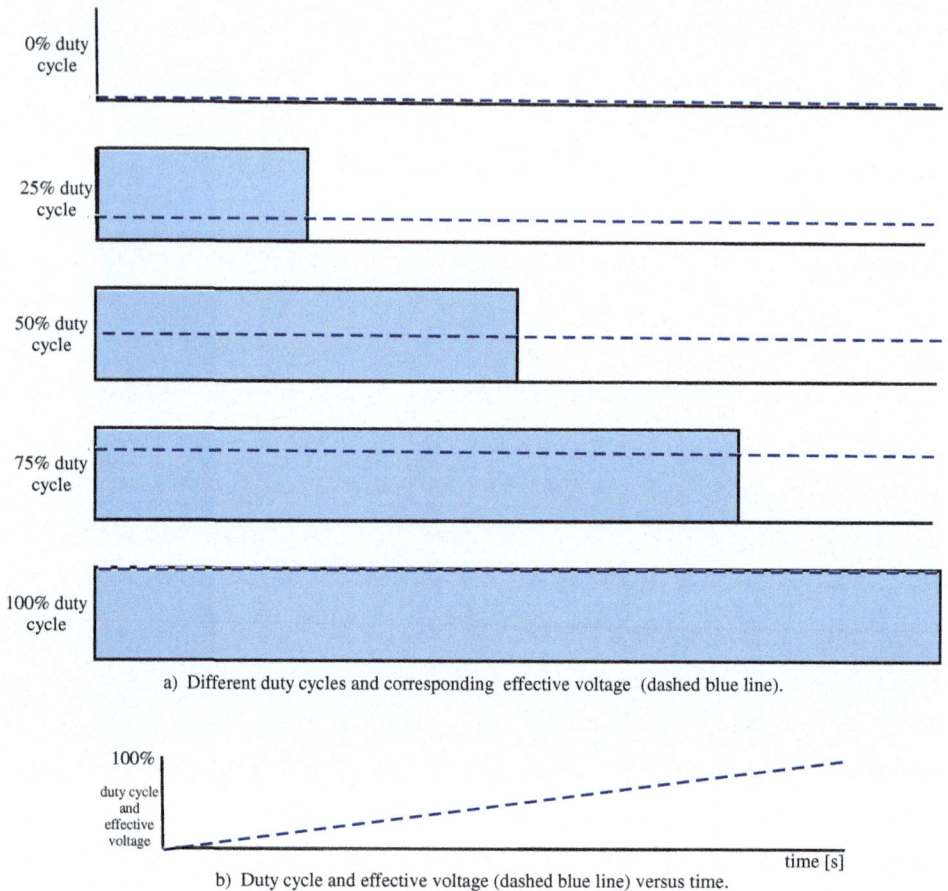

Fig. 1.11 Pulse width modulation

In the next sketch we vary the intensity of an LED by varying the PWM duty cycle.

```
//*******************************************************************
//PMC_PWM - generates a pulse width modulated signal to vary the
//          intensity of an LED
//
//Configuration:
//   - PMC connected to 24 VDC power supply
//   - Digital OUTPUTS 24V IN connected to 24 VDC power supply
//
//This example code is in the public domain.
//*******************************************************************

#include <Arduino_PortentaMachineControl.h>
```

```
unsigned int LED_int;                              //LED intensity

void setup()
{
MachineControl_DigitalOutputs.begin();             //init PMC digital out
}

void loop()
{
for(LED_int = 0; LED_int <= 100; LED_int++)
  {
  MachineControl_DigitalOutputs.write(0, HIGH); //digital out 00 on
  delayMicroseconds(LED_int);                      //delay in us
  MachineControl_DigitalOutputs.write(0, LOW);  //digital out 00 off
  delayMicroseconds(100 - LED_int);                //delay in us
  delay(5);
  }
}

//*******************************************************************
```

1.12 Application: Portable Lab Configuration

Provided in Fig. 1.12 is a layout diagram for a PMC panel. Raceway ducts are used to route wiring between components. The DIN compatible components are mounted on standard industrial DIN rails. A 24 VDC, 1.8A DIN rail power supply (Mean Well MDR–60–24) powers the PMC. A standard three conductor 115 VAC provides AC power to the 24 VDC supply. The AC input line is fused with a 2A fuse and the 24 VDC output is fused with a 2A fuse. The 10 x 38 mm fuses are housed within a DIN compatible BAOMIN fuse holder. The 24 VDC supply output is routed to an eight–channel block for distribution. The assembled panel is shown in Fig. 1.13.

1.13 Summary

In this chapter we began our exploration of the Operational Technology (OT) world. We started with a basic introduction to the Internet of Things (IoT). Within IoT there is a close relationship between Information Technology (IT) and Operational Technology (OT). We explored this relationship in some detail. The pervasiveness of IoT was then examined in industry or the Industrial Internet of Things (IIoT). We then shifted our focus to OT and basic PLC concepts. We concluded the chapter with an introduction to the Arduino Portenta Machine Control.

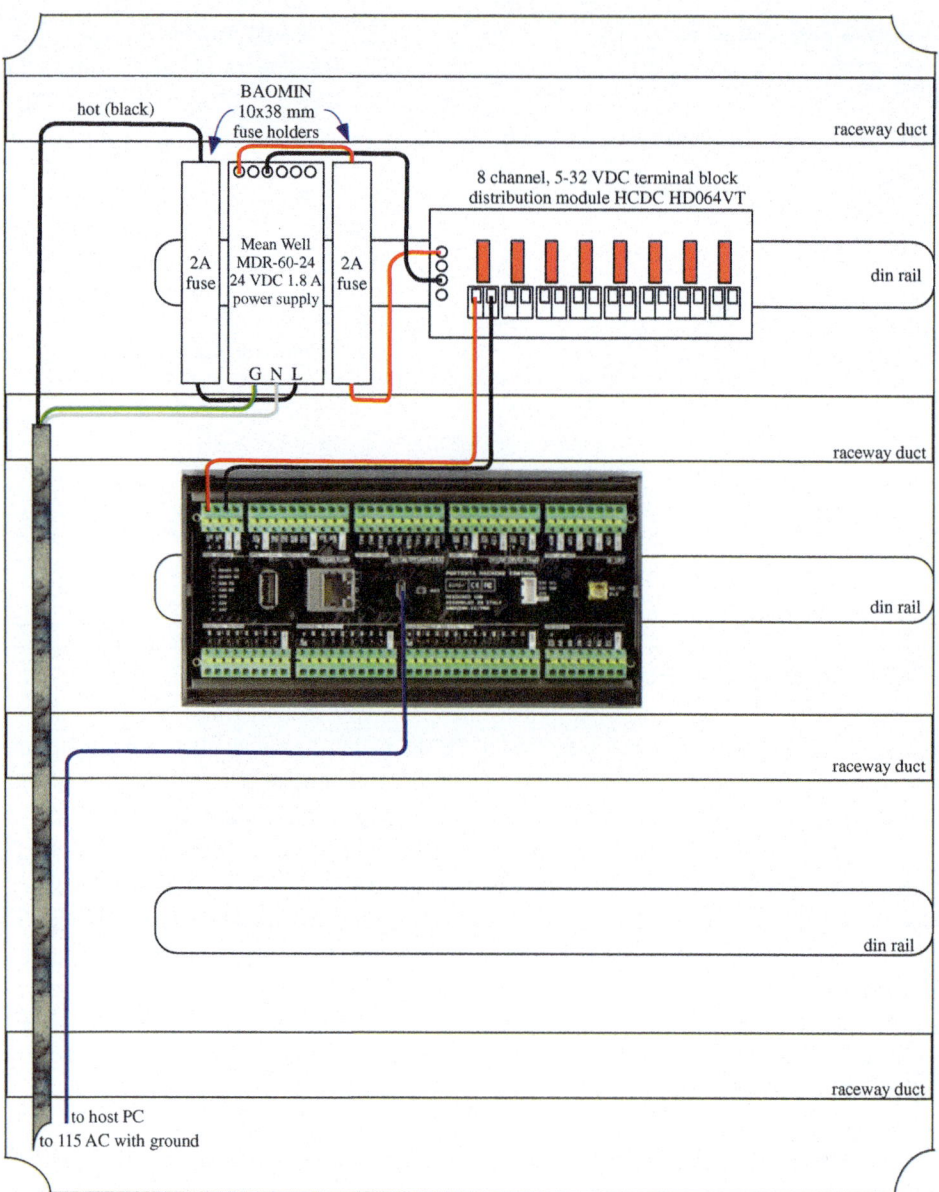

Fig. 1.12 PMC PLC panel. Images used courtesy of the Arduino Team (CC BY–NC–SA) (www.arduino.cc)

1.13 Summary

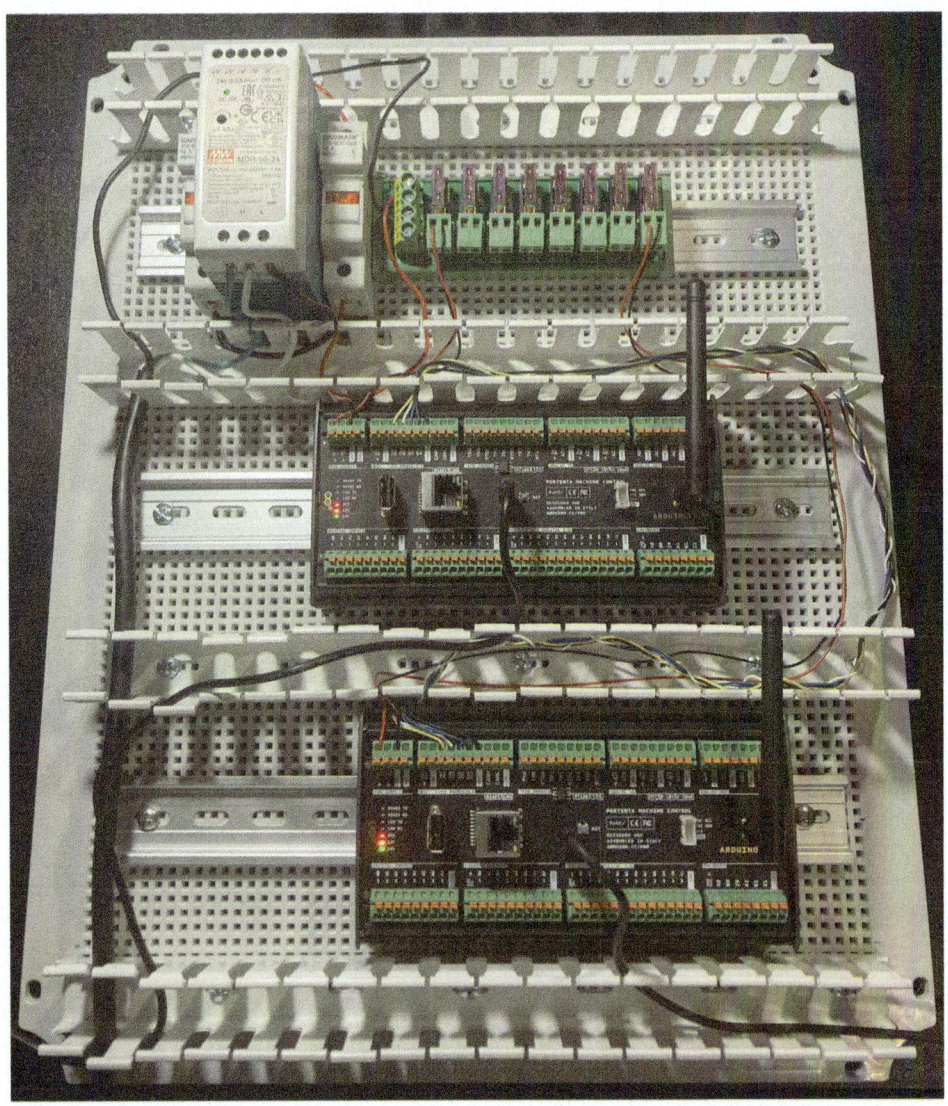

Fig. 1.13 Assembled PMC PLC panel

1.14 Chapter Problems

1. Describe different sources of cybersecurity threats.
2. Describe measures to counter cybersecurity threats.
3. Provide a working definition of IoT and IIoT.
4. What is the difference between IT and OT? How are the concepts related.
5. What is a PLC?
6. Describe the PLC scanning process.
7. Provide an IoT model. Describe the interaction between things and applications.
8. What is an "air gap?" Why is it essential for IIoT security?
9. Research and write a short paper on an IIoT security breach. How was the system penetrated? How could the situation been prevented?
10. Modify the sketch PMC_blink_sequence so the onboard PMC DIGITAL OUTPUT LEDs sequentially cycle left to right and then back right to left within a single loop pass.
11. In the above sketch, decrease the delay time. At what delay time do the LEDs appear to be always illuminated?
12. Modify the sketch PMC_PWM so the external LED goes from off to full intensity and then back to off.
13. Research and write a short paper on industrial control cybersecurity threats and countermeasures.
14. Describe how pulse width modulation techniques may be used to control LED intensity and DC motor speed.

References

Arduino homepage, www.arduino.cc.

Greer, C., M. Burns, D. Wollman, E. Griffor (2019) *Cyber–Physical Systems and Internet of Things*, NIST Special Publications 1900–202, National Institute of Standards and Technology, U.S. Department of Commerce.

Hanes D., G. Salgueiro, P. Grossetete, R. Barton, J. Henry (2017) *IoT Fundamentals–Networking Technologies, Protocols, and Use Cases for the Internet of Things*, Cisco Press.

Kurose, J. and K. Ross (1997) *Computer Networks–A Top–Down Approach*, 7th edition, Pearson Education, Inc.

Levine R. and M. Levine Young (2015) *The Internet for Dummies*, John Wiley and Sons Publishing, Inc.

Lowe, D. (2018) Networking All–In–One for dummies, 7th edition, John Wiley and Sons Publishing, Inc.

Rajkumar, R., I. Lee, L. Sha, J. Stankovic (2010) *Cyber–Physical Systems: The Next Coupling Revolution*, ACM Design Automation Conference, Anaheim, CA.

Stenerson, J. (2004) *Fundamentals of Programmable Logic Controllers, Sensors, and Communications*, Pearson Prentice Hall.

Z–Wave Alliance, *The Smart Home is Powered by Z–Wave*, www.z-wavealliance.org.

Arduino PMC Features 2

Objectives: After reading this chapter, the reader should be able to do the following:

- Describe features of the Arduino Portenta Machine Control (PMC);
- Design and implement basic control circuits and algorithms employing features of the Arduino PMC; and
- Design a control system employing an Arduino PMC.

2.1 Overview

In this chapter we become familiar with the features and subsystems onboard the Arduino Portenta Machine Control (PMC). For each subsystem we provide background theory, theory of operation, and detailed examples.

2.2 PMC and the Portenta H7 Processor

The PMC is designed as an industrial controller for equipment and machines. It couples a powerful dual processor microcontroller with industrial grade inputs and outputs. It also is equipped with a rich complement of onboard systems.

The Arduino PMC is equipped with the Portenta H7 processor. It is an ST STM32H747XI dual–core processor providing the features, subsystems and processing power for the PMC. The processor features an Arm Cortex–M7 core operating at 480 MHz and also an ARM 32–bit Cortex–M4 core operating at up to 240 MHz. The processor is equipped with multiple memory assets including: 1 MB of program memory and 2 MB of Flash memory. We explore STM32H747XI features as needed as we investigate PMC features.

Provided in Fig. 2.1 is a summary of PMC features. As shown in the figure the Arduino PMC is equipped with a rich complement of digital inputs and outputs, analog inputs and outputs, and specialized inputs for temperature and encoder measurements. We investigate each in turn. We begin by exploring the input and output features of the PMC.

2.3   Digital Output

In Chap. 1 we used a PMC digital output to illuminate an external LED and generate a pulse width modulated signal. The PMC Digital Outputs section is equipped with eight channels of digital output (00–07). Each output channel is rated at 24 VDC and 0.5 A. The Digital Output must be provided 24 VDC to the 24V IN pin.

2.4   Digital Input

The Arduino PMC DIGITAL INPUTS section is equipped with eight 24 VDC inputs. The threshold where an input transitions from logic low to logic high is approximately 11.3 VDC. The 24 VDC input is scaled to 3 VDC via a resistor dividing network as shown in Fig. 2.1.

Provided in Fig. 2.2 is a test circuit for digital input 00. A pushbutton tact switch with a series pull down resistor is used to introduce logic changes on input 00. When the pushbutton is pressed a logic one signal is provided to digital input 00.

In the following sketch the logic status of the switch is read and an external LED is illuminated for a logic one (switch pressed).

```
//*******************************************************************
//PMC_digital_in_out
//- Reads  Digital Input 00 to determine logic status
//- Updates LED status at Digital Output 00
//
//Adapted from:
//Library: Arduino Portenta Machine Control
//Sketch: Digital_output
//Authors: Riccardo Rizzo and Leonardo Cavagnis
//This code example is in the public domain.
//*******************************************************************

#include <Arduino_PortentaMachineControl.h>

uint16_t readings = 0;
```

2.4 Digital Input

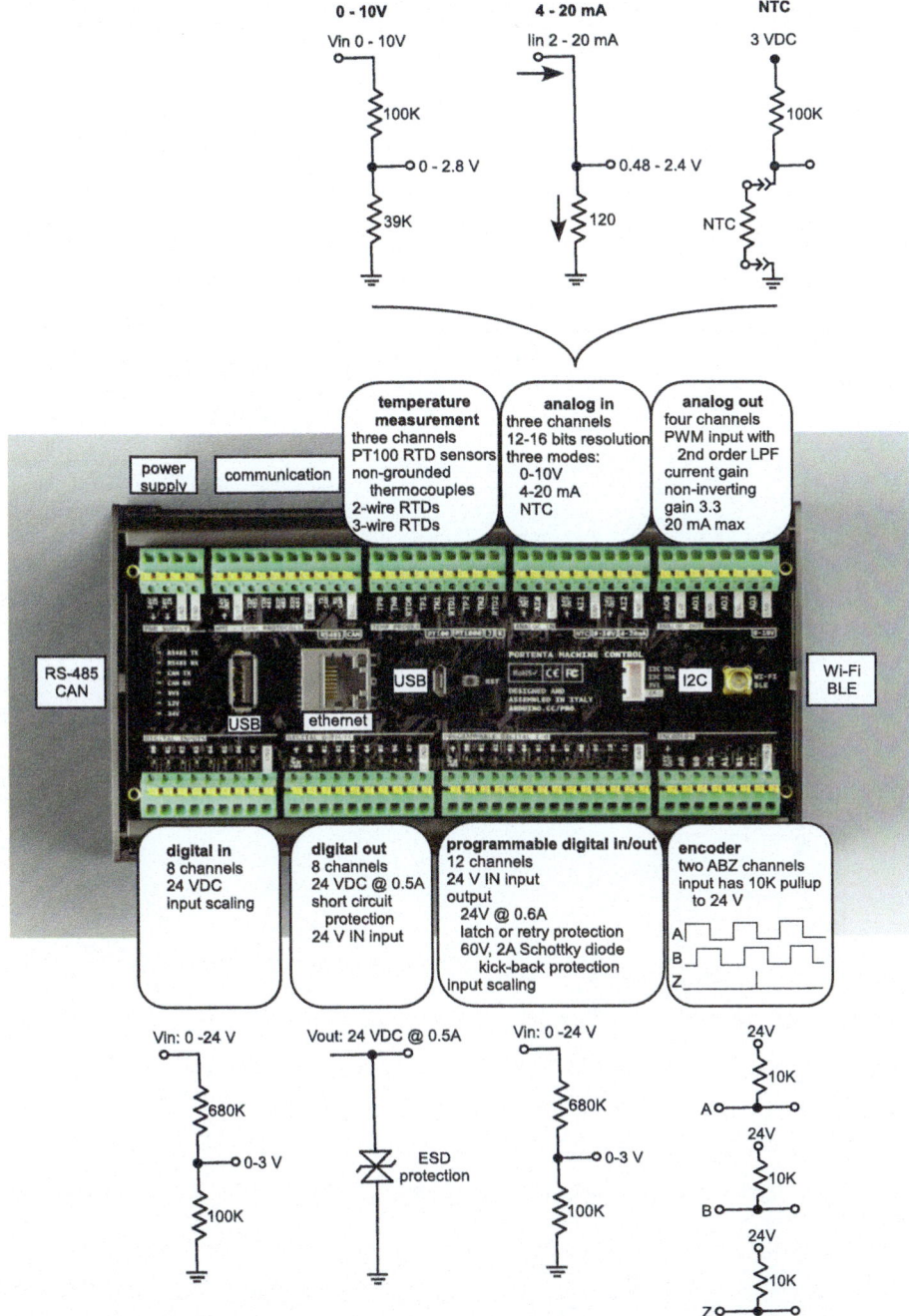

Fig. 2.1 PMC input/output summary. Images used courtesy of the Arduino Team (CC BY–NC–SA) (www.arduino.cc)

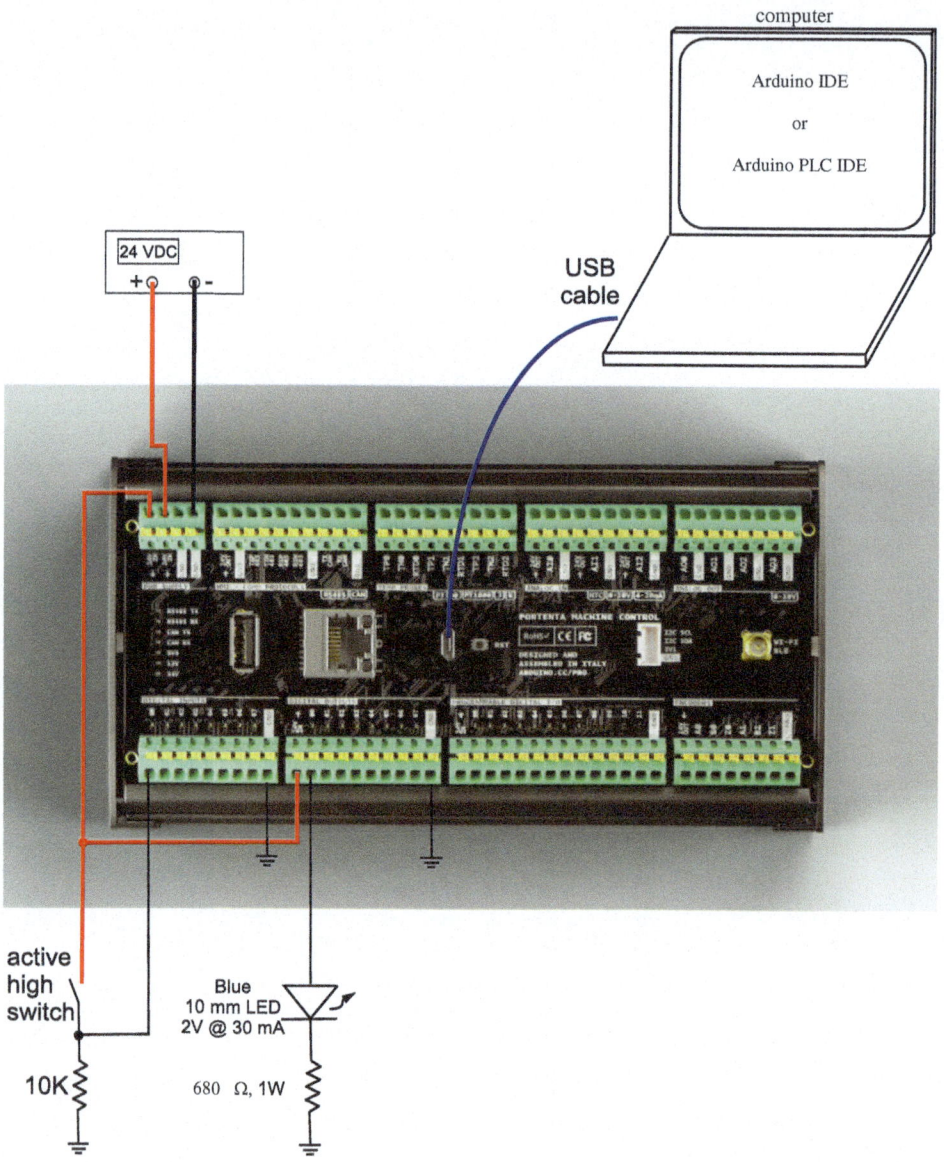

Fig. 2.2 PMC digital input test circuit. Images used courtesy of the Arduino Team (CC BY–NC–SA) (www.arduino.cc)

2.4 Digital Input

```
void setup()
{
Serial.begin(9600);
while(!Serial)
  {
  ; //wait for serial port to connect.
  }

  Wire.begin();
                                    //initialize PMC digital in
  if (!MachineControl_DigitalInputs.begin())
   {
    Serial.println("Digital input GPIO expander initialization fail!!");
  }
}

void loop()
{
//Read input status Digital Input 00
readings = MachineControl_DigitalInputs.read(DIN_READ_CH_PIN_00);

//Update LED status based on Digital Input 00 status
if(readings)
  {
  Serial.println(readings);
  MachineControl_DigitalOutputs.write(0, HIGH);
  }
else
  {
  Serial.println(readings);
  MachineControl_DigitalOutputs.write(0, LOW);
  }
Serial.println();

delay(100);
}

//*******************************************************************
```

2.4.1 Switch Debounce

Mechanical switches do not make a clean transition from one position (on) to another (off). When a switch is moved from one position to another, it may make and break contact multiple times. This activity may go on for tens of milliseconds. A processor such as the

PMC is relatively fast as compared to the action of the switch. Therefore, the processor is able to recognize each switch bounce as a separate and erroneous transition. To correct the switch bounce phenomena additional external hardware components may be used or software techniques may be employed.

2.4.1.1 Software Debounce

Software switch debouncing is accomplished by inserting a 30–50 ms lockout delay in the function responding to input changes. The delay prevents the processor from responding to the multiple switch transitions related to bouncing.

The following sketch scans for a change in input 00. When the switch is pressed, the switch count (switch_cnt) is incremented, and the corresponding LED is illuminated. Note the use of a 50 ms delay for switch debouncing. If the delay is removed (try this) the switch count (switch_cnt) may increment incorrectly. The corresponding circuit is provided in Fig. 2.3.

```
//*****************************************************************
//PMC_digital_in_out_four
//- Reads switch input at 00
//- Increments switch count variable for pushbutton press
//- Uses 50 ms software switch debounce
//- Illuminates appropriate LED based on switch count value
//
//Adapted from:
//Library: Arduino Portenta Machine Control
//Sketch: Digital_output
//Authors: Riccardo Rizzo and Leonardo Cavagnis
//This code example is in the public domain.
//*****************************************************************

#include <Arduino_PortentaMachineControl.h>

uint16_t    readings = 0;
unsigned int switch_cnt = 0;

void setup()
{
Wire.begin();
                                //initialize PMC digital in
if(!MachineControl_DigitalInputs.begin())
  {
  switch_cnt = 5;
  }
}

void loop()
```

2.4 Digital Input

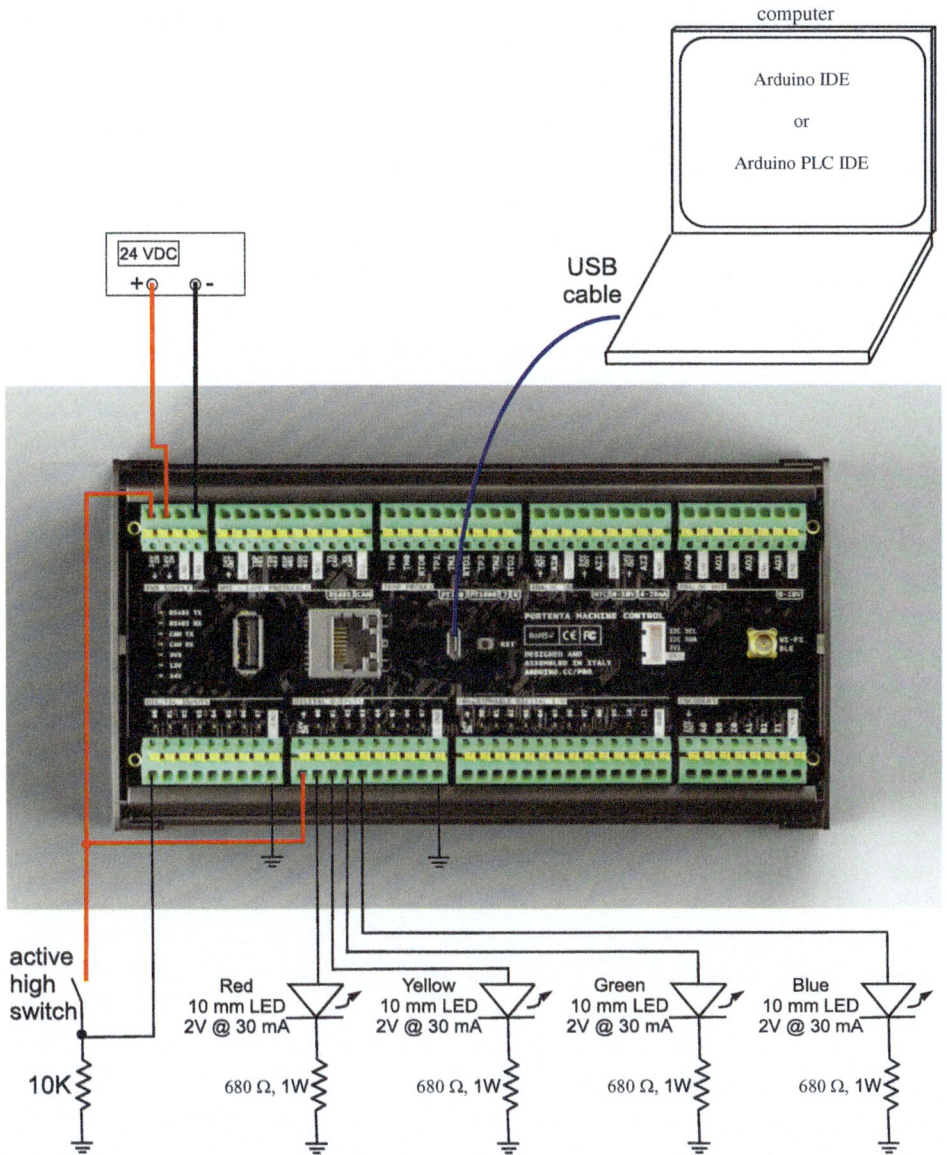

Fig. 2.3 PMC digital input test circuit for switch debounce. Images used courtesy of the Arduino Team (CC BY–NC–SA) (www.arduino.cc)

```
{                               //initialize PMC digital read
readings = MachineControl_DigitalInputs.read(DIN_READ_CH_PIN_00);
delay(50);                      //50 ms switch debounce delay

if(readings)
  {
  switch_cnt++;
  switch(switch_cnt)            //choose action based on switch_cnt
    {
    case 1: MachineControl_DigitalOutputs.write(0, HIGH);
            break;

    case 2: MachineControl_DigitalOutputs.write(1, HIGH);
            break;

    case 3: MachineControl_DigitalOutputs.write(2, HIGH);
            break;

    case 4: MachineControl_DigitalOutputs.write(3, HIGH);
            switch_cnt = 0;
            break;

    default:
            MachineControl_DigitalOutputs.write(0, LOW);
            MachineControl_DigitalOutputs.write(1, LOW);
            MachineControl_DigitalOutputs.write(2, LOW);
            MachineControl_DigitalOutputs.write(3, LOW);
            break;
    }//end switch
  }//end if
else
  {
  MachineControl_DigitalOutputs.write(0, LOW);
  MachineControl_DigitalOutputs.write(1, LOW);
  MachineControl_DigitalOutputs.write(2, LOW);
  MachineControl_DigitalOutputs.write(3, LOW);
  }
delay(100);
}

//****************************************************************
```

2.5 Programmable Digital I/O

The Arduino PMC's Programmable Digital I/O is equipped with 12 pins (00–11). Each pin may be configured as input or output.

- When configured as an input pin, the input is scaled using similar circuitry to the Digital Input pins.
- When configured as an output pin, the output is rated at 24 VDC with a maximum current of 500 mA.
- The output pin configuration includes 60V, 2A Schottky diodes for protection when driving an inductive load such as a motor, relay, etc.

In the sketch below, we duplicate the activity of the previous sketch. Note how Programmable Digital I/O 00 is configured for input while Programmable Digital I/O 04–07 is configured for output. The corresponding circuit is provided in Fig. 2.4.

```
//***************************************************************
//PMC_digital_in_out_four_prog
//
//Uses Programmable Digital I/O on the Portenta Machine Control
//- Reads switch input at 00
//- Increments switch count variable for pushbutton press
//- Uses 50 ms software switch debounce
//- Illuminates appropriate LED based on switch count value
//
//Adapted from:
//Library: Arduino Portenta Machine Control
//Sketch: GPIO_programmable
//Authors: Riccardo Rizzo and Leonardo Cavagni
//Adapted by: S. Barrett
//This code example is in the public domain.
//***************************************************************

#include <Arduino_PortentaMachineControl.h>

uint16_t      readings = 0;
unsigned int switch_cnt = 0;

void setup()
{
Wire.begin();
                              //initialize PMC I/O
if(!MachineControl_DigitalProgrammables.begin())
  {
  switch_cnt = 5;
  }
}

void loop()
{                             //read channel 00
readings = MachineControl_DigitalProgrammables.read(IO_READ_CH_PIN_00);
delay(50);                    //50 ms switch debounce delay
```

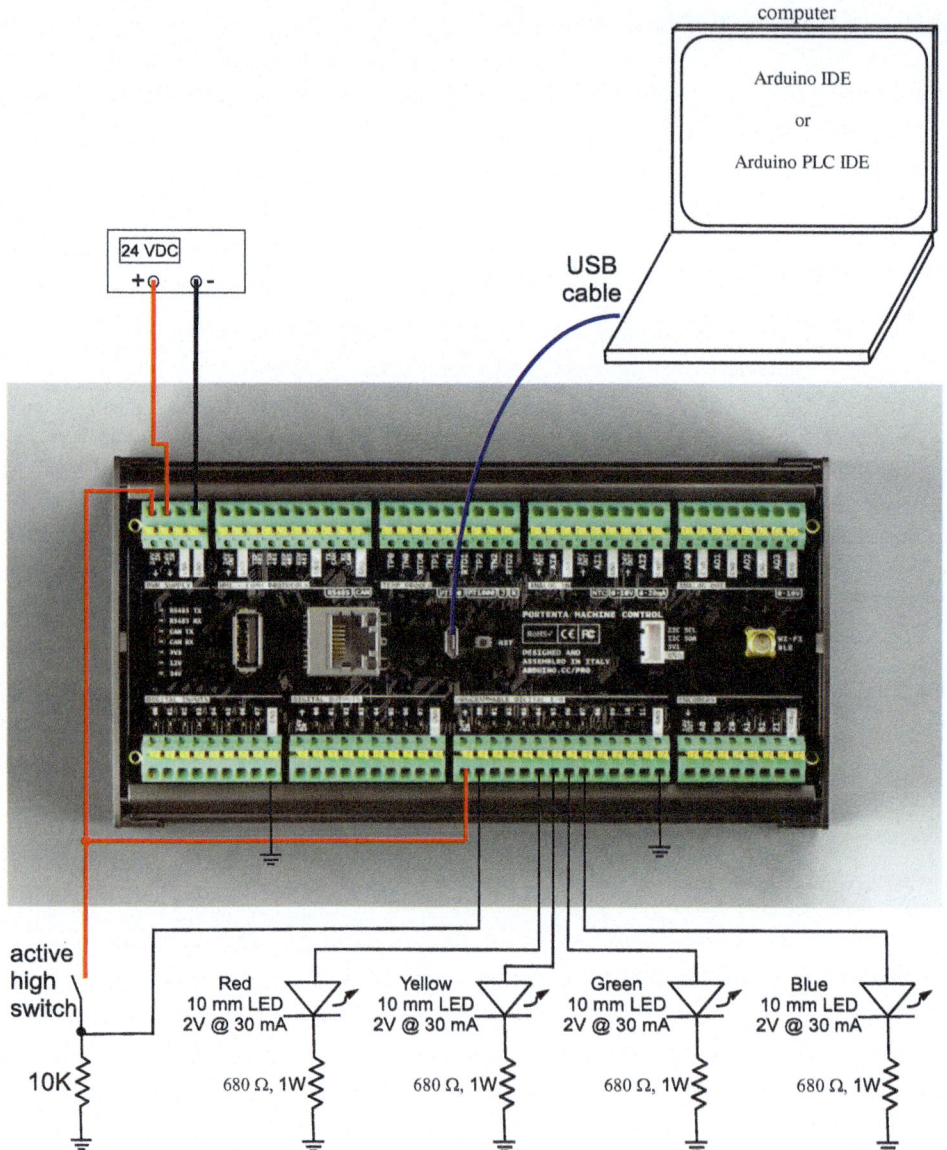

Fig. 2.4 PMC digital input test circuit for switch debounce. Images used courtesy of the Arduino Team (CC BY–NC–SA) (www.arduino.cc)

```
if(readings)
  {
  switch_cnt++;
  switch(switch_cnt)           //action based on switch_cnt
     {
     case 1: MachineControl_DigitalProgrammables.set(IO_WRITE_CH_PIN_04, SWITCH_ON);
             break;

     case 2: MachineControl_DigitalProgrammables.set(IO_WRITE_CH_PIN_05, SWITCH_ON);
             break;

     case 3: MachineControl_DigitalProgrammables.set(IO_WRITE_CH_PIN_06, SWITCH_ON);
             break;

     case 4: MachineControl_DigitalProgrammables.set(IO_WRITE_CH_PIN_07, SWITCH_ON);
             switch_cnt = 0;
             break;

     default:MachineControl_DigitalProgrammables.set(IO_WRITE_CH_PIN_04, SWITCH_OFF);
             MachineControl_DigitalProgrammables.set(IO_WRITE_CH_PIN_05, SWITCH_OFF);
             MachineControl_DigitalProgrammables.set(IO_WRITE_CH_PIN_06, SWITCH_OFF);
             MachineControl_DigitalProgrammables.set(IO_WRITE_CH_PIN_07, SWITCH_OFF);
             break;
     }//end switch
  }//end if
else
  {
  MachineControl_DigitalProgrammables.set(IO_WRITE_CH_PIN_04, SWITCH_OFF);
  MachineControl_DigitalProgrammables.set(IO_WRITE_CH_PIN_05, SWITCH_OFF);
  MachineControl_DigitalProgrammables.set(IO_WRITE_CH_PIN_06, SWITCH_OFF);
  MachineControl_DigitalProgrammables.set(IO_WRITE_CH_PIN_07, SWITCH_OFF);
  }
delay(100);
}

//**********************************************************************
```

2.6 Analog Input

The Arduino Portenta Machine Control is equipped with a flexible analog—to–digital converter (ADC) system. As shown in Fig. 2.5 the PMC has three analog channels. Each channel may be configured for conversion of a signal from 0–10 VDC, 4–20 mA, or a Negative Temperature Coefficient (NTC) signal. This allows the conversion of signals from a wide variety of sensors. The signal type for conversion is selected via a software setting. Once selected the signal is routed to the appropriate input circuitry as shown in Fig. 2.5.

The signal is routed to the ADC converter with a software configured resolution of 12–16 bits. A brief introduction to the analog–to–digital (ADC) conversion process follows.

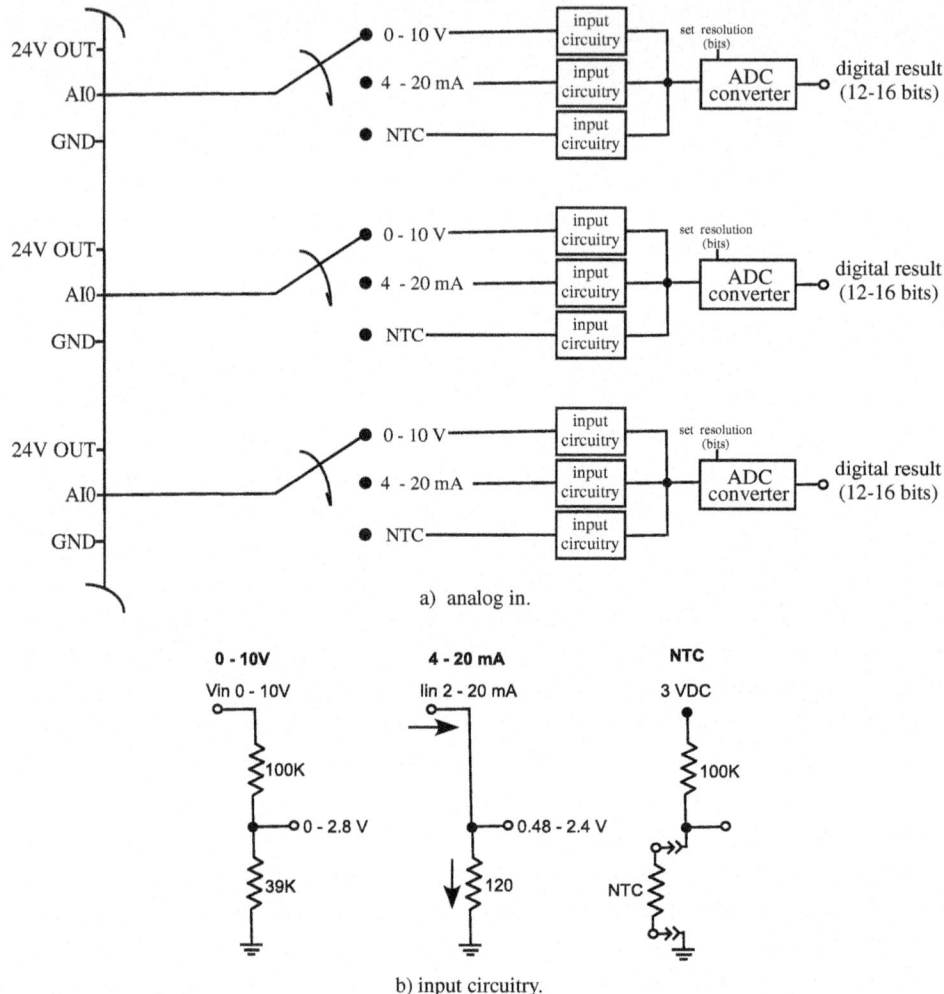

Fig. 2.5 PMC analog input channels. Images used courtesy of the Arduino Team (CC BY–NC–SA) (www.arduino.cc)

2.6.1 Analog-To-Digital Conversion (ADC)

A controller is used to process information from the natural world, use an algorithm to decide on a course of action based on the information collected, and then issue control signals to implement the decision.[1]

---

[1] This theory section is condensed and adapted with permission for the Arduino Portenta Machine Control from "Arduino II: Systems," S. Barrett, Morgan & Claypool Publishers, 2020.

2.6 Analog Input

Since the information from the natural world, is analog or continuous in nature, and the controller is a digital or discrete based processor, a method to convert an analog signal to a digital form is required. An ADC system performs this task while a digital–to–analog converter (DAC) performs the conversion in the opposite direction.

There are three important processes associated with the ADC process: sampling, quantization, and encoding.

Sampling. Sampling is the process of taking "snap shots" of a signal over time. When we sample a signal, we want to sample it in an optimal fashion such that we capture the essence of the signal while minimizing the use of memory resources. In essence, we want to minimize the number of samples while retaining the capability to faithfully reconstruct the original signal from the samples. Intuitively, the rate of change of a signal determines the number of samples required to faithfully reconstruct the signal.

Harry Nyquist from Bell Laboratory studied the sampling process and derived a criterion that determines the minimum sampling rate for a continuous analog signal. His, now famous, minimum sampling rate is known as the Nyquist sampling rate, which states that one must sample a signal at least twice as fast as the highest frequency content of the signal of interest.

For example, if we are dealing with a human voice signal that contains frequency components that span from about 20 Hz to 4 kHz, the Nyquist sample theorem requires that we must sample the signal at least at 8 kHz, 8000 "snap shots" every second. Figure 2.6 illustrates various sample rates.

When a signal is sampled a low pass anti–aliasing filter is employed to ensure the Nyquist sampling rate is not violated. In the voice example above, a low pass filter with a cutoff frequency of 4 kHz would be used before the sampling circuitry for this purpose.

Quantization. Each digital system has a number of bits it uses as the basic unit to represent data. A bit is the most basic unit where single binary information, one or zero, is represented.

Suppose you have a single bit to represent an incoming signal. You only have two different values, 0 and 1 for data representation. You may say that you can distinguish only low from high. Suppose you have two bits. You can represent four different levels, 00, 01, 10, and 11. What if you have three bits? You now can represent eight different levels: 000, 001, 010, 011, 100, 101, 110, and 111 as shown in Fig. 2.6. Similar discussion can lead us to conclude that given n bits, we have 2^n unique numbers or levels one can represent.

Figure 2.7 shows how n bits are used to quantize a range of values. In many digital systems, the incoming signals are voltage signals. The voltage signals are first obtained from physical signals (pressure, temperature, etc.) with the help of transducers, such as microphones, angle sensors, and infrared sensors.

The voltage signals are then conditioned to map their range with the input range of a digital system, typically 0–5 VDC for microcontrollers and 0–10 VDC for the Arduino PMC. In Fig. 2.7, n bits allow you to divide the input signal range of a digital system into 2^n different quantization levels. As can be seen from the figure, the more quantization levels means the better mapping of an incoming signal to its true value. As the number of bits

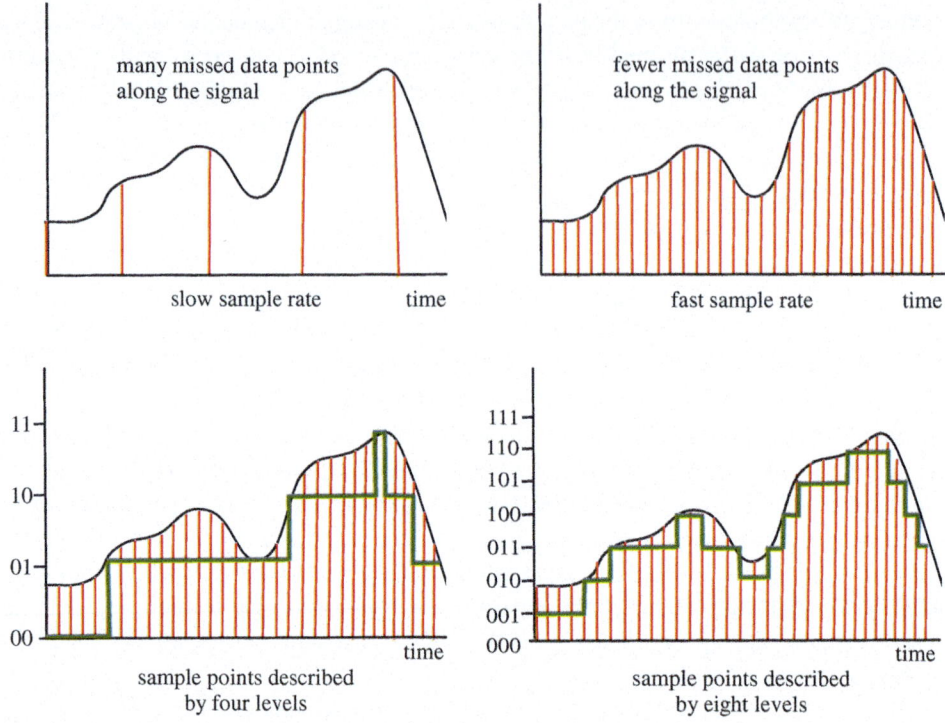

Fig. 2.6 Sampling rate

used for the quantization levels increases for a given input range the "distance" between two adjacent quantization levels decreases accordingly.

Encoding. Finally, the encoding process involves converting a quantized signal into a digital binary value. Suppose we are using eight bits to quantize a sampled analog signal. The quantization levels are determined by the eight bits and each sampled signal is quantized as one of 256 quantization levels. Consider the two sampled signals shown in Fig. 2.7. The first sample is mapped to quantization level two and the second one is mapped to quantization level 198. Note the amount of quantization error introduced for both samples. The quantization error is inversely proportional to the number of bits used to quantize the signal.

Once a sampled signal is quantized, the encoding process involves representing the quantization level with the available bits. Thus, for the first sample, the encoded sampled value is 0000_0010 (two), while the encoded sampled value for the second sample is 1100_0110 (198). As a result of the encoding process, sampled analog signals are now represented as a set of binary numbers. Thus, the encoding is the last necessary step to represent a sampled analog signal into its corresponding digital form, shown in Fig. 2.7.

2.6 Analog Input

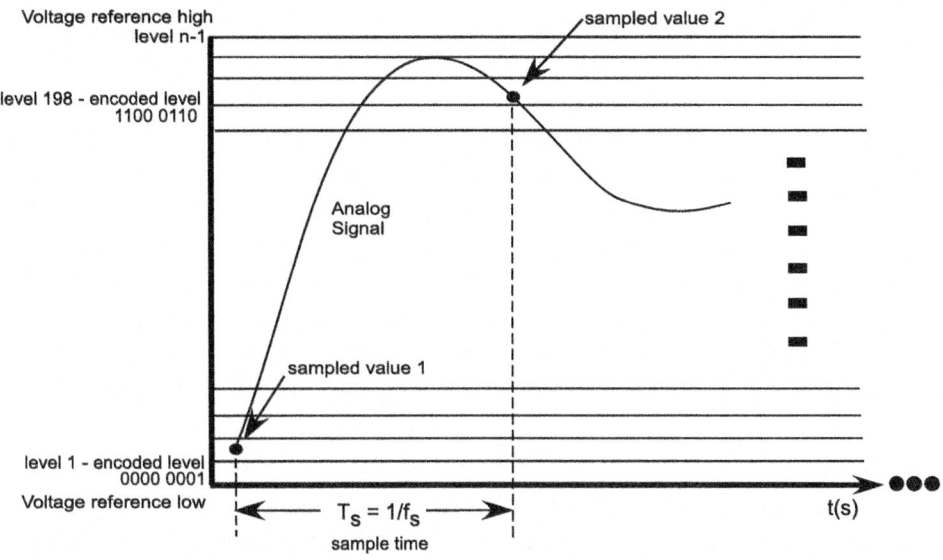

Fig. 2.7 Sampling, quantization, and encoding

Resolution. Resolution is a metric used to quantize an analog signal. Resolution is nothing more than the voltage "distance" between two adjacent quantization levels discussed earlier. The number of bits used for the quantization is directly proportional to the resolution of a system. In general, resolution may be defined as:

$$resolution = (voltage\ span)/2^b = (V_{ref\ high} - V_{ref\ low})/2^b$$

for the Arduino PMC, the best achievable resolution is:

$$resolution = (10 - 0)/2^{16} = 153\ \text{uV}$$

The desired resolution is chosen based on the specific requirements of the system. The ADC resolution for the Arduino PMC can be set from 12 to 16 bits.

2.6.2 Arduino PMC Examples

The Arduino PMC input pins are designated AI0, AI1, and AI2. Within the Arduino IDE environment they are designated 0, 1, and 2. The next three examples explore the ADC system used to convert a 0–10 VDC signal, a 4–20 mA signal, and an NTC based signal. In Chap. 4 we discuss a wide variety of input sensors. For now we use a DROK SG–03 Signal Generator to provide a 0–10 VDC or a 4–20 mA signal to the Arduino PMC analog input section. Figure 2.8 illustrates how each type of input is scaled by the PMC ADC circuitry.

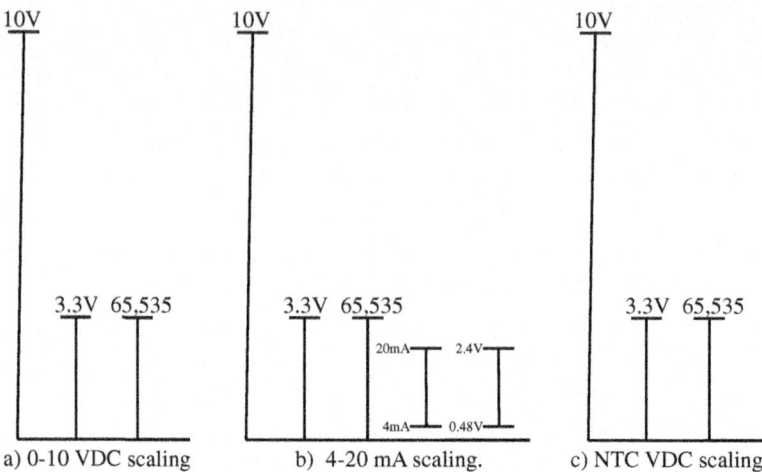

Fig. 2.8 PMC analog input channel scaling

2.6.2.1 Signal Conversion 0–10 VDC

In the sketch below an analog voltage between 0 and 10 VDC is converted via the onboard ADC. The resulting reading is converted to a voltage reading by the following equation:

$$measured\ voltage = ((ADC\ result)/(2^{16} - 1)) \times reference\ voltage$$

In the sketch the PMC ADC converter is set for 16 bit resolution. Note the additional 0.28 scaling factor to account for the resistive scaling network at the ADC input. Four sequential ADC conversions are taken and then averaged for noise reduction. The sketch is tested using a DROK SG–03 Signal Generator to provide a 0–10 VDC input to ADC channel 0 as shown in Fig. 2.9. The results are displayed on the serial monitor.

```
//*******************************************************************
//PMC_analog_input_0_10V_with_averaging
//
//- Each ANALOG IN channel is equipped with a 100k and 39k
//   resistor divider.  The input voltage is scaled by a 0.28 ratio
//- Maximum input voltage is 10V
//- Readings are averaged for noise reduction
//- DROK SG\---03 Signal Generator provides input to ADC channel 0
//
//Adapted from:
//Library: Arduino Portenta Machine Control
//Sketch: Analog_input_0_10V
//Authors: Riccardo Rizzo and Leonardo Cavagnis
//Adapted by: S. Barrett
//This code example is in the public domain.
//*******************************************************************
```

2.6 Analog Input

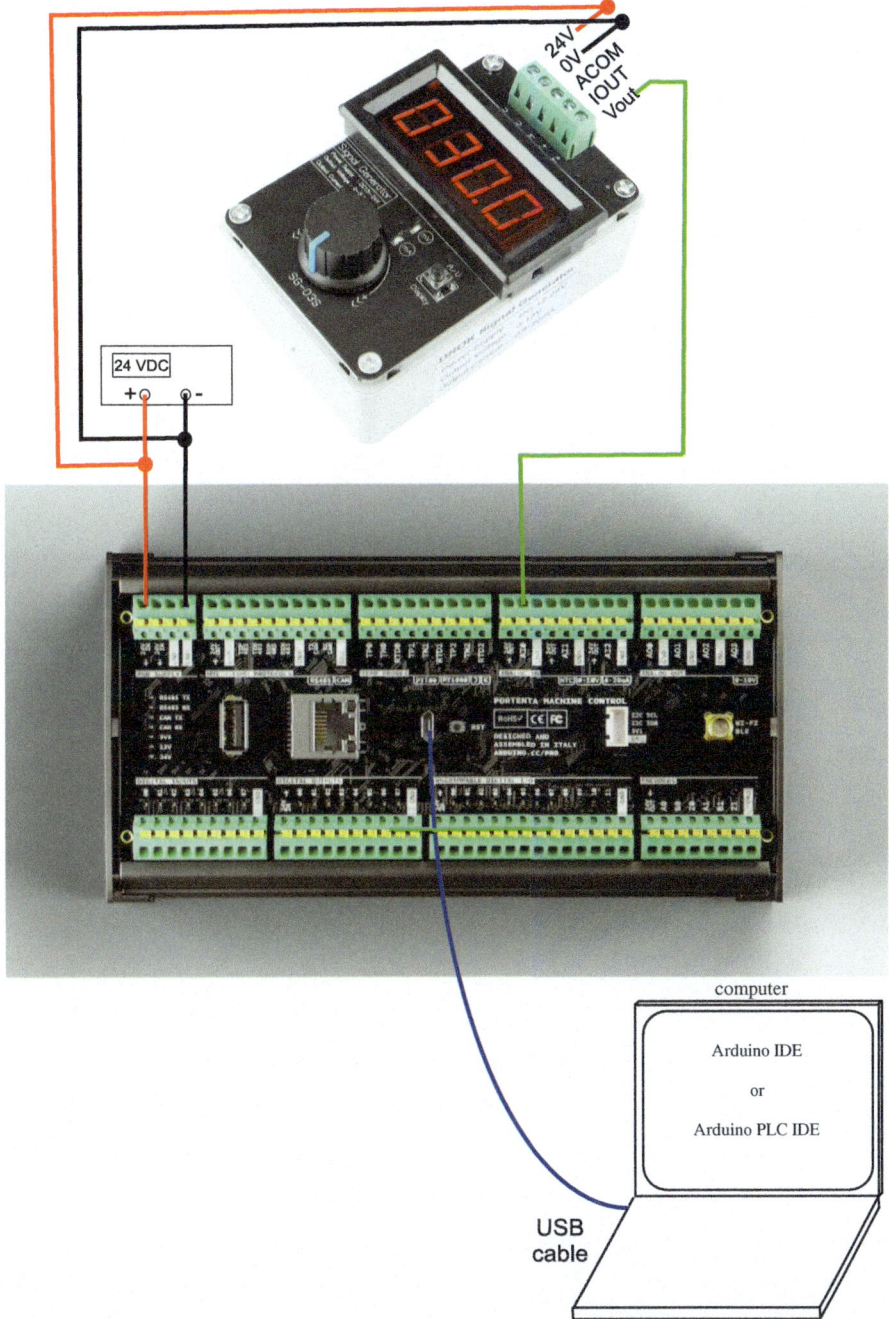

Fig. 2.9 Analog input with DROK signal generator. Images used courtesy of the Arduino Team (CC BY–NC–SA) (www.arduino.cc)

```
#include <Arduino_PortentaMachineControl.h>

const float RES_DIVIDER = 0.28057;
const float REFERENCE   = 3.0;
unsigned int analog_reading;

void setup()
{
Serial.begin(9600);
while(!Serial)
   {
   ; //wait for serial port connection
   }
                                         //set for 0-10 VDC in
MachineControl_AnalogIn.begin(SensorType::V_0_10);
}

void loop()
{
float raw_voltage_ch0 = MachineControl_AnalogIn.read(0);
float voltage_ch0 = (raw_voltage_ch0 * REFERENCE) / 65535 / RES_DIVIDER;
Serial.print("Voltage CH0: ");
Serial.print(voltage_ch0, 3);
Serial.println("V");

voltage_ch0 = 0.0;
for(analog_reading = 0; analog_reading <=3; analog_reading++)
   {
   float raw_voltage_ch0 = MachineControl_AnalogIn.read(0);
   float voltage_ch0_new = (raw_voltage_ch0 * REFERENCE) / 65535 / RES_DIVIDER;
   voltage_ch0 = voltage_ch0 + voltage_ch0_new;
   delay(5);
   }
voltage_ch0 = voltage_ch0/4.0;           //average results

Serial.print("Voltage CH0 average: ");
Serial.print(voltage_ch0, 3);
Serial.println("V");
Serial.println(" ");

delay(500);
}

//*********************************************************************
```

2.6.2.2 Signal Conversion 4–20 mA

Many industrial based sensors measure a physical variable and provide a corresponding signal from 4–20 mA. When the PMC ADC is configured for 4–20 mA reading, the signal current is routed through a 120 Ω resistor. The resulting voltage is routed to the ADC input for conversion. In the following sketch a DROK SG–03 Signal Generator provides a 4–20 mA input to ADC channel 0. The results are displayed on the Serial Monitor

2.6 Analog Input

```
//**********************************************************************
//PMC_analog_input_4_20mA
//
//- Each channel of ANALOG IN is equipped with 120 ohm resistor to GND.
//- Current from a 4-20mA sensor passes generates a voltage.
//- DROK SG\---03 Signal Generator provides input to ADC channel 0
//
//Adapted from:
//Library: Arduino Portenta Machine Control
//Sketch: Analog_input_0_20mA
//Authors: Riccardo Rizzo and Leonardo Cavagnis
//Adapted by: S. Barrett
//This code example is in the public domain.
//**********************************************************************

#include <Arduino_PortentaMachineControl.h>

#define SENSE_RES 120

const float REFERENCE = 3.0;

void setup()
{
Serial.begin(9600);
while (!Serial)
   {
   ; //wait for serial port connection
   }
                                          //4-20 mA mode
MachineControl_AnalogIn.begin(SensorType::MA_4_20);
}

void loop()
{
float raw_voltage_ch0 = MachineControl_AnalogIn.read(0);
float voltage_ch0 = (raw_voltage_ch0 * REFERENCE) / 65535;
float current_ch0 = (voltage_ch0 / SENSE_RES) * 1000;
Serial.print("Current CH0: ");
Serial.print(current_ch0);
Serial.println("mA");
Serial.println();
delay(250);
}

//**********************************************************************
```

2.6.2.3 Signal Conversion NTC

The PMC ADC negative temperature coefficient (NTC) mode is for thermistor measurements. We discuss the thermistor in Chapter 4. As its name implies the thermistor is a resistor whose value is determined by temperature. In this example we use a fixed 10 Kohm resistor as a stand in for the thermistor. The sketch below determines the resistor value by applying the voltage divider equation for series resistors. The resistor is connected between Analog Input AIO and ground. In Chap. 4 we discuss how to convert the resistance reading to a temperature.

```
//************************************************************
//PMS_analog_input_NTC
//
//- Provides the resistance value at ANALOG IN
//- 3V voltage REFERENCE connected to each channel of ANALOG IN
//- REFERENCE has 100k resistor in series
//- Voltage divider of 3V reference by 100K and unknown resistor
//- Unknown resistor value determined by solving voltage divider
//  equation for unknown resistor value
//
//Adapted from:
//Library: Arduino Portenta Machine Control
//Sketch: Analog_input_NTC
//Authors: Riccardo Rizzo and Leonardo Cavagnis
//This code example is in the public domain.
//************************************************************

#include <Arduino_PortentaMachineControl.h>

#define REFERENCE_RES 100000

const float REFERENCE = 3.0;
const float LOWEST_VOLTAGE = 2.7;

void setup()
{
Serial.begin(9600);
while(!Serial)
   {
   ; // wait for serial connection.
   }
                                            //set NTC mode
MachineControl_AnalogIn.begin(SensorType::NTC);
}

void loop()
{
float raw_voltage_ch0 = MachineControl_AnalogIn.read(0);
float voltage_ch0 = (raw_voltage_ch0 * REFERENCE) / 65535;
float resistance_ch0;
Serial.print("Resistance CH0: ");
if(voltage_ch0 < LOWEST_VOLTAGE)
   {
   resistance_ch0 = ((-REFERENCE_RES) * voltage_ch0) / (voltage_ch0 - REFERENCE);
   Serial.print(resistance_ch0);
   Serial.println(" ohm");
   }
else
   {
   resistance_ch0 = -1;
   Serial.println("NaN");
   }
```

```
Serial.println();
delay(250);
}

//*********************************************************************
```

2.7 Analog Output

As shown in Fig. 2.1, the Arduino PMC is equipped with four analog output channels designated ANALOG OUT AO0–AO3. Each channel may be configured for a 0–10 VDC signal with a maximum current of 20 mA. The DC output voltage is generated using a pulse width signal. The PWM signal is fed to an internal second order low pass filter with a voltage gain of 3.3.

In the next several sketches we use the test circuit of Fig. 2.10. Note how the LED output current is limited to 20 mA to comply with the maximum current rating of the Analog Output channels. In Chap. 4 we explore circuitry to boost the output current for specific examples.

In the following sketch four different analog output channels are set for four different voltages to illustrate the impact on LED intensity. The results captured with a DATAQ DI–1100 4–channel USB data acquisition kit is provided in Fig. 2.11.

```
//*********************************************************************
//PMC_analog_out
//
//- Configures four analog output channels with different voltages
//
//Adapted from:
//Library: Arduino Portenta Machine Control
//Sketch: analog_output
//Authors: Riccardo Rizzo and Leonardo Cavagnis
//Adapted by: S. Barrett
//This code example is in the public domain.
//*********************************************************************

#include <Arduino_PortentaMachineControl.h>

#define PERIOD_MS 4 /* 4ms - 250Hz */

float voltage = 2.0;

void setup()
{
Serial.begin(9600);
while (!Serial)
```

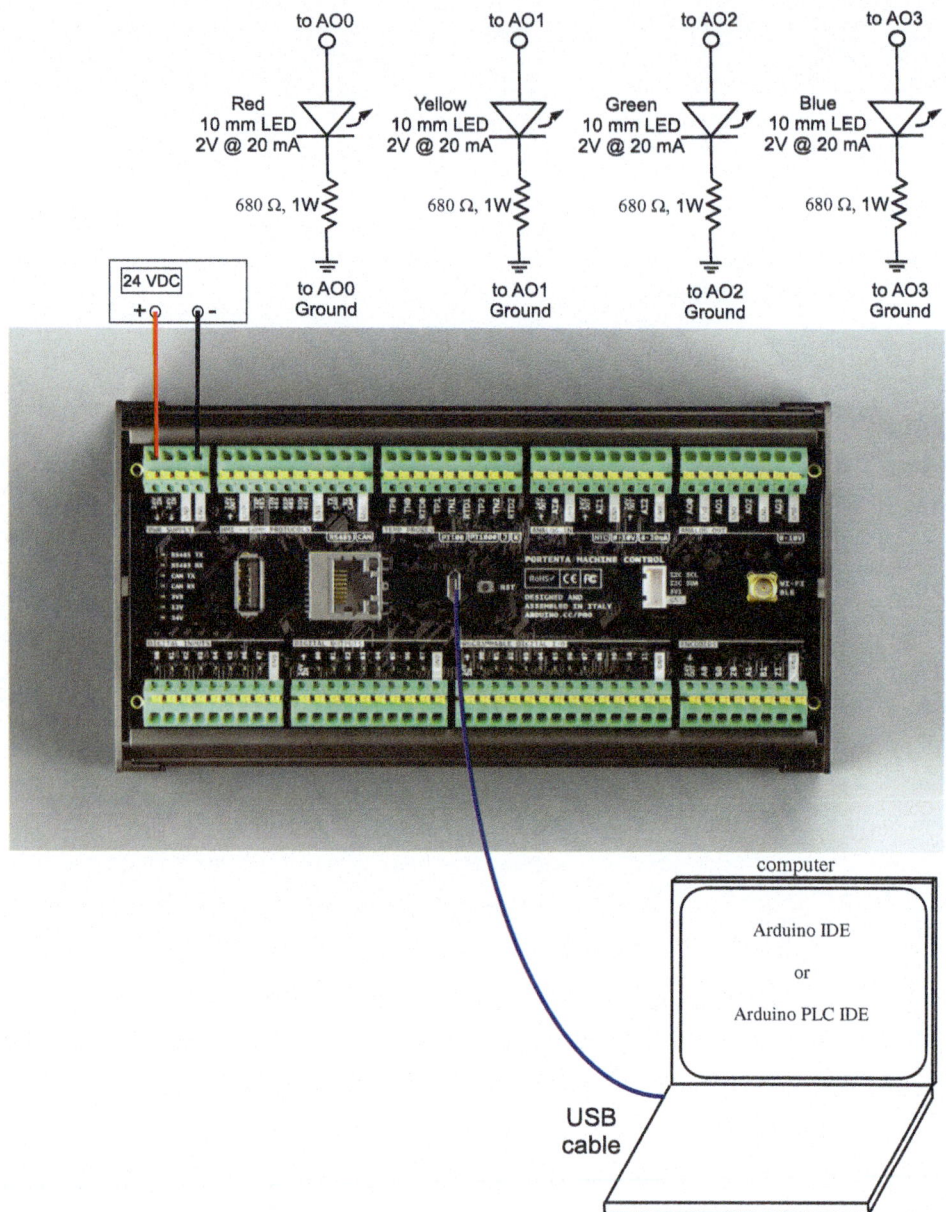

Fig. 2.10 PMC analog output channels. Images used courtesy of the Arduino Team (CC BY–NC–SA) (www.arduino.cc)

2.7 Analog Output

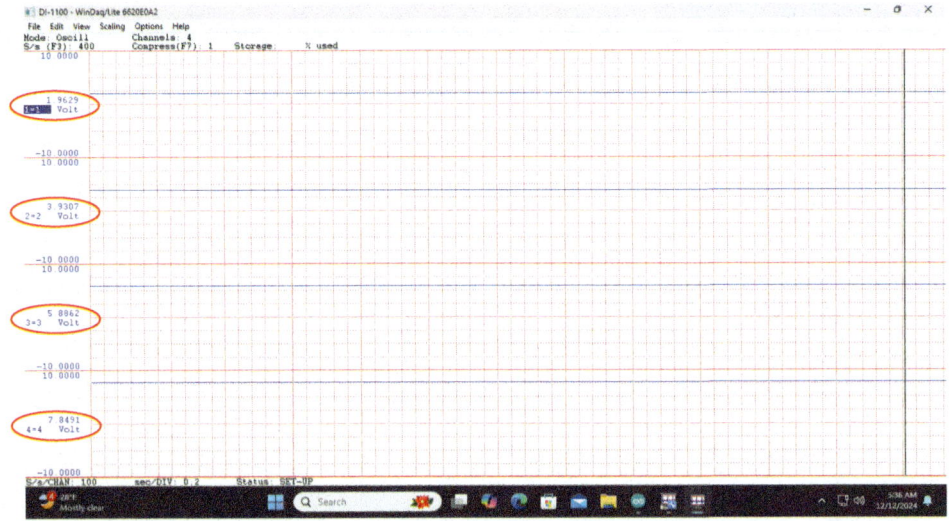

Fig. 2.11 PMC analog output results

```
  {
  ; //wait for serial port connection
  }
                      //initialize PMC analog out
MachineControl_AnalogOut.begin();

MachineControl_AnalogOut.setPeriod(0, PERIOD_MS);
MachineControl_AnalogOut.setPeriod(1, PERIOD_MS);
MachineControl_AnalogOut.setPeriod(2, PERIOD_MS);
MachineControl_AnalogOut.setPeriod(3, PERIOD_MS);
}

void loop()
{
MachineControl_AnalogOut.write(0, voltage +   0);
MachineControl_AnalogOut.write(1, voltage + 2.0);
MachineControl_AnalogOut.write(2, voltage + 4.0);
MachineControl_AnalogOut.write(3, voltage + 6.0);

Serial.println("CH0:" + String(voltage +   0) + "V");
Serial.println("CH1:" + String(voltage + 2.0) + "V");
Serial.println("CH2:" + String(voltage + 4.0) + "V");
Serial.println("CH3:" + String(voltage + 6.0) + "V");

delay(100);
}

//*************************************************************
```

In this sketch we generate a ramp voltage at Analog Out A0 to control the intensity of an external LED. This is accomplished by incrementing the output voltage by 0.1 V for each iteration of the for loop. The results captured with a DATAQ DI–1100 4–channel USB data acquisition kit is provided in Fig. 2.12.

```
//****************************************************************
//PMC_analog_out_ramp
//
//- Provides ramp output from 0 to 10V on ANALOG OUT A0 to vary
//LED intensity.
//
//Adapted from:
//Library: Arduino Portenta Machine Control
//Sketch: analog_output
//Authors: Riccardo Rizzo and Leonardo Cavagnis
//Adapted by: S. Barrett
//This code example is in the public domain.
//****************************************************************

#include <Arduino_PortentaMachineControl.h>

#define PERIOD_MS 4 /* 4ms - 250Hz */

float voltage = 0.0;
float voltage_step = 0.1;
unsigned int time_incr;

void setup()
{
Serial.begin(9600);
while (!Serial)
   {
   ; //wait for serial port connection
   }

MachineControl_AnalogOut.begin();

MachineControl_AnalogOut.setPeriod(0, PERIOD_MS);
}

void loop()
{
for(time_incr = 0; time_incr <= 100; time_incr++)
   {
   voltage = voltage + voltage_step;
   MachineControl_AnalogOut.write(0, voltage);
   Serial.println("CH0:" + String(voltage) + "V");
   delay(5);
```

2.7 Analog Output

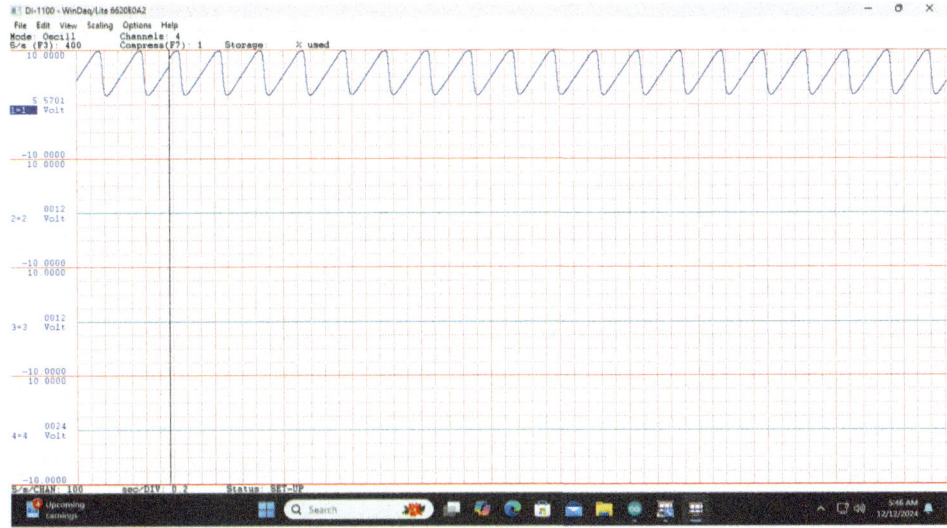

Fig. 2.12 PMC analog output results

```
   }
   voltage = 0.0;                        //reset for next loop
   delay(100);
}

//*******************************************************************
```

In the next sketch we send a ramp voltage sequentially to four different LEDs to illustrate how pulse width modulation techniques (PWM) may be used to control LED intensity as shown in Fig. 2.13a. The results captured with a DATAQ DI–1100 4–channel USB data acquisition kit is provided in Fig. 2.14.

```
//*******************************************************************
//PMC_analog_out_ramp_four
//
//- Provides ramp output for four LEDs from 0 to 10V on ANALOG OUT A0-A3
//
//Adapted from:
//Library: Arduino Portenta Machine Control
//Sketch: analog_output
//Authors: Riccardo Rizzo and Leonardo Cavagnis
//Adapted by: S. Barrett
//This code example is in the public domain
//*******************************************************************

#include <Arduino_PortentaMachineControl.h>

#define PERIOD_MS 4 /* 4ms - 250Hz */
```

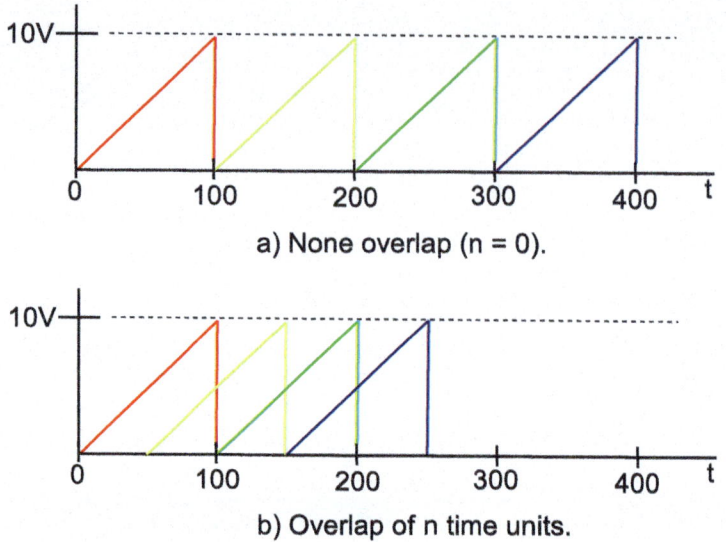

- When overlap set to n:
    - First signal (red) is stationary
    - Second signal (yellow) starts n time units early
    - Third signal (green) starts 2n time units early
    - Fourth signal (blue) starts 3n time units early

Fig. 2.13 PMC analog output ramp to four different LEDs

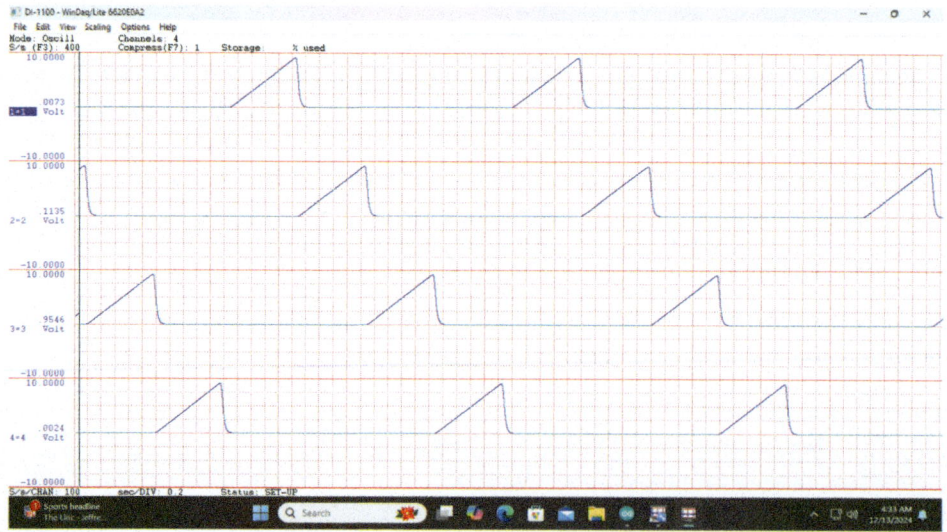

Fig. 2.14 PMC analog output results

2.7 Analog Output

```
float red_voltage=0.0, yellow_voltage=0.0, green_voltage=0.0, blue_voltage=0.0;
float voltage_step = 0.1;
unsigned int time_incr;

void setup()
{
Serial.begin(9600);
while (!Serial)
  {
  ; //wait for serial port connection
  }

MachineControl_AnalogOut.begin();

MachineControl_AnalogOut.setPeriod(0, PERIOD_MS);
MachineControl_AnalogOut.setPeriod(1, PERIOD_MS);
MachineControl_AnalogOut.setPeriod(2, PERIOD_MS);
MachineControl_AnalogOut.setPeriod(3, PERIOD_MS);
}

void loop()
{
for(time_incr = 1; time_incr <= 400; time_incr++)
  {
  if((time_incr >= 1)&&(time_incr <= 100))
    {
    red_voltage = red_voltage + voltage_step;
    if(time_incr == 100) red_voltage = 0.0;   //reset for next loop
    MachineControl_AnalogOut.write(0, red_voltage);
    Serial.println("CH0 (red):" + String(red_voltage) + "V");
    }
  if((time_incr >= 1)&&(time_incr <= 100))
    {
    red_voltage = red_voltage + voltage_step;
    if(time_incr == 100) red_voltage = 0.0;   //reset for next loop
    MachineControl_AnalogOut.write(0, red_voltage);
    Serial.println("CH0 (red):" + String(red_voltage) + "V");
    }
  if((time_incr >= 101)&&(time_incr <= 200))
    {
    yellow_voltage = yellow_voltage + voltage_step;
    if(time_incr == 200) yellow_voltage = 0.0;   //reset for next loop
    MachineControl_AnalogOut.write(1, yellow_voltage);
    Serial.println("CH1 (yellow):" + String(yellow_voltage) + "V");
    }
  if((time_incr >= 201)&&(time_incr <= 300))
    {
    green_voltage = green_voltage + voltage_step;
    if(time_incr == 300) green_voltage = 0.0;   //reset for next loop
    MachineControl_AnalogOut.write(2, green_voltage);
    Serial.println("CH2 (green):" + String(green_voltage) + "V");
    }
  if((time_incr >= 301)&&(time_incr <= 400))
    {
    blue_voltage = blue_voltage + voltage_step;
    if(time_incr == 400) blue_voltage = 0.0;   //reset for next loop
    MachineControl_AnalogOut.write(3, blue_voltage);
    Serial.println("CH3 (blue):" + String(blue_voltage) + "V");
```

```
    }
  delay(10);
  }
delay(100);
}
//******************************************************************
```

2.8 Real Time Clock

The Arduino PMC is equipped with a PCF8563T/F4 real time clock. This provides the PMC with the capability to keep calendar time in year, month, day, hour, etc. A 32,768 kHz crystal oscillator is the time base for the RTC. The oscillator frequency provides 1 Hz (one pulse per second) time base when passed through a 15 stage flip flop network. Each flip flop stage divides the input frequency by a factor of two. The RTC is equipped with a 100 mF capacitor to provide backup power for up to 48 h.

In the sketch below the RTC is set for a specified time and then reports time to the Serial Monitor every second. In an upcoming example we equip the PMC with an I2C based liquid crystal display (LCD) to display time.

```
//******************************************************************
//PMC_RTC
//Portenta Machine Control - RTC Example
//
//Demonstrates PMC RTC PCF8563T configuration and operation
//
//Adapted from:
//Library: Arduino Portenta Machine Control
//Sketch: RTC
//Authors: Riccardo Rizzo and Leonardo Cavagnis
//Adapted by: S. Barrett
//This code example is in the public domain.
//******************************************************************

#include <Arduino_PortentaMachineControl.h>

int year = 24;
int month = 4;
int day = 2;
int hours = 4;
int minutes = 30;
int seconds = 00;

void setup()
{
Serial.begin(9600);
while(!Serial)                    //wait for serial monitor
  {
  ;
```

2.9 Interrupts

```
    }
Serial.print("RTC Initialization");
if(!MachineControl_RTCController.begin())
    {
    Serial.println(" fail!");
    }
Serial.println(" done!");

//APIs to set date's fields: years, months, days, hours, minutes and seconds
//The RTC time can be set as epoch, using one of the following two options:
// - Calendar time: MachineControl_RTCController.setEpoch
//   (years,  months,  days, hours, minutes, seconds);
// - UTC time: MachineControl_RTCController.setEpoch(date_in_seconds);
MachineControl_RTCController.setYear(year);
MachineControl_RTCController.setMonth(month);
MachineControl_RTCController.setDay(day);
MachineControl_RTCController.setHours(hours);
MachineControl_RTCController.setMinutes(minutes);
MachineControl_RTCController.setSeconds(seconds);
MachineControl_RTCController.setEpoch();
}

void loop()
{
// APIs to get date's fields
Serial.print("Date: ");
Serial.print(MachineControl_RTCController.getYear());
Serial.print("/");
Serial.print(MachineControl_RTCController.getMonth());
Serial.print("/");
Serial.print(MachineControl_RTCController.getDay());
Serial.print(" - ");
Serial.print(MachineControl_RTCController.getHours());
Serial.print(":");
Serial.print(MachineControl_RTCController.getMinutes());
Serial.print(":");
Serial.println(MachineControl_RTCController.getSeconds());

time_t utc_time = time(NULL);
Serial.print("Date as UTC time: ");
Serial.println(utc_time);
Serial.println();

delay(1000);
}

//**********************************************************************
```

2.9 Interrupts

The interrupt system onboard a processor allows it to respond to higher priority events. Appropriate responses to these events are planned, but we do not know when these events will occur. When an interrupt event occurs, the processor will normally complete the instruction

Fig. 2.15 Processor interrupt response

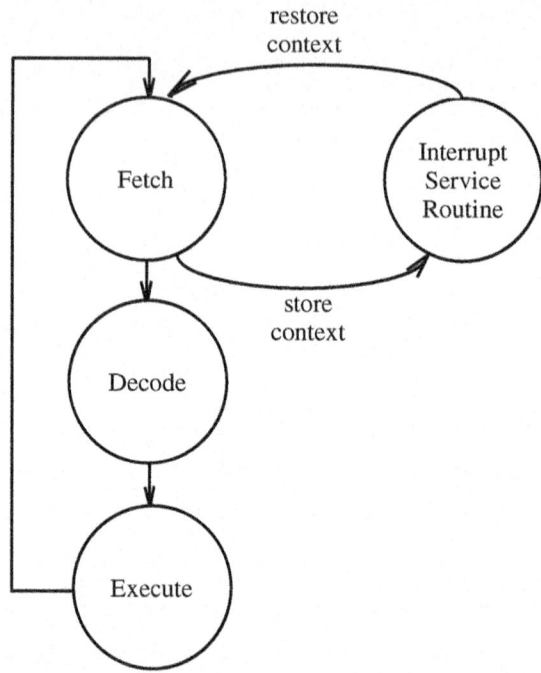

it is currently executing and then transition program control to interrupt event specific tasks. These tasks, which resolve the interrupt event, are organized into a function called an interrupt service routine (ISR). Each interrupt will normally have its own interrupt specific ISR. Once the ISR is complete, the processor will return to the main program where it left off before the interrupt event occurred (Fig. 2.15).

The Arduino Development Environment has four built–in functions to support external interrupts (www.arduino.cc).

These are the four functions:

- **interrupts()**. This function enables interrupts.
- **noInterrupts()**. This function disables interrupts.
- **attachInterrupt(interrupt, function, mode)**. This function links the interrupt to the appropriate interrupt service routine.
- **detachInterrupt(interrupt)**. This function turns off the specified interrupt.

The Arduino PMC is equipped with a complement of different interrupts including edge transition interrupts on the DIGITAL IN pins. The **attachInterrupt(interrupt, function, mode)** function is used to link the hardware pin to the appropriate interrupt service pin. The three arguments of the function are configured as follows:

2.9 Interrupts

- **interrupt**. Interrupt specifies the interrupt pin.
- **function**. Function specifies the name of the interrupt service routine.
- **mode**. Mode specifies what activity on the interrupt pin will initiate the interrupt: **LOW** level on pin, **CHANGE** in pin level, **RISING** edge, or **FALLING** edge.

In the following example a rising edge interrupt is configured to detect a transition on DIGITAL IN 00. Note the sketch interrupt configuration notes. The circuit provided in Fig. 2.2 is used in the next two sketches.

```
//***********************************************************************
//PMC_input_interrupt:  This example demonstrates the configuration
//of a Digital Input interrupt.
//
//Some notes:
// - PMC inputs are connected to an I2C IO expander (TC6424ARGJR).
// - In TOP level schematic note how the interrupt pin on the IO
//   expander (/DIG_IN_EXPANDER_INT) is connected to pin PB4.
// - Pin goes LOW when change (RISING or FALLING) occurs on an input.
// - Perform read operation to reset interrupt status on IO expander.
// - MBED does not allow read operation in interrupt callback.
// - Allowable interrupt response provided below for rising edge on
//   Digital In 00.
//
//Created by:   rreignier, Sep 20, 2022, Arduino Forum posting
//Adapted by:   S. Barrett, updated with Arduino_PortentaMachineControl.h
//This code example is in the public domain.
//***********************************************************************

#include <Arduino_PortentaMachineControl.h>
#include "Wire.h"

uint16_t  readings = 0;

volatile bool interruptOccured = false;

void inputCallback()
{
interruptOccured = true;
}

void setup()
{
Serial.begin(9600);
while(!Serial);                    //wait for serial monitor
Wire.begin();
attachInterrupt(PB_4, inputCallback, FALLING);

if(!MachineControl_DigitalInputs.begin())
  {
  Serial.println("Digital input GPIO expander initialization fail!");
  }
```

```
}
void loop()
{
if(interruptOccured)
   {
   interruptOccured = false;

   //Reading digital input resets the interrupt flag on the IO expander
   //Check the value to see if this is a RISING or FALLING edge
   readings = MachineControl_DigitalInputs.read(DIN_READ_CH_PIN_00);
   if(!readings)
      {
      Serial.println("Interrupt occured");
      }
   }
delay(1);
}

//*********************************************************************
```

2.9.1 Foreground and Background Processing

A sequential processor can only execute a single instruction at a time. It processes instructions in a fetch–decode–execute sequence as determined by the program and its response to external events. In many cases, a processor has to process multiple events seemingly simultaneously. How is this possible with a single sequential processor?

Normal processing accomplished by the processor is called foreground processing. An interrupt may be used to periodically break into foreground processing, 'steal' some clock cycles to accomplish another event called background processing, and then return processor control back to the foreground process.

As an example, a processor controlling access for an electronic door must monitor input commands from a user and generate the appropriate pulse width modulation (PWM) signals to open and close the door. Once the door is in motion, the controller must monitor door motor operation for obstructions, malfunctions, and other safety related parameters. This may be accomplished using interrupts. In this scenario, the processor is responding to user input status in the foreground while monitoring safety related status in the background using interrupts as illustrated in Fig. 2.16.

Example: As an example, we configure DIGITAL INPUT 00 as an interrupt. During normal operation PMC DIGITAL OUTPUT 00 has an external LED flashing at 100 ms intervals. When the button is pressed, an interrupt service routine (ISR) is called and sequentially illuminates LEDs on PMC DIGITAL OUTPUTS 2, 3, and 4 at 100 ms intervals. Within the

2.9 Interrupts

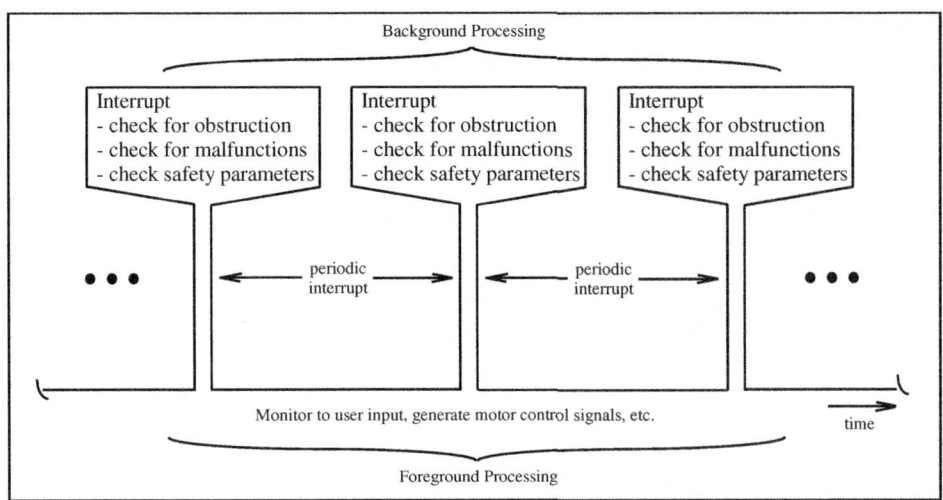

Fig. 2.16 Interrupt used for background processing. The processor responds to user input status in the foreground while monitoring safety related status in the background using interrupts

ISR multiple NOP (no operation) instructions are used to generate time delays. Each NOP requires a single clock cycle for execution.

```
//*****************************************************************
//PMC_foreground_background:  This example illustrates foreground
//versus background processing for an interrupt.  Arduino PMC
//Digital Input 00 is configured as a rising edge interrupt.
//In main foreground program, LED connected to Digital Output 00
//flashes at 1 second interval.  When button is pressed, ISR
//executes to sequentially illuminate LEDs on PMC Digital
//Output 01, 02, 03.
//
//Some notes:
// - PMC inputs are connected to an I2C IO expander (TC6424ARGJR).
// - In TOP level schematic note how the interrupt pin on the IO
//    expander (/DIG_IN_EXPANDER_INT) is connected to pin PB4.
// - Pin goes LOW when change (RISING or FALLING) occurs on an input.
// - Perform read operation to reset interrupt status on IO expander.
// - MBED does not allow read operation in interrupt callback.
// - Allowable interrupt response provided below for rising edge on
//    Digital In 00.
//
//Created by:  rreignier, Sep 20, 2022, Arduino Forum posting
//Adapted by:  S. Barrett
//This code example is in the public domain,
//*****************************************************************
```

```
#include <Arduino_PortentaMachineControl.h>
#include "Wire.h"

uint16_t    readings = 0;
unsigned long int i;

volatile bool interruptOccured = false;

void inputCallback()
{
interruptOccured = true;
}

void setup()
{
Serial.begin(9600);
while(!Serial);
Wire.begin();
attachInterrupt(PB_4, inputCallback, FALLING);

if(!MachineControl_DigitalInputs.begin())
   {
   Serial.println("Digital input GPIO expander initialization fail!");
   }
}

void loop()
{                                   //foreground processing
MachineControl_DigitalOutputs.write(0, HIGH);
for (i=0; i<=24000000; i++)         //0.1s delay 240e5 clock cycles
   {
   asm("nop");                      //1 clock cycle
   }

MachineControl_DigitalOutputs.write(0, LOW);
for (i=0; i<=24000000; i++)         //0.1s delay 240e5 clock cycles
   {
   asm("nop");                      //1 clock cycle
   }

if(interruptOccured)
   {
   interruptOccured = false;

   //Reading digital input resets the interrupt flag on the IO expander
   //Check the value to see if this is a RISING or FALLING edge

   readings = MachineControl_DigitalInputs.read(DIN_READ_CH_PIN_00);
```

2.9 Interrupts

```
    if(!readings)
      {
      Serial.println("Interrupt occured");
      }

    //background processing
    MachineControl_DigitalOutputs.write(1, HIGH);
    for(i=0; i<=24000000; i++)        //0.1s delay 240e5 clock cycles
      {
      asm("nop");                     //1 clock cycle
      }

    MachineControl_DigitalOutputs.write(1, LOW);
    for(i=0; i<=24000000; i++)        //0.1s delay 240e5 clock cycles
      {
      asm("nop");                     //1 clock cycle
      }

    MachineControl_DigitalOutputs.write(2, HIGH);
    for(i=0; i<=24000000; i++)        //0.1s delay 240e5 clock cycles
      {
      asm("nop");                     //1 clock cycle
      }

    MachineControl_DigitalOutputs.write(2, LOW);
    for(i=0; i<=24000000; i++)        //0.1s delay 240e5 clock cycles
      {
      asm("nop");                     //1 clock cycle
      }

     MachineControl_DigitalOutputs.write(3, HIGH);
     for(i=0; i<=24000000; i++)       //0.1s delay 240e5 clock cycles
      {
      asm("nop");                     //1 clock cycle
      }

     MachineControl_DigitalOutputs.write(3, LOW);
     for(i=0; i<=24000000; i++)       //0.1s delay 240e5 clock cycles
      {
      asm("nop");                     //1 clock cycle
      }
    }
delay(1);
}
//****************************************************************
```

2.10 System and Data Integrity

The PMC may be used in an industrial environment to control complex processes and machinery. It is essential the PMC operates correctly within potentially noisy environments. Cybersecurity techniques presented here are an introduction to a wide variety of techniques to maintain system integrity within harsh environments. The same processes may be used to prevent system tampering from a nefarious actor.[2]

2.10.1 Cyclic Redundancy Check (CRC)

In a previous professional life, the author served as a missileer in the United States Air Force. On a routine basis, the guidance set aboard an assigned missile was updated with critical data to ensure the missile would serve its intended mission. Maintenance crews from a nearby support base would transport information tapes out to the missile site where the onboard missile guidance set was updated. A CRC checksum was generated on the information tapes before they left the support base. After the information was loaded from the tapes to the missile guidance set, a CRC checksum was performed. If the checksum generated by the missile guidance set matched the checksum generated at the support base, the missile was designated as properly updated.

This scenario illustrates the application and importance of using a Cyclic Redundancy Check or CRC to maintain data integrity. This technique is often performed to ensure the integrity of transmitted or stored data.

The basic concept behind generating a CRC checksum is binary division. The basic operation of division can be defined as (FIPS–197):

$$Dividend/divisor = quotient + remainder$$

The block of data to be protected via the checksum is considered the dividend. The dividend is divided by a pre-selected CRC polynomial which serves as the divisor. At the completion of the division operation, a quotient and a remainder result. The remainder of the operation serves as the CRC checksum.

Generation of a checksum is based on the concept that when a given block of data is divided by a specific polynomial initialized with the same value (seed), the same checksum will result every time the operation is performed. Similarly, if the input data is different or in a different order, the polynomial is changed, or seeded with a different initial value, a

---

[2] This section was adapted with permission from "Embedded Systems Design with the Texas Instruments MSP432 32–bit Processor," D. Dang, D. J. Pack, and S. F. Barrett, Morgan and Claypool Publishers, 2017.

2.10 System and Data Integrity

different checksum will result. A number of common polynomials have been developed to support CRC checksum generation. Two common ones include (SLAU367O) :

- CRC16–CCITT defined as $f(x) = X^{15} + X^{12} + X^5 + 1$
- CRC32–IS3309 defined as

$$f(x) = X^{32} + X^{26} + X^{23} + X^{22} + \\ X^{16} + X^{12} + X^{11} + X^{10} + X^8 + \\ X^7 + X^5 + X^4 + X^2 + X + 1 \quad (2.1)$$

A hardware CRC generator uses a linear feedback shift register (LFSR) to generate the checksum. The polynomial divisor chosen to generate the checksum specifies the hardware connection for the LFSR. For example, the LFSR configuration for the CRC16–CCITT polynomial is shown in Fig. 2.17. Note how the polynomial terms specify the output connections of certain flip–flops within the LFSR. To generate the checksum, the LFSR is initially configured to the seed value. The data block is fed in as a serial data stream. The resulting remainder is used as the checksum and appended to the original data block for transmission. The operation of the hardware based checksum generator can also be implemented with a software algorithm (SLAU367O).

In the following sketch we use a CRC 16 example from the Arduino CRC Library.

```
//*****************************************************************
//FILE: CRC16_test.ino
//AUTHOR: Rob Tillaart
//PURPOSE: demo
//URL: \url{https://github.com/RobTillaart/CRC}
//Source: Arduino CRC Library
//This code example is in the public domain
//*****************************************************************

#include "CRC16.h"
#include "CRC.h"

char str[24] =  "123456789";

CRC16 crc;

void setup()
{
Serial.begin(115200);                   //initialize serial monitor
while (!Serial)                         //wait for serial monitor to open
  {
  ;
  }

//Serial.println(__FILE__);
//Serial.println("Verified with - \url{http://zorc.breitbandkatze.de/crc.html} \n");

test();
```

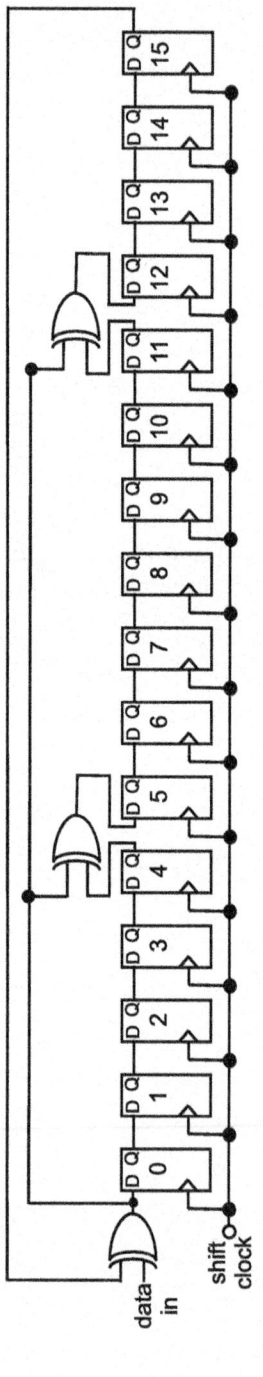

Fig. 2.17 CRC16–CCITT polynomial and LFSR configuration (SLAU367O)

```
}

void loop()
{
}

//*********************************************************************
//void test - performs CRC16 calculations using three different
//            approaches.
//*********************************************************************

void test()
{                                    //CRC calculation 1
                                     //contains default CRC polynomial
Serial.print("CRC calc 1: ");
Serial.println(calcCRC16((uint8_t *) str, 9), HEX);
Serial.println(" ");

                                     //CRC calculation 2
                                     //add single value to CRC calculation
crc.add((uint8_t*)str, 9);           //performed 9 times
Serial.print("CRC calc 2: ");
Serial.println(crc.calc(), HEX);     //report result to serial monitor
Serial.println(" ");

crc.restart();                       //reset internal CRC count

                                     //CRC calculation 3
                                     //incremental steps report to serial
                                     //monitor
for(int i = 0; i < 9; i++)           //perform operation nine times
  {
  crc.add(str[i]);                   //add single value to CRC calculation
  Serial.print("CRC calc iteration: ");
  Serial.println(crc.count());
  Serial.print("CRC calc: ");
  Serial.println(crc.calc(), HEX);

  }
}

//   -- END OF FILE --
//*********************************************************************
```

Provided in Fig. 2.18 provides a CRC16 calculation using three different approaches.

2.11 Advanced Encryption Standard (AES)

For transmitting and receiving data, data may be encrypted at the transmitter and decrypted at the receiver. AES uses the Rijndael cryptographic algorithm. The algorithm allows the encryption of a 128–bit plain text data block into a corresponding size cipher text block. The

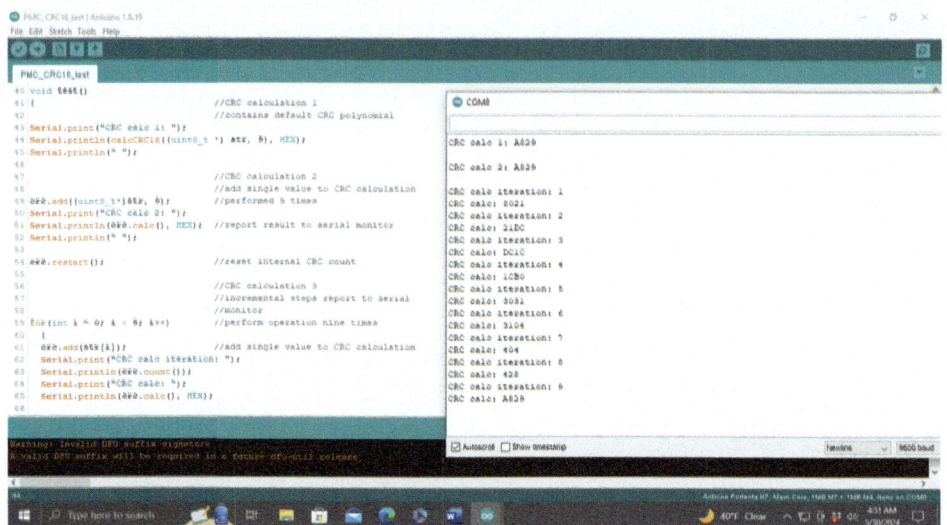

Fig. 2.18 CRC16 results

data may then be transmitted in an encrypted format and decrypted using a similar algorithm at the receiving end (FIPS–197, SLAU367O).

The data algorithm uses a 128, 192, or 256–bit cipher key to encrypt the plain text data block. The length of the cipher key determines the number of rounds (10, 12, or 14 respectively) of encryption performed on the plain text data to transform it into the cipher text block. The basic encryption process is shown in Figure 2.19 a, b. The plain text 128-–bit block is formatted into a state block. The state block then goes through a series of transformation rounds including an initial round, the sub–byte round, the shift rows round, the mix columns round, the add key round, and the final round to encrypt the data. As shown in Fig. 2.19b, a specific round key is derived from the original cipher key and used in a given round (FIPS–197, SLAU367O).

In the following sketch we use a AES128_Basics example from the Arduino CryptoAES_CBC Library.

```
//****************************************************************************
//PMC_AES128_Basics
//
//Adapted from:
//Library: Arduino CryptoAES_CBC
//Sketch: AES128_Basics
//
//This example explains basic AES128 implementation.
//
//AES128_Basics Copyright (C) 2015 Southern Storm Software, Pty Ltd.
//Permission is hereby granted, free of charge, to any person obtaining a
//copy of this software and associated documentation files (the "Software"),
//to deal in the Software without restriction, including without limitation
```

2.11 Advanced Encryption Standard (AES)

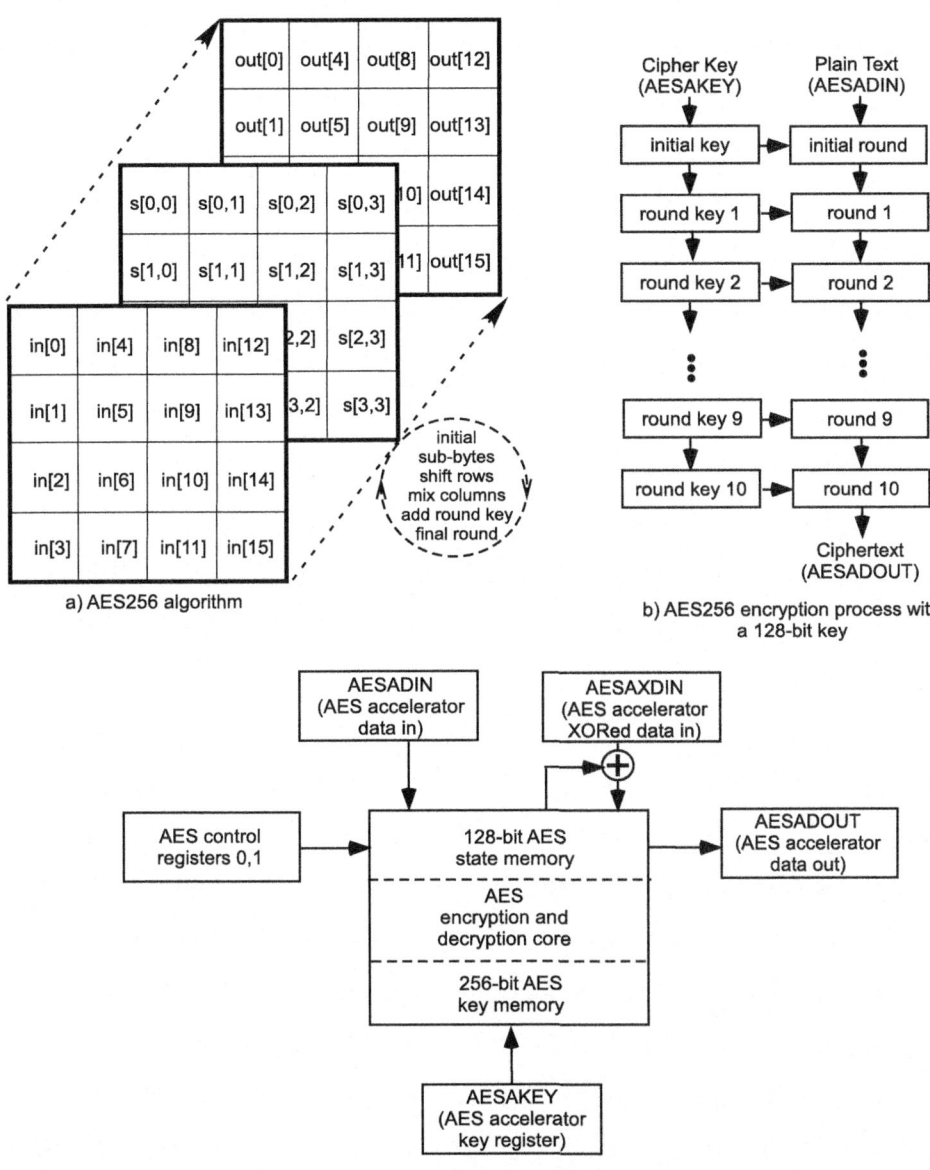

Fig. 2.19 AES256 encryption process. **a** AES256 algorithm. **b** AES256 encryption process with 128–bit key. (SLAU367O)

```
//the rights to use, copy, modify, merge, publish, distribute, sublicense,
//and/or sell copies of the Software, and to permit persons to whom the
//Software is furnished to do so, subject to the following conditions:
//- The above copyright notice and this permission notice shall be included
//  in all copies or substantial portions of the Software.
//- THE SOFTWARE IS PROVIDED "AS IS", WITHOUT WARRANTY OF ANY KIND, EXPRESS
//  OR IMPLIED, INCLUDING BUT NOT LIMITED TO THE WARRANTIES OF MERCHANTABILITY,
//  FITNESS FOR A PARTICULAR PURPOSE AND NONINFRINGEMENT. IN NO EVENT SHALL THE
//  AUTHORS OR COPYRIGHT HOLDERS BE LIABLE FOR ANY CLAIM, DAMAGES OR OTHER
//  LIABILITY, WHETHER IN AN ACTION OF CONTRACT, TORT OR OTHERWISE, ARISING
//  FROM, OUT OF OR IN CONNECTION WITH THE SOFTWARE OR THE USE OR OTHER
//  DEALINGS IN THE SOFTWARE.
//This code example is in the public domain.
//****************************************************************************

#include <CryptoAES_CBC.h>
#include <AES.h>
#include <string.h>

//key[16] contains 16 byte key(128 bit) for encryption
byte key[16]={0x00, 0x01, 0x02, 0x03, 0x04, 0x05, 0x06, 0x07,
              0x08, 0x09, 0x0A, 0x0B, 0x0C, 0x0D, 0x0E, 0x0F};
//plaintext[16] contains the text we need to encrypt
byte plaintext[16]={0x00, 0x11, 0x22, 0x33, 0x44, 0x55, 0x66,
              0x77,0x88, 0x99, 0xAA, 0xBB, 0xCC, 0xDD, 0xEE, 0xFF};
//cypher[16] stores the encrypted text
byte cypher[16];
//decryptedtext[16] stores decrypted text after decryption
byte decryptedtext[16];

//creating an object of AES128 class
AES128 aes128;

void setup()
{
Serial.begin(9600);
aes128.setKey(key,16);              //Setting Key for AES

while(!Serial)                      //wait to open serial monitor
  {
  ;
  }

Serial.print("Before Encryption:");
for(int i=0; i<sizeof(plaintext); i++)
  {
  Serial.print(plaintext[i]);
  Serial.print("\t");
  }
                                    //cypher->output block and plaintext->input block
aes128.encryptBlock(cypher,plaintext);
Serial.println();
Serial.print("After Encryption:");
for(int j=0;j<sizeof(cypher);j++)
  {
  Serial.print(cypher[j]);
  Serial.print("\t");
  }

aes128.decryptBlock(decryptedtext,cypher);
Serial.println();
Serial.print("After Dencryption:");
for(int i=0; i<sizeof(decryptedtext); i++)
```

2.12 Application I: Programming Practice

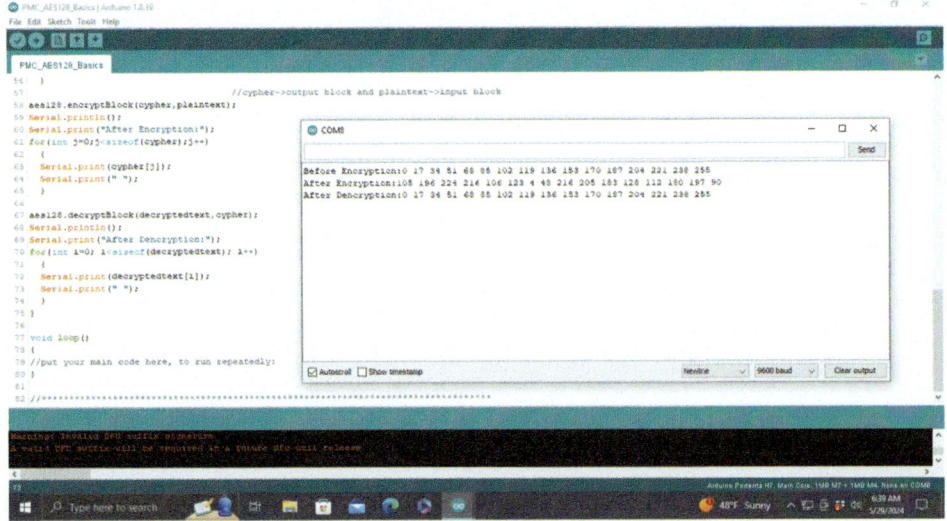

Fig. 2.20 AES128 basic results

```
  {
  Serial.print(decryptedtext[i]);
  Serial.print("\t");
  }
}

void loop()
{
//put your main code here, to run repeatedly:
}
//*******************************************************************
```

Provided in Fig. 2.20 shows the data before encryption, after encryption, and after decryption.

2.12 Application I: Programming Practice

Just for fun we provide another sketch where the ramp voltages and hence the LED intensity variation overlaps as shown in Fig. 2.13b. The results captured with a DATAQ DI–1100 4–channel USB data acquisition kit is provided in Fig. 2.21.

```
//*******************************************************************
//PMC_analog_out_ramp_four_overlap
//
//- Provides ramp output for four LEDs from 0 to 10V on ANALOG OUT A0-A3
//
```

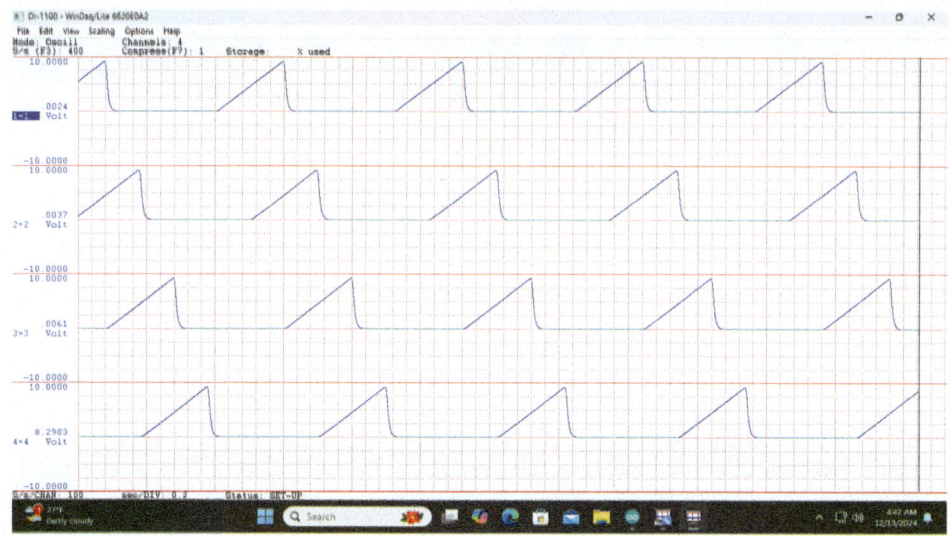

Fig. 2.21 PMC analog output results

```
//Adapted from Arduino Portenta Machine Control Library
//Sketch: analog_output
//Authors: Riccardo Rizzo and Leonardo Cavagnis
//Adapted by: S. Barrett
//This code example is in the public domain.
//**********************************************************************

#include <Arduino_PortentaMachineControl.h>

#define PERIOD_MS 4 /* 4ms - 250Hz */

float red_voltage=0.0, yellow_voltage=0.0, green_voltage=0.0, blue_voltage=0.0;
float voltage_step = 0.1;
unsigned int time_incr;
unsigned int overlap = 50;

void setup()
{
Serial.begin(9600);
while (!Serial)
  {
  ; //wait for serial port connection
  }

MachineControl_AnalogOut.begin();

MachineControl_AnalogOut.setPeriod(0, PERIOD_MS);
MachineControl_AnalogOut.setPeriod(1, PERIOD_MS);
MachineControl_AnalogOut.setPeriod(2, PERIOD_MS);
MachineControl_AnalogOut.setPeriod(3, PERIOD_MS);
}
```

```
void loop()
{
for(time_incr = 1; time_incr <= (400 - (3 * overlap)); time_incr++)
  {
  if((time_incr >= 1)&&(time_incr <= 100))
    {
    red_voltage = red_voltage + voltage_step;
    if(time_incr == 100) red_voltage = 0.0;    //reset for next loop
    MachineControl_AnalogOut.write(0, red_voltage);
    Serial.println("CH0 (red):" + String(red_voltage) + "V");
    }
    if((time_incr >= (101 - overlap))&&(time_incr <= (200 - overlap)))
    {
    yellow_voltage = yellow_voltage + voltage_step;
    if(time_incr == (200 - overlap)) yellow_voltage = 0.0;   //reset for next loop
    MachineControl_AnalogOut.write(1, yellow_voltage);
    Serial.println("CH1 (yellow):" + String(yellow_voltage) + "V");
    }
  if((time_incr >= (201 - (2*overlap)))&&(time_incr <= (300- (2*overlap))))
    {
    green_voltage = green_voltage + voltage_step;
    if(time_incr == 300 - (2 * overlap)) green_voltage = 0.0;    //reset for next loop
    MachineControl_AnalogOut.write(2, green_voltage);
    Serial.println("CH2 (green):" + String(green_voltage) + "V");
    }
  if((time_incr >= (301- (3*overlap)))&&(time_incr <= (400 - (3*overlap))))
    {
    blue_voltage = blue_voltage + voltage_step;
    if(time_incr == 400 - (3*overlap)) blue_voltage = 0.0;   //reset for next loop
    MachineControl_AnalogOut.write(3, blue_voltage);
    Serial.println("CH3 (blue):" + String(blue_voltage) + "V");
    }
  delay(10);
  }
delay(100);
}
//*****************************************************************
```

Modify the sketch such that the amount of overlap is determined by a potentiometer setting.

2.13 Application II: RGB Color

The RGB color wheel shown in Fig. 2.22a) provides for multiple colors. This is accomplished by mixing different amounts of the three primary colors (Red, Green, Blue). In this exercise we use three LEDs (R, G, B) and a ping pong ball as a light diffuser to mix different color components to obtain other colors.

Write a sketch and develop a circuit to demonstrate RGB color mixing.

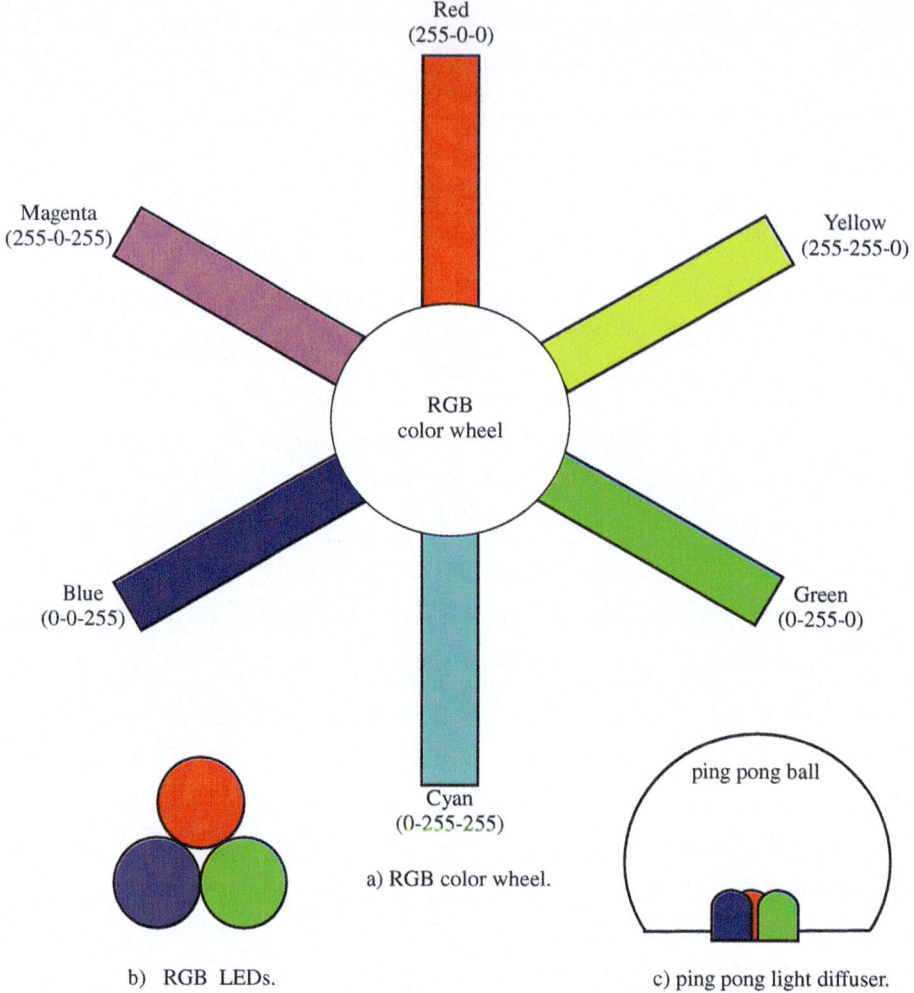

Fig. 2.22 RGB color

2.14 Summary

We began the chapter with a brief review of the Arduino Programmable Machine Control. We explored PMC features and employed them to explore fundamental input/output control concepts. Throughout the chapter we provided illustrative examples.

2.15 Chapter Problems

1. What is the difference between a microcontroller and a programmable logic controller?
2. Describe the PMC host processor features.
3. Construct a feature table for the Arduino PMC.
4. What are the voltage and current characteristics of the PMC digital inputs and outputs.
5. What is a DIN rail? How is it used in industrial control applications?
6. What is switch bouncing? How is it corrected in PLC applications?
7. Create UML activity diagrams for all chapter sketches.
8. What is switch bouncing? How is it countered in a PMC application? What impact is there on a system/algorithm if a switch input is not debounced?
9. An analog input signal is routed to an ADC. The result is then provided to a DAC. Will the resulting analog voltage equal the original input analog signal? Explain.
10. What is the effect on ADC resolution for increasing the number of bits per sample? Sampling the analog input signal more frequently?
11. Discuss the voltage and current specification of the PMC analog output features. Provide examples of devices that can be directly driven from the output.
12. What is the purpose of a Real Time Clock (RTC)?
13. Discuss the tradeoffs of using polling versus interrupt techniques.
14. Provide several scenarios where it would be helpful to use foreground/background processing techniques.
15. What is the difference between CRC and AES techniques. Provide examples where each and a combination of both would be helpful.
16. Discuss the three different ADC input configurations.

References

Federal Information Processing Standards Publication 197 (FIPS–197). November 26, 2001.
M. Fowler with K. Scott "UML Distilled– A Brief Guide to the Standard Object Modeling Language," 2nd edition. Boston: Addison–Wesley, 2000. hero.
Hanes D., G. Salgueiro, P. Grossetete, R. Barton, J. Henry (2017) *IoT Fundamentals–Networking Technologies, Protocols, and Use Cases for the Internet of Things*, Cisco Press.
Horowitz P, Hill W (2015) The Art of Electronics, third edition, Cambridge University Press.
MSP430FR58xx, MSP430FR59xx, and MSP430FR6xx Family User's (SLAU367O). Texas Instruments: 2017.
Stenerson, J. (2004) *Fundamentals of Programmable Logic Controllers, Sensors, and Communications*, Pearson Prentice Hall.

Connectivity 3

Objectives: After reading this chapter, the reader should be able to do the following:

- Describe the differences between serial and parallel communication;
- Provide definitions for serial communications terminology;
- Describe the purpose of the Inter–Integrated Circuit (I2C) protocol;
- Describe the basic concepts supporting Bluetooth communications;
- Describe the operation of the Universal Synchronous and Asynchronous Receiver and Transmitter (UART) and related RS–232 concepts;
- Describe the basic concepts supporting Ethernet and Wi–Fi;
- Describe the importance of and basic concepts supporting the Controller Area Network (CAN);
- Describe the basic concepts and unique features of RS–485; and
- Implement a communication link for a specific technology.

3.1 Overview

This chapter describes methods of connecting the Portenta Machine Control (PMC) to external peripheral devices and other processors via different communication links. Figure 3.1 provides a summary of communication technology supported by the PMC. For each range[1] we provide the communication network type and also implementation technologies. The

---

[1] Note the range scale is not linear but a logarithm scale.

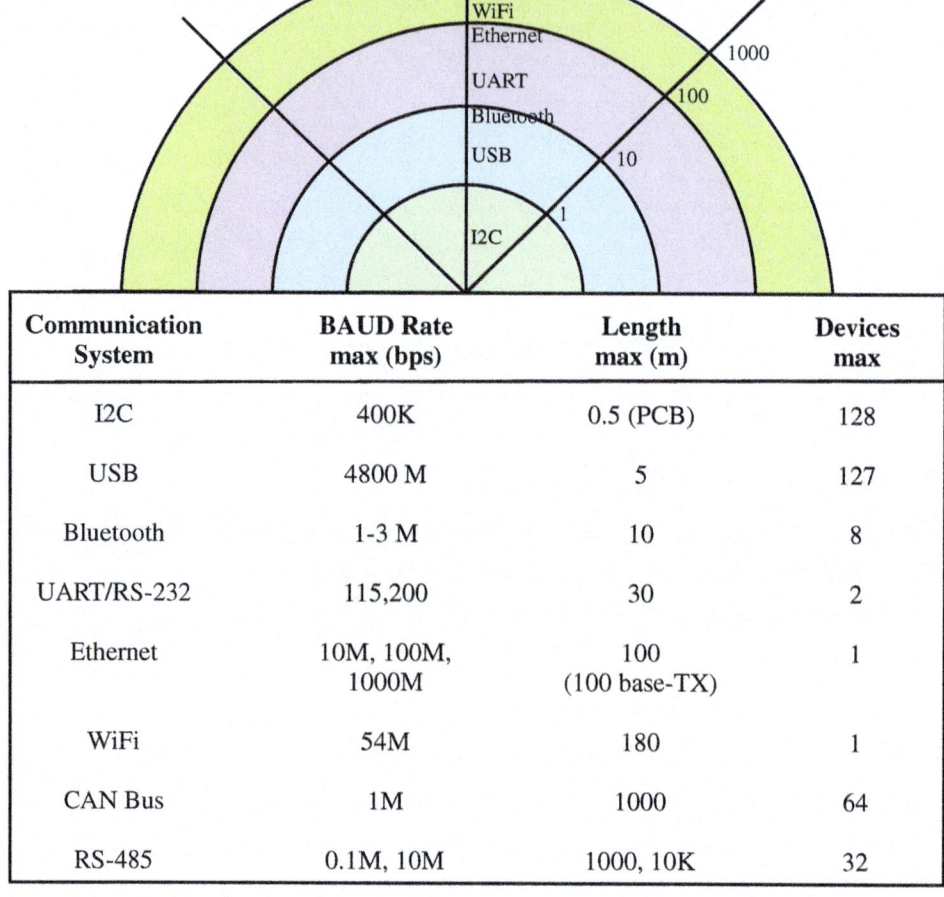

Fig. 3.1 Connectivity (Hanes, Horowitz and Hill)

chapter begins with a brief description of serial communications and related serial communication terminology. We then review different connectivity technologies beginning with close range technologies, then mid–range technologies, and conclude with long range technologies.

3.2 Serial Communications

Serial communication techniques provide a vital link between a processor and certain input devices, output devices, and other processors. In this chapter, we investigate the serial communication features beginning with a review of serial communication concepts and terminology.[2]

Processors must often exchange data with other processors or peripheral devices. Data may be exchanged by using parallel or serial techniques. With parallel techniques, an entire byte of data is typically sent simultaneously from the transmitting device to the receiver device. While this is efficient from a time point of view, it requires eight separate lines (or more) for the data transfer.

In serial transmission, a byte of data is sent a single bit at a time. Once eight bits have been received at the receiver, the data byte is reconstructed. While this is inefficient from a time point of view, it only requires a line (or two) to transmit the data. Before discussing the different serial communication features aboard the PMC, we review serial communication terminology.

3.3 Serial Communication Terminology

In this section, we review common terminology associated with serial communication.

Asynchronous versus Synchronous Serial Transmission: In serial communications, the transmitting and receiving device must be synchronized to one another and use a common data rate and protocol. Synchronization allows both the transmitter and receiver to be expecting data transmission/reception at the same time. There are two basic methods of maintaining "sync" between the transmitter and receiver: asynchronous and synchronous.

In an asynchronous serial communication system, framing bits are used at the beginning and end of a data byte. These framing bits alert the receiver that an incoming data byte has arrived and also signals the completion of the data byte reception. The data rate for an asynchronous serial system is typically much slower than the synchronous system, but it only requires a single wire (or two) between the transmitter and receiver.

A synchronous serial communication system maintains "sync" between the transmitter and receiver by employing a common clock between the two devices. Data bits are sent and received on the edge of the clock. This allows data transfer rates higher than with asynchronous techniques but requires two lines (or more), data and clock, to connect the receiver and transmitter.

Baud rate: Data transmission rates are typically specified as a Baud or bits per second rate. For example, 9600 Baud indicates the data is being transferred at 9600 bits per second.

---

[2] The serial communication theory was adapted for the Arduino Portenta Machine Control with permission from "Microcontroller Fundamentals for Engineers and Scientists," S. F. Barrett and D. J. Pack, Morgan and Claypool Publishers, 2006.

Full Duplex: Often serial communication systems must both transmit and receive data. To do both transmission and reception, simultaneously, requires separate hardware for transmission and reception. A half duplex system has a single complement of hardware that must be switched from transmission to reception configuration. A full duplex serial communication system has separate hardware for transmission and reception.

Non–return to Zero (NRZ) Coding Format: There are many different coding standards used within serial communications. The important point is the transmitter and receiver must use a common coding standard so data may be interpreted correctly at the receiving end. In NRZ coding a logic one is signaled by a logic high during the entire time slot allocated for a single bit; whereas, a logic zero is signaled by a logic low during the entire time slot allocated for a single bit.

Parity: To further enhance data integrity during transmission, parity techniques may be used. Parity is an additional bit (or bits) that may be transmitted with the data byte. With a single parity bit, a single bit error may be detected. Parity may be even or odd. In even parity, the parity bit is set to one or zero such that the number of ones in the data byte including the parity bit is even. In odd parity, the parity bit is set to one or zero such that the number of ones in the data byte including the parity bit is odd. At the receiver, the number of bits within a data byte including the parity bit are counted to ensure that parity has not changed, indicating an error, during transmission. Additional parity bits may be used to detect and correct errors.

ASCII: The American Standard Code for Information Interchange or ASCII is a standardized, seven bit method of encoding alphanumeric data. It has been in use for many decades, so some of the characters and actions listed in the ASCII table are not in common use today. However, ASCII is still the most common method of encoding alphanumeric data. The ASCII code is provided in Fig. 3.2. For example, the capital letter "G" is encoded in ASCII as 0x47. The "0x" symbol indicates the hexadecimal number representation. Unicode is the international counterpart of ASCII. It provides standardized 16–bit encoding format for the written languages of the world. ASCII is a subset of Unicode. The interested reader is referred to the Unicode home page website, www.unicode.org, for additional information on this standardized encoding format.

In the remainder of this chapter we examine different methods the PMC can communicate with external devices.

3.4 Inter–Integrated Circuit (I2C)

The I2C subsystem allows the system designer to connect a number of I2C configured devices (microcontrollers, transducers, displays, etc.) together into a system using a two–wire interconnecting scheme. The I2C allows a maximum of 128 devices to be connected together. Each device has its own unique address and may both transmit and receive over the two–wire bus at frequencies up to 400 kHz. This allows the device to freely exchange

3.4 Inter-Integrated Circuit (I2C)

		Most significant digit							
		0x0_	0x1_	0x2_	0x3_	0x4_	0x5_	0x6_	0x7_
Least significant digit	0x_0	NUL	DLE	SP	0	@	P	`	p
	0x_1	SOH	DC1	!	1	A	Q	a	q
	0x_2	STX	DC2	"	2	B	R	b	r
	0x_3	ETX	DC3	#	3	C	S	c	s
	0x_4	EOT	DC4	$	4	D	T	d	t
	0x_5	ENQ	NAK	%	5	E	U	e	u
	0x_6	ACK	SYN	&	6	F	V	f	v
	0x_7	BEL	ETB	'	7	G	W	g	w
	0x_8	BS	CAN	(	8	H	X	h	x
	0x_9	HT	EM	)	9	I	Y	i	y
	0x_A	LF	SUB	*	:	J	Z	j	z
	0x_B	VT	ESC	+	;	K	[	k	{
	0x_C	FF	FS	,	<	L	\	l	|
	0x_D	CR	GS	-	=	M	]	m	}
	0x_E	SO	RS	.	>	N	^	n	~
	0x_F	SI	US	/	?	O	_	o	DEL

Fig. 3.2 ASCII code. The ASCII code is used to encode alphanumeric characters. The "0x" indicates hexadecimal notation in the C programming language

information with other devices in a small area network. The I2C is alternately known as the Two Wire Interface (TWI) protocol (Philips). I2C was originally developed for connecting multiple devices within the area of approximately 12 by 12 in. (25 by 25 cm). Its range may be extended up to one to two meters by reducing the data rate.

An overview of the I2C system is shown in Fig. 3.3. Devices within the small area network are connected by two wires to share data (SDA) and a common clock (SCL). Pullup resistors are required on each of the lines. I2C compatible devices are connected to the SCL and SDA lines as shown.

The I2C system is a state machine to control the "hand shaking" protocol between the I2C master(s) and the multiple slave devices on the I2C bus. If the system contains more than one master designated device, arbitration detection and resolution protocols prevent bus contention. Each slave device has a unique seven–bit address to allow one–to–one communication using the Address Match Unit and address comparator. The I2C bus frequency should not exceed 400 kHz. The bus frequency is derived from the microcontroller clock signal and scaling hardware. The I2C system also includes signal conditioning features for the SCL and SDA pins including slew rate and spike control. The I2C system is configured and controlled using a series of registers.

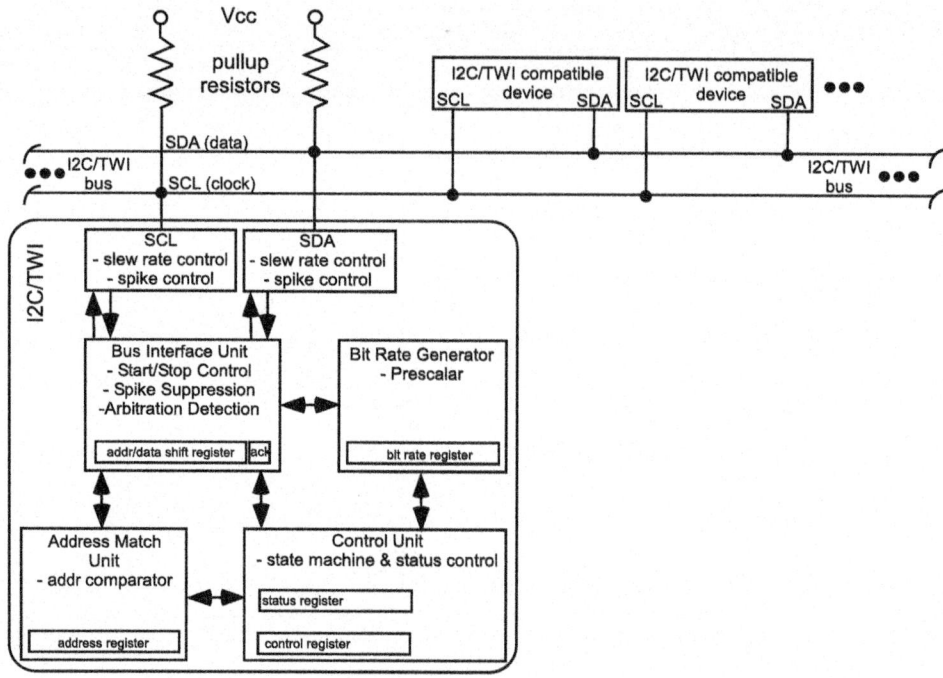

Fig. 3.3 I2C/TWI system overview (Microchip)

3.4.1 Programming the I2C Subsystem with the Arduino IDE

The Arduino IDE uses the "Wire" instruction set for I2C/TWI communications. The command set includes:

- Wire.begin(address): allows I2C/TWI compatible controller or peripheral device to join the I2C/TWI bus.
- Wire.end(): electronically disconnects the I2C/TWI compatible controller or peripheral device from the I2C/TWI bus.
- Wire.beginTransmission(address): begins a transmission to the I2C/I2C peripheral device with the specified address.
- Wire.endTransmission(): ends a transmission to the peripheral device that was begun by Wire.beginTransmission(address).
- Wire.write(): writes data to a peripheral device.

3.4 Inter-Integrated Circuit (I2C)

3.4.2 I2C LCD

To connect a microcontroller to a standard parallel configured LCD requires multiple pins. Typically, eight data lines and several control lines are required. The LCD2004 (20 characters x 4 lines) provides an I2C backpack to provide for a serial link between the Arduino PMC and the LCD and hence conserve precious digital I/O pins. The connections between the Arduino PMC and the LCD2004 is shown in Fig. 3.4.

There are four connections required for the backpack as shown in the figure:

- 5 V to 5 VDC from the voltage regulator circuit.
- PMC GND to LCD ground via the Adafruit Grove connector (#5244).
- PMC I2C SDA to LCD SDA via the Adafruit Grove connector (#5244).
- PMC I2C SCL to LCD SCL via the Adafruit Grove connector (#5244).

The Arduino Library supporting the backpack, HD44870 by Bill Perry, may be downloaded using the Library Manager from within the Arduino IDE. With hardware connections complete, download the library, and upload "HelloWorld_I2C." Adjust the contrast control as necessary to visualize the LCD characters.

```
//************************************************************************
//HelloWorld - simple demonstration of lcd
//Created by Bill Perry 2016-07-02
//bperrybap@opensource.billsworld.billandterrie.com
//
//This example code is unlicensed and is released into the public domain.
//************************************************************************

#include <Wire.h>
#include <hd44780.h>                      //main hd44780 header
#include <hd44780ioClass/hd44780_I2Cexp.h> //i2c expander i/o class header

hd44780_I2Cexp lcd2(0x27); //declare lcd object: auto locate & auto config expander chip

//If you wish to use an i/o expander at a specific address, you can specify the
//i2c address and let the library auto configure it. If you don't specify
//the address, or use an address of zero, the library will search for the
//i2c address of the device.
//hd4480_I2Cexp lcd2;
//hd44780_I2Cexp lcd2(0x27);

// LCD geometry
const int LCD_COLS = 20;
const int LCD_ROWS = 4;

void setup()
{
int status;

status = lcd2.begin(LCD_COLS, LCD_ROWS);
if(status) // non zero status means it was unsuccesful
  {
  hd44780::fatalError(status); // does not return
```

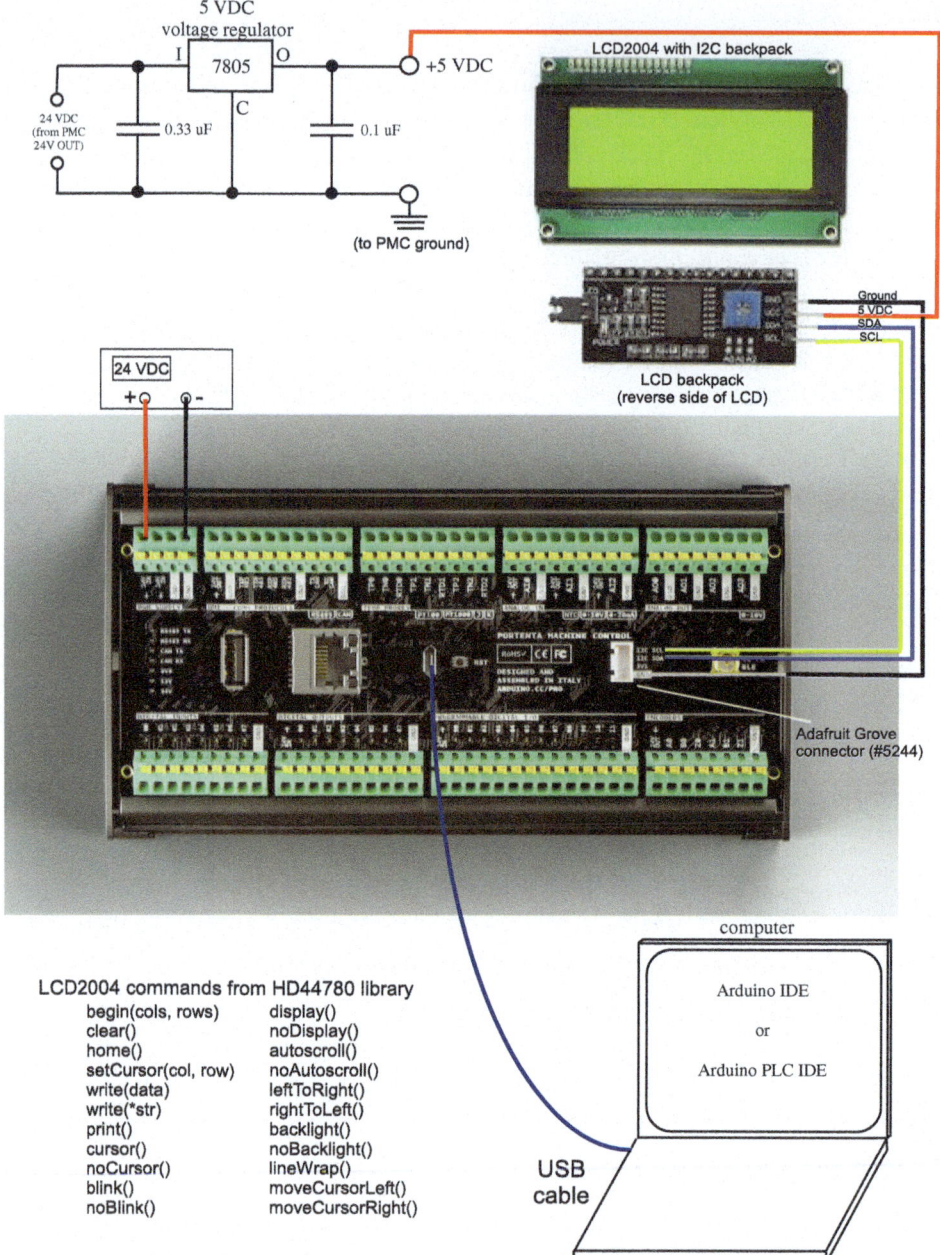

Fig. 3.4 PMC to LCD2004 connection via I2C. Images used courtesy of the Arduino Team (CC BY–NC–SA) (www.arduino.cc)

3.4 Inter-Integrated Circuit (I2C)

```
}

//Print a message to the LCD
lcd2.print("Hello, World!");
}

void loop()
{

}
//*****************************************************************
```

In Chap. 2 we discussed the Real Time Clock features of the PMC. In the following sketch we display Real Time Clock values to the LCD2004.

```
//*****************************************************************
//PMC_RTC_LCD_I2C
//Portenta Machine Control - RTC Example
//
//Demonstrates PMC RTC PCF8563T configuration and operation
//Displays curren time on LCD2004
//
//Adapted from:
//Library: Arduino Portenta Machine Control
//Sketch: RTC
//Authors:
//- RTC: Riccardo Rizzo and Leonardo Cavagnis
//- LCD: Code created by Bill Perry 2016-07-02
//   bperrybap@opensource.billsworld.billandterrie.com
//Adapted by: S. Barrett
//This code example is in the public domain.
//*****************************************************************

#include <Arduino_PortentaMachineControl.h>
#include <Wire.h>
#include <hd44780.h>                      //main hd44780 header
#include <hd44780ioClass/hd44780_I2Cexp.h> //i2c expander i/o class header

hd44780_I2Cexp lcd2(0x27);                //declare lcd object
                                          //auto locate, config expander

int year = 24;
int month = 4;
int day = 2;
int hours = 5;
int minutes = 24;
int seconds = 00;

// LCD geometry
const int LCD_COLS = 20;
const int LCD_ROWS = 4;

void setup()
{
int status;

status = lcd2.begin(LCD_COLS, LCD_ROWS);
if(status) // non zero status means it was unsuccesful
  {
  hd44780::fatalError(status); // does not return
  }


Serial.begin(9600);
```

```
Serial.print("RTC Initialization");
if(!MachineControl_RTCController.begin())
  {
  Serial.println(" fail!");
  }
Serial.println(" done!");

//APIs to set date's fields: years, months, days, hours, minutes and seconds
//The RTC time can be set as epoch, using one of the following two options:
// - Calendar time: MachineControl_RTCController.setEpoch(years, months, days, hours, minutes, seconds);
// - UTC time: MachineControl_RTCController.setEpoch(date_in_seconds);
MachineControl_RTCController.setYear(year);
MachineControl_RTCController.setMonth(month);
MachineControl_RTCController.setDay(day);
MachineControl_RTCController.setHours(hours);
MachineControl_RTCController.setMinutes(minutes);
MachineControl_RTCController.setSeconds(seconds);
MachineControl_RTCController.setEpoch();
}

void loop()
{
//APIs to get date's fields
Serial.print("Date: ");
Serial.print(MachineControl_RTCController.getMonth());
Serial.print("/");
Serial.print(MachineControl_RTCController.getDay());
Serial.print("/");
Serial.print(MachineControl_RTCController.getYear());
Serial.print(" - ");
Serial.print(MachineControl_RTCController.getHours());
Serial.print(":");
Serial.print(MachineControl_RTCController.getMinutes());
Serial.print(":");
Serial.println(MachineControl_RTCController.getSeconds());

//Display on LCD2004
lcd2.setCursor(2,0);
lcd2.print("Real Time Clock");
lcd2.setCursor(0,1);
lcd2.print("Date:");
lcd2.print(MachineControl_RTCController.getMonth());
lcd2.print("/");
lcd2.print(MachineControl_RTCController.getDay());
lcd2.print("/");
lcd2.print(MachineControl_RTCController.getYear());
lcd2.print("-");
lcd2.print(MachineControl_RTCController.getHours());
lcd2.print(":");
lcd2.print(MachineControl_RTCController.getMinutes());
lcd2.print(":");
lcd2.print(MachineControl_RTCController.getSeconds());
delay(1000);
lcd2.clear();
}

//*******************************************************************
```

3.5 Universal Serial Bus (USB)

The Portenta Machine Control typically serves as a USB client (peripheral) in most applications. In these application the support laptop/PC serves as the USB host (controller). In this application the PMC is configured as the USB host with a keyboard as the client.

The configuration for this example is provided in Fig. 3.5. In the following sketch when either the R/r, G/g, or B/b keyboard keys are pressed; the corresponding LED is toggled on and off. The sketch is available from the Arduino Pro Tutorials Library in Portenta H7 as a USB Host → LED Keyboard Controller. The sketch must be slightly modified for use with the PMC as shown below. After testing the sketch you may need to press the PMC reset button (RST) to return the PMC to client configuration. This basic example can be extended to control complex PMC operations from a keyboard.

```
//****************************************************************************
//
//           _____                  _   _ ____  ____
//          |_   _|                | | | / ___|| __ \
//            | |  ___   ___  _ __ | | | \___ \| |_) |
//            | | / _ \ / _ \| '_ \| |_| |___) |  _ <
//            | ||  __/|  __/| | | |\___/|____/| |_) |
//           |_| \___| \___||_| |_|       \____/|____/
//                      _/  |
//                     |___/
//
// TeenyUSB - light weight usb stack for STM32 micro controllers
//
// Copyright (c) 2019 XToolBox    - admin@xtoolbox.org
//                       \url{www.tusb.org}
//
// Permission is hereby granted, free of charge, to any person obtaining a copy
// of this software and associated documentation files (the "Software"), to deal
// in the Software without restriction, including without limitation the rights
// to use, copy, modify, merge, publish, distribute, sublicense, and/or sell
// copies of the Software, and to permit persons to whom the Software is
// furnished to do so, subject to the following conditions:
//
// The above copyright notice and this permission notice shall be included in all
// copies or substantial portions of the Software.
//
// THE SOFTWARE IS PROVIDED "AS IS", WITHOUT WARRANTY OF ANY KIND, EXPRESS OR
// IMPLIED, INCLUDING BUT NOT LIMITED TO THE WARRANTIES OF MERCHANTABILITY,
// FITNESS FOR A PARTICULAR PURPOSE AND NONINFRINGEMENT. IN NO EVENT SHALL THE
// AUTHORS OR COPYRIGHT HOLDERS BE LIABLE FOR ANY CLAIM, DAMAGES OR OTHER
// LIABILITY, WHETHER IN AN ACTION OF CONTRACT, TORT OR OTHERWISE, ARISING FROM,
// OUT OF OR IN CONNECTION WITH THE SOFTWARE OR THE USE OR OTHER DEALINGS IN THE
// SOFTWARE.
//
//Adapted for Portenta Machine Control:  S. Barrett, June 1, 2024
//This code example is in the public domain.
//****************************************************************************

USBHost  usb;
#include "USBHost.h"
#include <Arduino_PortentaMachineControl.h>

static int process_key(tusbh_ep_info_t* ep, const uint8_t* key);

static const tusbh_boot_key_class_t cls_boot_key = {
  .backend = &tusbh_boot_keyboard_backend,
  .on_key = process_key
```

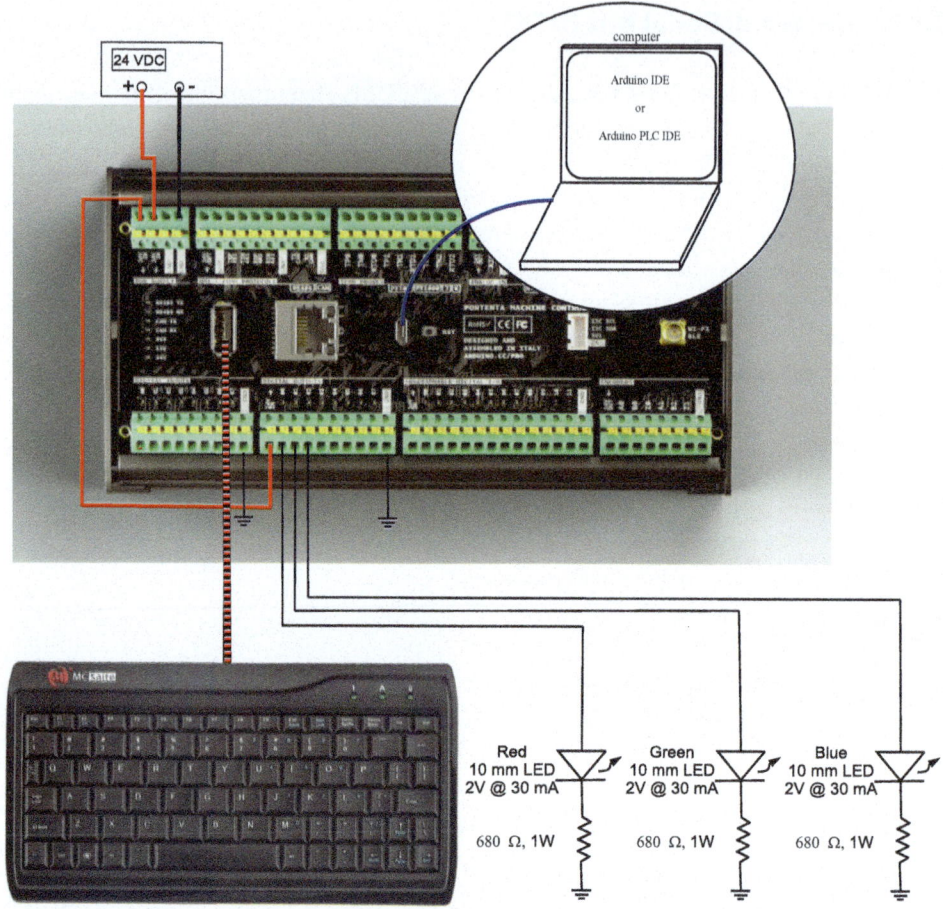

Fig. 3.5 PMC as USB host to keyboard client. Images used courtesy of the Arduino Team (CC BY–NC–SA) (www.arduino.cc)

```
};

static const tusbh_boot_mouse_class_t cls_boot_mouse = {
  .backend = &tusbh_boot_mouse_backend,
  // .on_mouse = process_mouse
};

static const tusbh_hid_class_t cls_hid = {
  .backend = &tusbh_hid_backend,
  //.on_recv_data = process_hid_recv,
  //.on_send_done = process_hid_sent,
};

static const tusbh_hub_class_t cls_hub = {
  .backend = &tusbh_hub_backend,
};
```

3.5 Universal Serial Bus (USB)

```
static const tusbh_vendor_class_t cls_vendor = {
  .backend = &tusbh_vendor_backend,
  //.transfer_done = process_vendor_xfer_done
};

int msc_ff_mount(tusbh_interface_t* interface, int max_lun, const tusbh_block_info_t* blocks);
int msc_ff_unmount(tusbh_interface_t* interface);

static const tusbh_msc_class_t cls_msc_bot = {
  .backend = &tusbh_msc_bot_backend,
  //   .mount = msc_ff_mount,
  //   .unmount = msc_ff_unmount,
};

static const tusbh_cdc_acm_class_t cls_cdc_acm = {
  .backend = &tusbh_cdc_acm_backend,
};

static const tusbh_cdc_rndis_class_t cls_cdc_rndis = {
  .backend = &tusbh_cdc_rndis_backend,
};

static const tusbh_class_reg_t class_table[] = {
  (tusbh_class_reg_t)&cls_boot_key,
  (tusbh_class_reg_t)&cls_boot_mouse,
  (tusbh_class_reg_t)&cls_hub,
  (tusbh_class_reg_t)&cls_msc_bot,
  (tusbh_class_reg_t)&cls_cdc_acm,
  (tusbh_class_reg_t)&cls_cdc_rndis,
  (tusbh_class_reg_t)&cls_hid,
  (tusbh_class_reg_t)&cls_vendor,
  0,
};

bool ledRstate = false;
bool ledGstate = false;
bool ledBstate = false;

void setup()
{
Serial1.begin(115200);
MachineControl_USBController.begin();

usb.Init(USB_CORE_ID_FS, class_table);
//usb.Init(USB_CORE_ID_HS, class_table);

pinMode(LEDR, OUTPUT);
pinMode(LEDG, OUTPUT);
pinMode(LEDB, OUTPUT);
MachineControl_DigitalOutputs.begin();

//Turn off the LEDs
//Onboard Portenta H7
digitalWrite(LEDR, HIGH);
digitalWrite(LEDG, HIGH);
digitalWrite(LEDB, HIGH);

//External LEDs
MachineControl_DigitalOutputs.write(0, LOW);
MachineControl_DigitalOutputs.write(1, LOW);
MachineControl_DigitalOutputs.write(2, LOW);
}

void loop()
{
usb.Task();
}
```

```c
#define MOD_CTRL      (0x01 | 0x10)
#define MOD_SHIFT     (0x02 | 0x20)
#define MOD_ALT       (0x04 | 0x40)
#define MOD_WIN       (0x08 | 0x80)

#define LED_NUM_LOCK     1
#define LED_CAPS_LOCK    2
#define LED_SCROLL_LOCK  4

#define stdin_recvchar  Serial1.write

static uint8_t key_leds;
static const char knum[] = "1234567890";
static const char ksign[] = "!@#$%^&*()";
static const char tabA[] = "\t -=[]\\#;',./";
static const char tabB[] = "\t _+{}|~:\"~<>?";

// route the key event to stdin
static int process_key(tusbh_ep_info_t* ep, const uint8_t* keys)
{
  printf("\n");
  uint8_t modify = keys[0];
  uint8_t key = keys[2];
  uint8_t last_leds = key_leds;
  if (key >= KEY_A && key <= KEY_Z) {
    char ch = 'A' + key - KEY_A;
    if ( (!!(modify & MOD_SHIFT)) == (!!(key_leds & LED_CAPS_LOCK)) ) {
      ch += 'a' - 'A';
    }
    stdin_recvchar(ch);
    if (ch == 'r' || ch == 'R')
    {
      ledRstate = !ledRstate;
      if (ledRstate)
        {
          digitalWrite(LEDR, LOW);
          MachineControl_DigitalOutputs.write(0, LOW);
        }
      else
        {
          digitalWrite(LEDR, HIGH);
          MachineControl_DigitalOutputs.write(0, HIGH);
        }
    }

    if (ch == 'g' || ch == 'G')
    {
      ledGstate = !ledGstate;
      if (ledGstate)
        {
          digitalWrite(LEDG, LOW);
          MachineControl_DigitalOutputs.write(1, LOW);
        }
      else
        {
          digitalWrite(LEDG, HIGH);
          MachineControl_DigitalOutputs.write(1, HIGH);
        }
    }

    if (ch == 'b' || ch == 'B')
    {
      ledBstate = !ledBstate;
      if (ledBstate)
        {
          digitalWrite(LEDB, LOW);
```

3.6 Bluetooth Low Energy (BLE)

```
        MachineControl_DigitalOutputs.write(2, LOW);
       }
     else
       {
       digitalWrite(LEDB, HIGH);
       MachineControl_DigitalOutputs.write(2, HIGH);
       }
  }
  } else if (key >= KEY_1 && key <= KEY_0) {
    if (modify & MOD_SHIFT) {
      stdin_recvchar(ksign[key - KEY_1]);
    } else {
      stdin_recvchar(knum[key - KEY_1]);
    }
  } else if (key >= KEY_TAB && key <= KEY_SLASH) {
    if (modify & MOD_SHIFT) {
      stdin_recvchar(tabB[key - KEY_TAB]);
    } else {
      stdin_recvchar(tabA[key - KEY_TAB]);
    }
  } else if (key == KEY_ENTER) {
    stdin_recvchar('\r');
  } else if (key == KEY_CAPSLOCK) {
    key_leds ^= LED_CAPS_LOCK;
  } else if (key == KEY_NUMLOCK) {
    key_leds ^= LED_NUM_LOCK;
  } else if (key == KEY_SCROLLLOCK) {
    key_leds ^= LED_SCROLL_LOCK;
  }

  if (key_leds != last_leds) {
    tusbh_set_keyboard_led(ep, key_leds);
  }
  return 0;
}
//*****************************************************************
```

In the Application section of this chapter we configure the PMC to serve as USB data logger.

3.6 Bluetooth Low Energy (BLE)

The Arduino Portenta Machine Control is equipped with Bluetooth features. The Classic form of Bluetooth was designed to provide a wireless replacement for the common RS–232 serial connection standard. The Arduino PMC is equipped with Bluetooth Low Energy (BLE) features. It is important to note that Bluetooth Classic and BLE features are not compatible with one another.[3]

Bluetooth BLE provides for low transmit power (10 mW), short (maximum 100 m) range RF connections to replace wires. It uses the crowded Industrial, Scientific, and Medical (ISM) frequency band from 2.40 to approximately 2.50 GHz. The BLE band is divided into

---

[3] Portions of this section are provided with permission from "Arduino V: Machine Learning," S. Barrett, Springer, 2022.

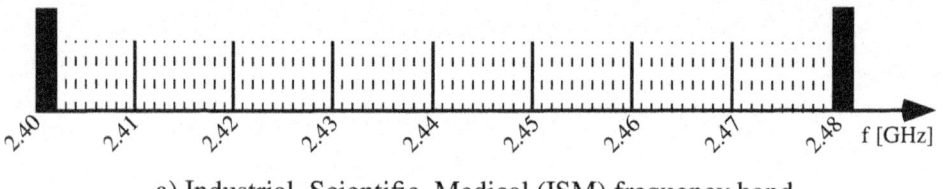
a) Industrial, Scientific, Medical (ISM) frequency band.

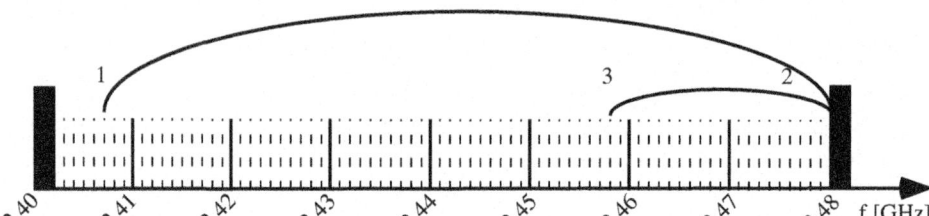
b) Frequency Hopping Spread Spectrum (FHSS) (R and S).

Fig. 3.6 Bluetooth BLE communication concepts

40 different, 2 MHz channels as shown in Fig. 3.6. BT BLE employs an interesting frequency hopping technique to communicate. Data for transmission is divided into packets at data rates from 125 to 2 Mb/s. The device transmits a packet of data at the first carrier frequency. It then hops to a different carrier frequency for the next packet and so on until the entire message is transmitted as shown in Fig. 3.6b. Formally the BT BLE modulation technique is called Direct Sequence Spread Spectrum (DSSS) (www.bluetooth.com).

BLE uses the Generic Attribute (GATT) Profile to establish two different primary roles for a BLE connection:

- The peripheral or server role provides bulletin board features where data is posted for reading.
- The central or client role can read and interact with the posted data.

In Fig. 3.7 we use the PMC in a peripheral server role to collect important greenhouse information such as external temperature, internal temperature, humidity, and soil moisture content. The greenhouse related data is collected and organized into a BLE service. The service related data is provided as BLE configured characteristics. To allow ease of access to the information from an external central client device, the BLE service and characteristics are each assigned a universally unique identifier (UUID) (www.bluetooth.com). If we were to expand the features of the project with additional services, we could group them into a profile.

There are a number of 16 bit pre–assigned UUIDs. The UUIDs represent different manufacturers and technology companies employing Bluetooth–based technologies. Also, UUIDs

3.6 Bluetooth Low Energy (BLE)

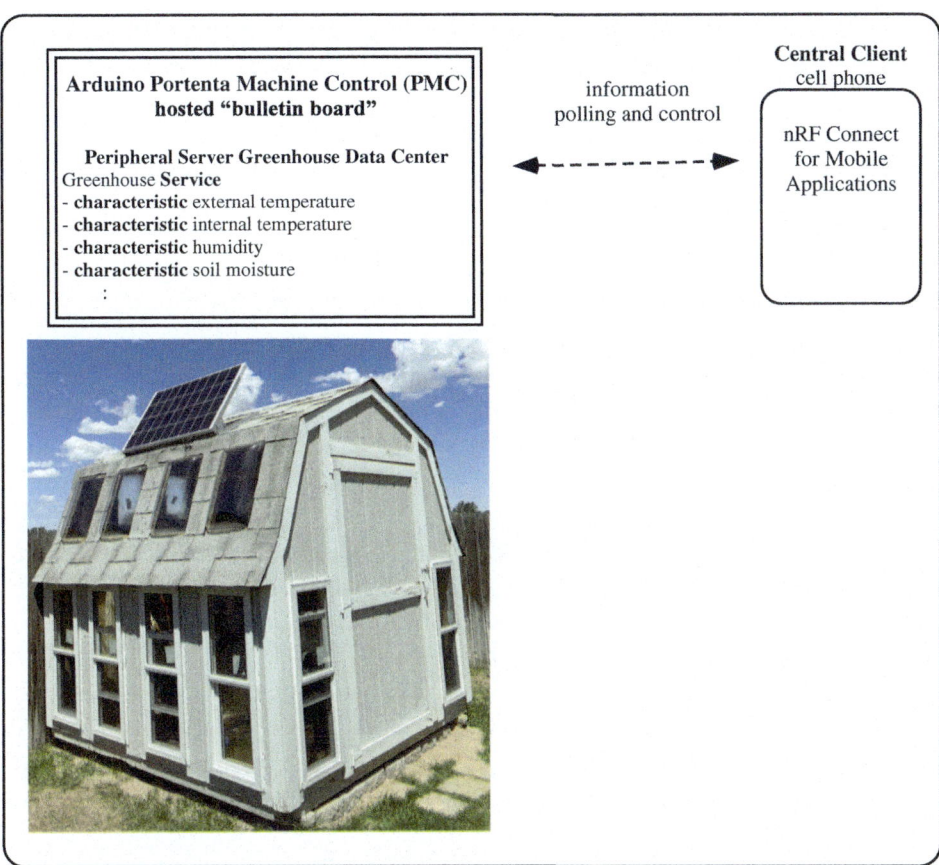

Fig. 3.7 Bluetooth BLE equipped greenhouse

have been pre-assigned to common Bluetooth features and common pre-assigned data types (e.g. temperature, pressure, etc.) (www.bluetooth.com):

- Bluetooth members: 0xFxxx
- GATT characteristic and object type: 0x2xxx
- GATT declarations: 0x28xx and 0x29xx
- GATT service: 0x18xx
- GATT unit: 0x27xx
- protocol identifier: 0x00xx
- SDO GATT service: 0XFFFx
- service classes and profiles: 0x10xx and 0x11xx

For BLE services and characteristics without a 16–bit pre–assigned UUID, a unique 128 bit UUID code is used. A Bluetooth unique UUID may be obtained using a number of online UUID generators.

In the greenhouse example, a cell phone is configured as a BLE central or client. Through the BLE wireless radio interconnect, the cell phone can read and interact with the greenhouse data and features.

3.6.1 Arduino BLE Library

The Arduino BLE Library provides for a wide variety of BLE configurations. The library is downloaded from within the Arduino IDE using the Library Manager. The library is organized into different classes including the (www.arduino.cc):

- BLE Class used to enable the BLE module,
- BLE Device Class to get information about connected devices,
- BLE Service Class to enable services and interaction with services,
- BLE Characteristic Class to enable characteristics and interaction with them, and
- BLE Descriptor Class to describe characteristics.

To get acquainted with the library we continue with a series of examples. The examples are adapted from the Arduino BLE Library. A cell phone is configured as a client to poll and interact with the data. The cell phone is equipped with a BLE compatible app to interact with the PLC (e.g. nRF Connect, LightBlue).

Example: In this first example "LED," from the Arduino BLE Library, a cell phone serves as a central client to control an onboard PMC LED. The PMC is configured as a server. For BLE operation the PMC requires a 2.4 GHz antenna (e.g. dual band Wi–Fi 2.4 GHz, 5 GHz, SMA male antenna). The antenna is connected to the PMC via the SMA connector.

To get better acquainted with the sketch, we study the Bluetooth configuration related code steps. In Fig. 3.8 we detail these steps in a UML activity diagram (Fowler and Scott).

```
//**************************************************************
//PMC_BLE_LED: This example creates a BLE peripheral with
//service that contains a characteristic to control an
//onboard PMC LED (DIGITAL OUTPUT LED 00).
//
//A generic BLE central phone app, like LightBlue or
//nRF Connect is used to interact with the PMC hosted BLE
//services and characteristics created in this sketch.
//
//Adapted from:
//Library: Arduino BLE Library
//This example code is in the public domain.
//**************************************************************
```

3.6 Bluetooth Low Energy (BLE) 89

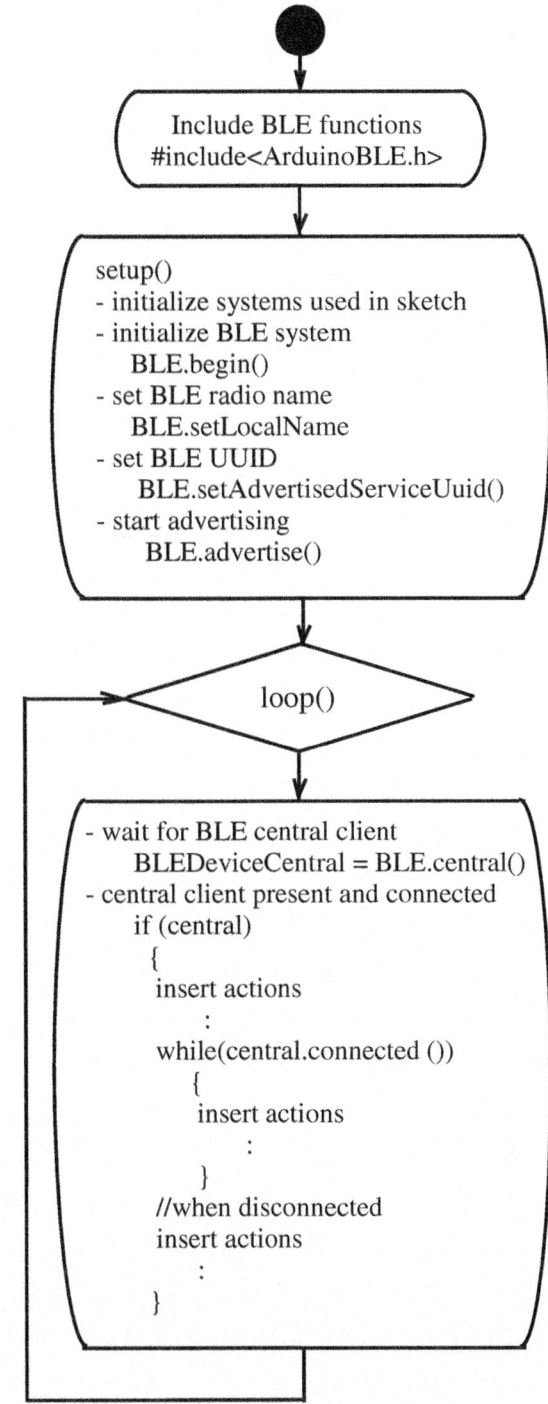

Fig. 3.8 Bluetooth BLE configuration

```cpp
#include <Arduino_PortentaMachineControl.h>
#include <ArduinoBLE.h>
                                   //Declare BLE LED Service
                                   //Link to 128 bit UUID
BLEService ledService("19B10000-E8F2-537E-4F6C-D104768A1214");

//BLE LED Switch Characteristic - custom 128-bit UUID, read and
//writable by central client device (cell phone)
BLEByteCharacteristic switchCharacteristic
   ("19B10001-E8F2-537E-4F6C-D104768A1214", BLERead | BLEWrite);

void setup()
{
MachineControl_DigitalOutputs.begin();       //init PMC digital outputs
Serial.begin(9600);                 //status to serial monitor
while (!Serial);

if(!BLE.begin())                    //BLE initialization
   {
   Serial.println("starting BLE failed!");
   while (1);
   }

   //set advertised local name and service UUID:
   BLE.setLocalName("LED");
   BLE.setAdvertisedService(ledService);

   //add the characteristic to the service
   ledService.addCharacteristic(switchCharacteristic);

   //add service
   BLE.addService(ledService);

   //set the initial value for the characeristic:
   switchCharacteristic.writeValue(0);

   //start advertising
   BLE.advertise();

   Serial.println("BLE LED Peripheral");
}

void loop()
{
//listen for BLE clients (central) to connect:
BLEDevice central = BLE.central();

//if a client (central) is connected to peripheral:
if(central)
   {
   Serial.print("Connected to client: ");

   //print the client's MAC address:
   Serial.println(central.address());

   //while the client (central) is still connected to
   //the server (peripheral):
   while (central.connected())
      {
```

3.6 Bluetooth Low Energy (BLE)

```
    //if the remote client device wrote to the
    //server characteristic, use the
    //value to control the LED:
      if(switchCharacteristic.written())
        {
        if(switchCharacteristic.value())
          {
          Serial.println("LED on");      //any value other than 0
          MachineControl_DigitalOutputs.write(0, HIGH); //digital out 00 on
          }
        else
          {
          Serial.println(F("LED off")); //a 0 value
          MachineControl_DigitalOutputs.write(0, LOW); //digital out 00 off
          }
        }
    }//end while
  //when the central disconnects, print it out:
  Serial.print(F("Disconnected from central: "));
  Serial.println(central.address());
  }//end if(central)
}

//*************************************************************
```

The sketch may be compiled and uploaded to the Arduino PMC. Once uploaded, the sketch may be tested:

- Open the Serial Monitor in the Arduino IDE to monitor sketch status.
- Using a cell phone as a client, open "nRF Connect" to establish Bluetooth BLE connection with the PMC based server.[4]
- Find "LED" in the nRF scanner list.
- Tap "Connect" to connect the client (cell phone) to the server (PMC).
- By selecting "Client" and the up arrow, values may be sent from the client to the server to control the LED.
- Select "Write Value" and "Unsigned."
- Sending a non–zero turns the LED on while sending zero turns the LED off.

The results of the interaction between the cell phone and the PMC is provided in 3.9.
Example: In this example, "battery monitor," adapted from the Arduino BLE Library, the Arduino PMC is configured as a server. The Arduino PMC monitors the analog signal on ANALOG IN AI0 and posts this characteristic to the server based bulletin board. A cell phone based client equipped with BLE compatible app is used to poll the posted data.

To simulate a battery, the wiper arm of a 1M Ohm potentiometer is connected to PMC ANALOG IN pin AI0. The potentiometer is in series with a 1.4 M Ohm equivalent resistance to limit the AI0 input voltage to values below 10 VDC as shown in Fig. 3.10. For BLE

---

[4] BLE applications such as nRF connect or LightBlue are available from your cell phone app store.

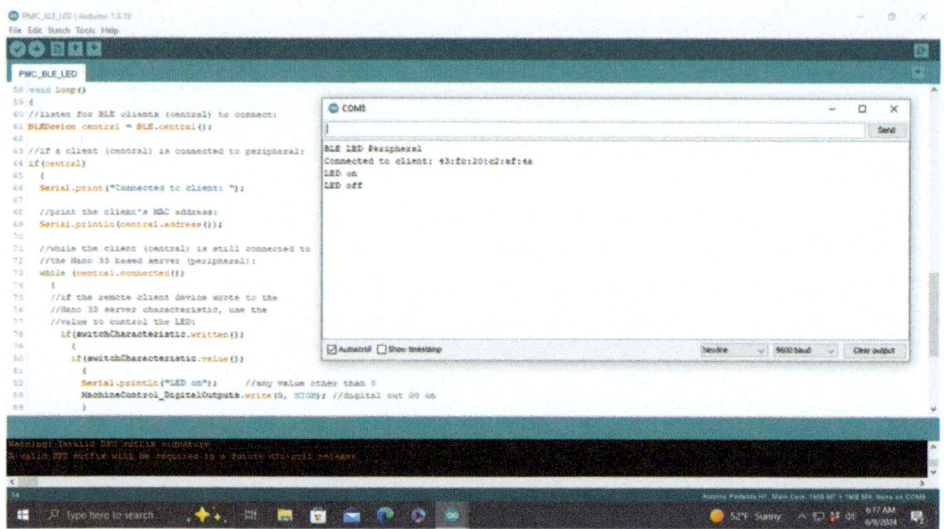

Fig. 3.9 Bluetooth BLE link between PMC and cell phone

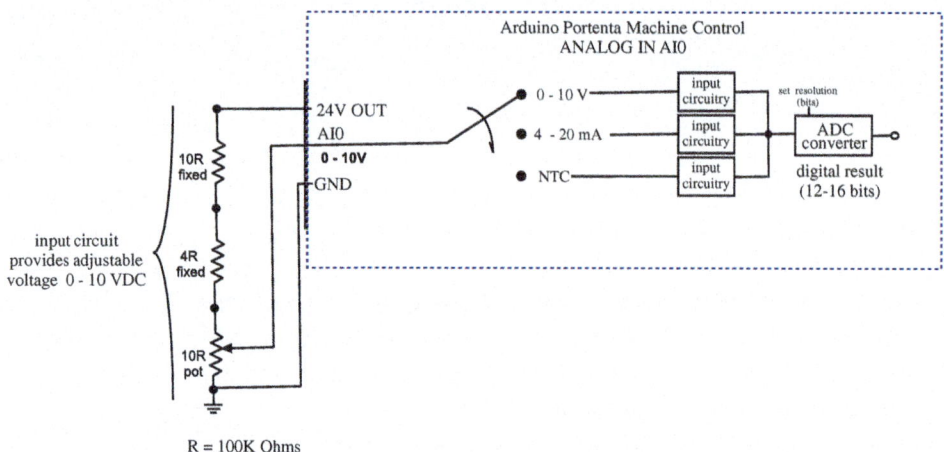

Fig. 3.10 Bluetooth BLE as battery monitor

operation the PMC requires a 2.4 GHz antenna (e.g. dual band Wi–Fi 2.4 GHz, 5 GHz, SMA male antenna). The antenna is connected to the PMC via the SMA connector.

The client/server connection is tested using techniques similar to those provided in the previous example.

```
//***************************************************************
//PMC_BLE_battery_monitor - This example creates a BLE server
//(peripheral) with the standard battery service and level
```

3.6 Bluetooth Low Energy (BLE)

```
//characteristic. The ANALOG IN (AI0) is used to monitor the
//battery level.
//
//A generic BLE central phone app, like LightBlue or
//nRF Connect is used to interact with the Arduino PMC
//hosted BLE services and characteristics created in this
//sketch.
//
//Adapted from:
//Library: Arduino BLE
//This example code is in the public domain.
//**************************************************************

#include <ArduinoBLE.h>
#include <Arduino_PortentaMachineControl.h>

const float RES_DIVIDER = 0.28057;
const float REFERENCE   = 3.0;
unsigned int analog_reading;


BLEService batteryService("180F");          //BLE Battery Service

//BLE Battery Level Characteristic
//  - standard 16-bit characteristic UUID
//  - remote clients get notifications if characteristic changes
BLEUnsignedCharCharacteristic batteryLevelChar("2A19",
                                    BLERead | BLENotify);
long battery;                               //battery reading
int oldBatteryLevel = 0;                    //last battery level reading from A0
long previousMillis = 0;                    //last time battery level checked (ms)

void setup()
{
MachineControl_DigitalOutputs.begin();
MachineControl_AnalogIn.begin(SensorType::V_0_10);  //set for 0-10 VDC in
Serial.begin(9600);                         //initialize serial communication
while (!Serial);
                                            //init built-in LED pin
pinMode(LED_BUILTIN,OUTPUT);                //indicates when central is connected
//analogReadResolution(12);                   //set 12 to 16 bits


if(!BLE.begin())                            //initialize Bluetooth BLE device
  {
  Serial.println("starting BLE failed!");
  while (1);
  }

//Set a local name for the BLE device. This name appears
//in advertising packets. Name used by remote devices to
//identify this BLE device.
BLE.setLocalName("BatteryMonitor");
BLE.setAdvertisedService(batteryService);   //add the service UUID
                                            //add the battery level characteristic
batteryService.addCharacteristic(batteryLevelChar);
BLE.addService(batteryService);             //add battery service
batteryLevelChar.writeValue(oldBatteryLevel); //set initial value
```

```
//Start advertising BLE.  Continuously transmits BLE advertising
//packets. Advertising will be visible to remote BLE central devices.
BLE.advertise();
Serial.println("Bluetooth device active, waiting for connections...");
}

void loop()
{
BLEDevice central = BLE.central();           //wait for a BLE central
                                             //if a central client
                                             //is connected to peripheral
if(central)
{
Serial.print("Connected to central: ");
Serial.println(central.address());           //print the central's BT address
digitalWrite(LED_BUILTIN, HIGH);             //LED on when client connected
MachineControl_DigitalOutputs.write(0, HIGH); //DIGITAL OUTPUTS LED 00

                                             //while client connected
                                             //check battery level every 200ms
while(central.connected())
  {
  long currentMillis = millis();
                                             //if 200ms have passed,
                                             //check the battery level
  if(currentMillis - previousMillis >= 200)
    {
    previousMillis = currentMillis;
    updateBatteryLevel();
    }
  }

  //when the central client disconnects, turn off the LED
  digitalWrite(LED_BUILTIN, LOW);
  MachineControl_DigitalOutputs.write(0, LOW);//DIGITAL OUTPUTS LED 00
  Serial.print("Disconnected from central: ");
  Serial.println(central.address());
  }
}

//*************************************************************
//void updateBatteryLevel() - Read the current voltage level
//on the AI0 analog input pin. This is used here to simulate
//the battery level.
//*************************************************************

void updateBatteryLevel()
{
//Read the input on analog input AI0
float raw_voltage_ch0 = MachineControl_AnalogIn.read(0);
Serial.print("Raw Voltage CH0: ");
Serial.print(raw_voltage_ch0, 3);

float voltage_ch0 = (raw_voltage_ch0 * REFERENCE) / 65535 / RES_DIVIDER;
Serial.print("Voltage CH0: ");
Serial.print(voltage_ch0, 3);
Serial.println("V");

int batteryLevel = map(raw_voltage_ch0, 0, 65535, 0, 100);
```

```
if(batteryLevel != oldBatteryLevel)
  {                                   //if the battery level has changed
  Serial.print("Battery Level %: ");  // print it
  Serial.println(batteryLevel);
  batteryLevelChar.writeValue(batteryLevel);  //update battery level
                                      //characteristic
  oldBatteryLevel = batteryLevel;     //save level for comparison
  }
}

//**************************************************************
```

3.7 UART/RS–232 Communication Protocol

When serial transmission occurs over a long distance additional techniques may be used to insure data integrity. Over long distances logic levels degrade and may be corrupted by noise. At the receiving end, it is difficult to discern a logic high from a logic low.

The RS–232 standard has been around for some time. With the RS–232 standard (EIA–232), a logic one is represented with a negative 12 VDC level while a logic zero is represented by a + 12 VDC level. Level shifter chips are commonly available (e.g. MAX232) that convert the 5 and 0 V output levels from a processor based Universal Asynchronous Receiver Transmitter (UART) to RS–232 compatible levels and convert back to 5V and 0V levels at the receiver. The RS–232 standard also specifies other features for this communication protocol.

The Arduino PMC is equipped with a TJA1049T/3J RS–232 transceiver as shown in Fig. 3.11. The transceiver provides RS–232 compatible signals for communication to other RS232 compatible peripheral devices and processors.

The output from a UART system such as the Arduino UNO R3 is shown Fig. 3.11. The signal idles at logic high. The transmission of a single character begins with a logic low start bit. The start bit falling edge provides for synchronization between the transmitter and receiver. The data bits (7 or 8) is then transmitted. In the example shown, the ASCII letter "G" is transmitted least significant bit first. The data is then followed by a parity bit, stop bit(s), and then goes back to idle. To maintain synchronization between the transmitter and receiver, a common Baud rate and parity must be shared.

Example: Arduino PMC to UNO R3 RS–232 connection. In this example the Arduino PMC is connected to an Arduino UNO R3 using an RS–232 link. The UNO R3 does not have a resident RS–232 transceiver. The UNO R3 is equipped with a TinySine RS232/RS485 transceiver shield as shown in Fig. 3.34. The UNO R3 communicates with the shield via a Software Serial UART connection.

The following PMC sketch is available in the Portenta Machine Control Library. It sends "Hello" with an incrementing count from the PMC to the UNO R3. The UNO R3 receives

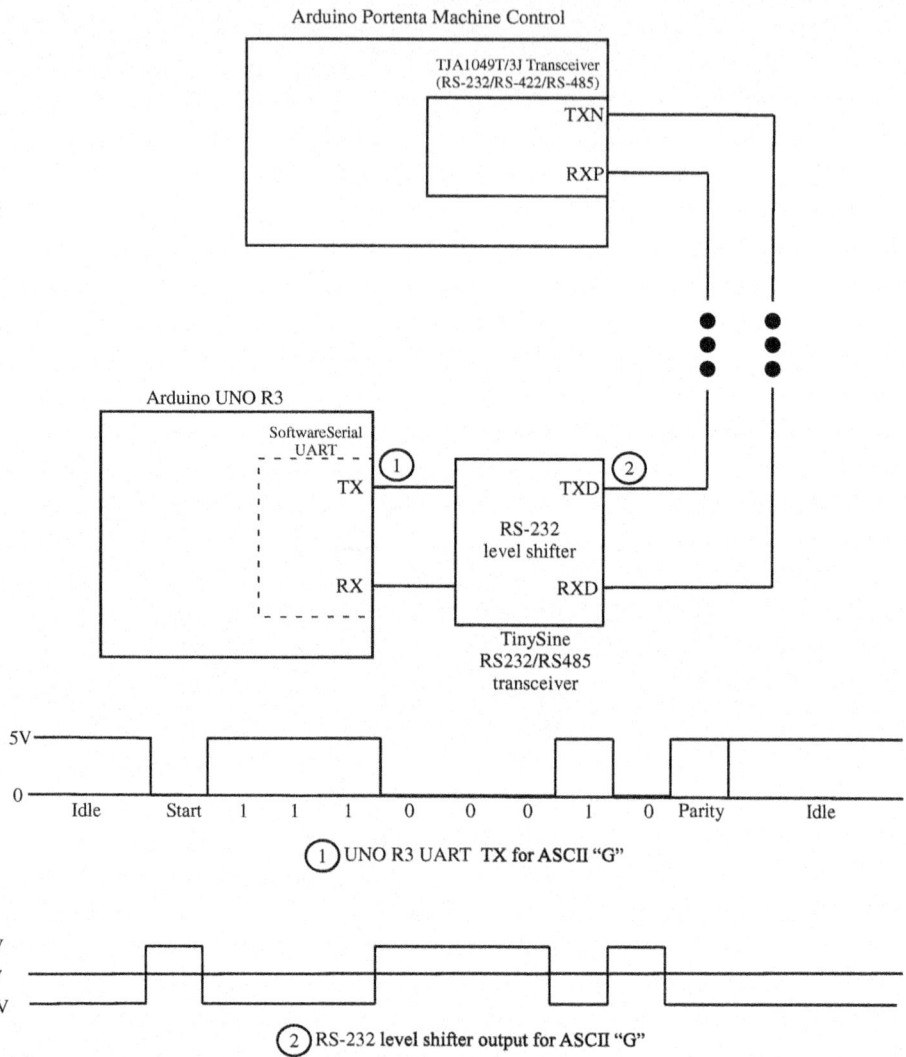

Fig. 3.11 Arduino PMC to UNO R3 communication via RS–232 link

and reports the message to the Serial Monitor. The UNO R3 responds with a "Got it!" message to the PMC (Fig. 3.12).

```
//*************************************************************
//Portenta Machine Control RS-232 Communication Example
//
//This sketch shows the usage of the SP335ECR1 on the Machine
//Control as an RS-232 interface. It demonstrates how to periodically
//send a string on the RS-232 TX channel and how to receive data from
```

3.7 UART/RS–232 Communication Protocol

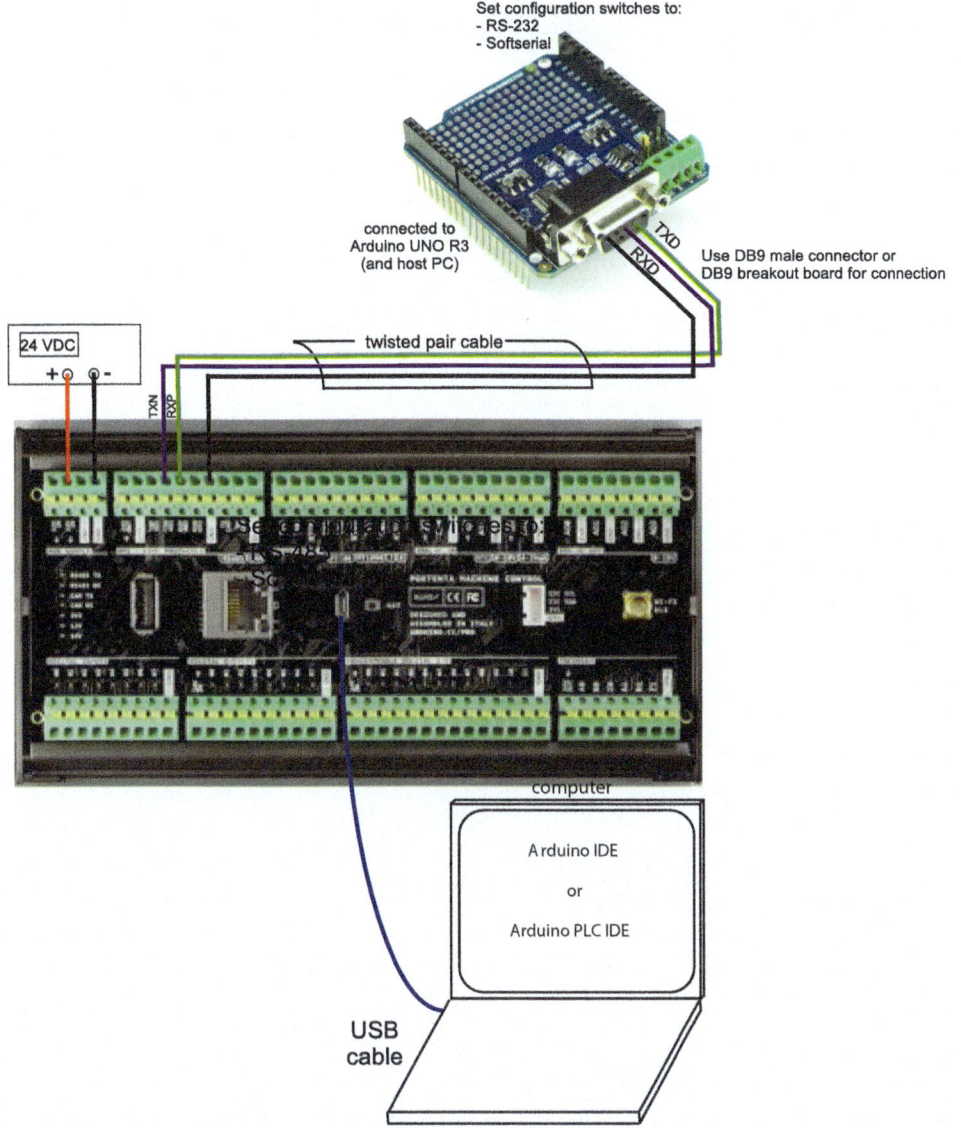

Fig. 3.12 Arduino PMC to UNO R3 communication via RS–232 link. The UNO R3 is equipped with a TinySine RS232/RS485 transceiver shield. Images used courtesy of the Arduino Team (CC BY–NC–SA) (www.arduino.cc)

```cpp
//the interface RX channel.
//
//Circuit:
//- Portenta H7
//- Portenta Machine Control
//- Device with RS-232 interface
//- Connect PMC TXN to RS-232 Device RXD
//- Connect PMC RXP to RS-232 Device TXD
//- Connect PMC GND to RS-232 Device GND
//
//Adapted from:
//Library: Arduino Portenta Machine Control
//Sketch: RS232
//Initial author: Riccardo Rizzo @Rocketct
//Authors who have contributed to updates:
//- Leonardo Cavagnis @leonardocavagnis
//The example code is in the public domain.
//**************************************************************

#include <Arduino_PortentaMachineControl.h>

constexpr unsigned long sendInterval { 5000 };
unsigned long sendNow { 0 };
unsigned long counter { 0 };
unsigned char count = 0;

void setup()
{
Serial.begin(9600);
while (!Serial)
  {
  ;                             //wait for serial monitor to open
  }
delay(2500);
Serial.println("Start RS232 initialization");

//Set the PMC Communication Protocols to default config and
//enable the RS485/RS232 system.
//RS485/RS232 default config is:
//- RS485/RS232 system disabled
//- RS485 mode
//- Half Duplex
//- No A/B and Y/Z 120 Ohm termination enabled

//Specify baudrate for the communication
MachineControl_RS485Comm.begin(9600);

//Enable RS232 mode
MachineControl_RS485Comm.setModeRS232(true);

//Start in receive mode
MachineControl_RS485Comm.receive();

Serial.println("Initialization done!");
}

void loop()
{
if(MachineControl_RS485Comm.available())
```

3.7 UART/RS–232 Communication Protocol

```
Serial.write(MachineControl_RS485Comm.read());

if(millis() > sendNow)
{
String log = "[";
            log += sendNow;
            log += "] ";
String msg = "hello ";
msg += counter++;
log += msg;
//Serial.println(log);
Serial.println(msg);

//Disable receive mode before transmission
MachineControl_RS485Comm.noReceive();

//Transmit message
MachineControl_RS485Comm.beginTransmission();
MachineControl_RS485Comm.println(msg);
MachineControl_RS485Comm.endTransmission();

//Re-enable receive mode after transmission
MachineControl_RS485Comm.receive();

sendNow = millis() + sendInterval;
  }
}

//**************************************************************
```

Sketch for Arduino UNO R3.

```
//**************************************************************
//UART_RX
//
//S. Barrett
//This code example is in the public domain.
//**************************************************************

unsigned char incomingByte;        //for incoming serial data

#include <SoftwareSerial.h>

#define RS232_RX_pin   2
#define RS232_TX_pin   3

SoftwareSerial RS232(RS232_RX_pin, RS232_TX_pin);

bool sent_one = 0;

void setup()
{
Serial.begin(9600);                //initialize serial comm 9600 baud rate
while(!Serial)                     //wait for serial monitor to open
  {
  ;
  }
RS232.begin(9600);                 //start RS-485 comm, BAUD rate 9600
```

```
delay(100);
}

void loop()
{
if(RS232.available() > 0)
  {
  incomingByte = RS232.read();     //read the incoming byte:
    Serial.write(incomingByte);
    delay(100);
    sent_one = 0;
  }
else if(sent_one == 0)
  {
  RS232.write("Got it!   ");
  sent_one = 1;
  }
}

//*************************************************************
```

The result of the RS–232 transmission between the PMC and the Arduino UNO R3 is provided in Fig. 3.13.

3.8 Internet Concepts

In this section we begin with fundamental internet concepts to support upcoming Ethernet and Wi–Fi discussions. As discussed in Chap. 1, the internet is used extensively in Operational Technology (OT), the Internet of Things (IoT), and the Industrial Internet of Things (IIoT).

3.8.1 A Big Picture of the Internet

From its early beginnings in the late 1960s to today, the internet has become ubiquitous (found everywhere) in every facet of our lives. A few examples where the internet has become prevalent include industry, agriculture, energy production, education, healthcare, entertainment, manufacturing, retail, communications, and many other areas. References for the following internet sections are from (Hanes, Lowe, and Null) and others as cited.[5]

Many (including the author) take a safe, secure, and reliable internet for granted. Portions of the internet, consisting of a global network of interconnected computers, are referred to as the "cloud." In this section we examine connections between computers that comprise the internet in a home and work environment and then examine what is inside the cloud.

---

[5] Portions of this section are adapted with permission from "Arduino III: Internet of Things," S. Barrett, Springer, 2021.

3.8 Internet Concepts

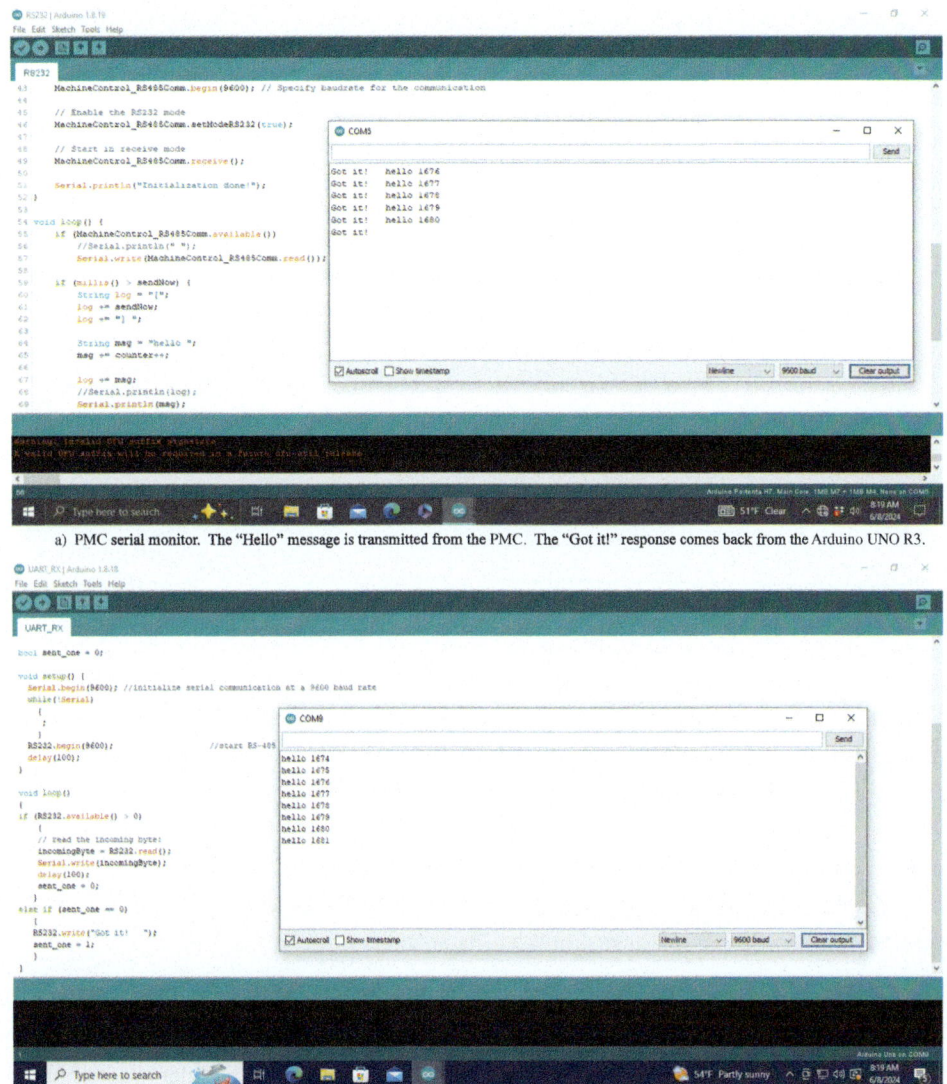

a) PMC serial monitor. The "Hello" message is transmitted from the PMC. The "Got it!" response comes back from the Arduino UNO R3.

b) The "Hello" message comes from the PCM. The Arduino UNO R3 responds with "Got it!".

Fig. 3.13 Arduino PMC to UNO R3 communication via RS–232 link

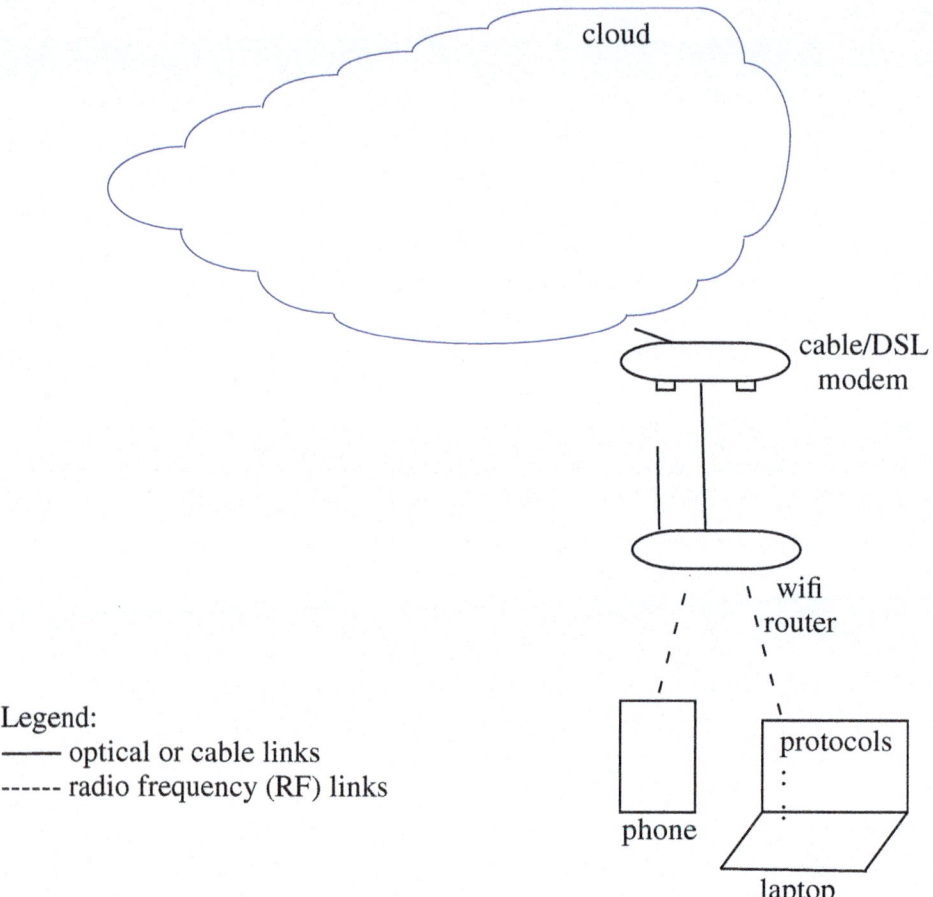

Fig. 3.14 Internet at home or a small business configuration

Figure 3.14 provides a typical internet connection found in a home or small business such as a cafe or small store. A cable or digital subscriber line (DSL) modulator/demodulator (modem) provides a connection to the internet. The cable/DSL modem is provided by an Internet Service Provider (ISP) when you subscribe to their internet connection service. The connection between the cable/DSL modem and the ISP provider may be a combination of copper cable, optical fiber, and wireless radio frequency connection links.

With internet service available via the cable/DSL modem, a Wi–Fi router is used to establish a wireless local area network (WLAN) within your home or small business. The WLAN serves as an internet access point to a broader area using radio frequency (RF) signals operating at 2.5 or 5 GHz. The link between the cable/DSL modem to the Wi–Fi router is via a cable. Wireless devices such as a cell phone or a laptop within range of the Wi–Fi

3.8 Internet Concepts

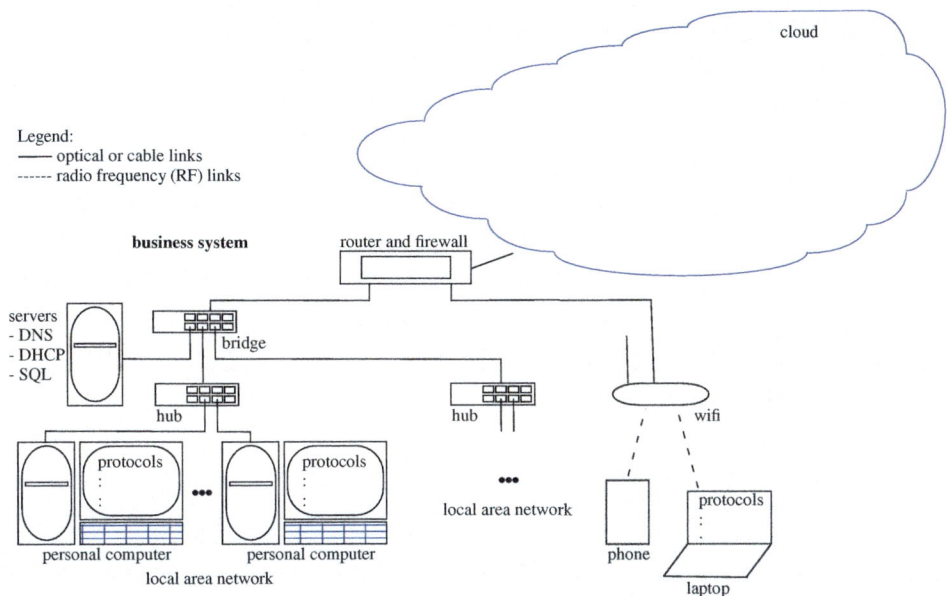

Fig. 3.15 Internet at a large business or university

router are able to access the internet with proper credentials (i.e. WLAN password and ISP subscription). The Wi–Fi router range may be extended using a Wi–Fi range extender.

Internet service typically found in a larger business or a university is shown in 3.15. Internet service is provided to the organization via an ISP. Connection is made to the ISP via a firewall. The firewall provides protection from internet hazards outside the organization. It may also be used to limit outgoing information from the organization (e.g. sensitive company information, classified material, etc.). There is also a router at the organization portal. It is used to route internet message traffic to its next destination.

Individual computers are provided access to the internet via a cabled connection to a hub. The hub is used to connect computers into a local area network (LAN) sharing the same location or similar function such as an academic department or company section. Multiple LANs are then connected via a bridge. The bridge will also provide connection to a series of local servers such as a Domain Name System (DNS) server, a Dynamic Host Configuration Protocol (DHCP) server, a Structured Query Language (SQL) database server, and a mail server (among others). Wi–Fi access may also be provided by a Wi–Fi router as previously described.

3.8.2 Internet Cloud

Figure 3.16 shows the configuration of the internet cloud. The cloud contains the global connection of multiple internet service providers. Regional internet service providers share internet traffic via a metropolitan area exchange (MAE). The regional network ISPs connect to a network service provider who are in turn connected to other network service providers via network access points. Connectivity across the globe is provided by submarine optical fiber cables spanning the oceans. Internet access may be provided to remote areas via balloon borne network access points. The overall result is a global network of interconnected computers for the open exchange of information.

Aside from the internet hardware components, there are internet protocols and applications used to ensure reliable and compatible communications from one location to another. As an example, when you are checking your favorite news website or sending an e–mail to

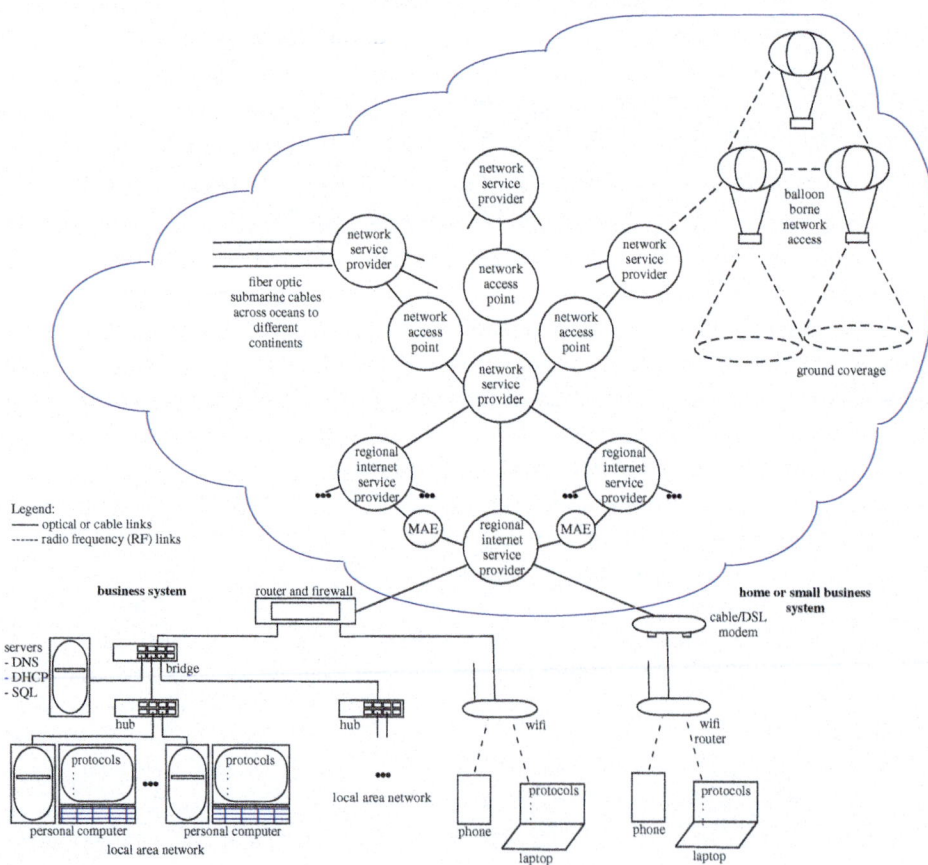

Fig. 3.16 The cloud

3.8 Internet Concepts

a friend, the specific application you are using along with the computer's operating system has built in features to interact with the internet to accomplish your desired task.

The application and the operating system apply the different layer activities to transmit information over the internet. For example, the e-mail message you are sending is broken into packets of information, provided source and destination IP addresses, and converted to an electronic signal. The packets are then routed to the destination computer via a series of internet hops directed by routers along the path. At the destination the received packets are reassembled and the protocol steps are applied in reverse order.

3.8.3 Internet Protocol Models

There are two different models of protocol stacks commonly used within the internet community. A protocol is a standardized set of rules and procedures. The layered protocol models provide guidelines on how data is processed within applications and prepared for transmission over the internet. Both protocols were developed in the early 1980s.

The International Organization for Standardization (ISO) developed the seven layered Open Systems Internet protocol stack ISO/OSI reference model shown in Fig. 3.17a. The adjacent layers interact with one another in a given system while similar layers interact with one another in different systems.

The transmission control protocol/internet (TCP/IP) maps into several layers of the OSI model as shown in Fig. 3.17a. Figure 3.17b shows the four layers of the TCP/IP model: application, transport, internet, and link layer.

The data to be shared with another computer resides within the application layer. The data is divided into packets. As the data packet is processed through each layer of the sending computer, additional header and footer information is appended to the data payload. The IP address allows the sender and receiver to find one another on the internet (Leiden, Lowe, Null).

3.8.4 Internet Addressing Techniques

In this section we discuss the importance of and techniques used to address network assets. We begin with IP addressing and packet headers. There are two different versions of the IP header: IPv4 and IPv6 as shown in Fig. 3.18.

3.8.4.1 IPv4 Header
The earlier IPv4 header version consists of 24 bytes followed by the data payload. The overall packet datagram must be at least 40 bytes. The IPv4 header consists of the following fields:

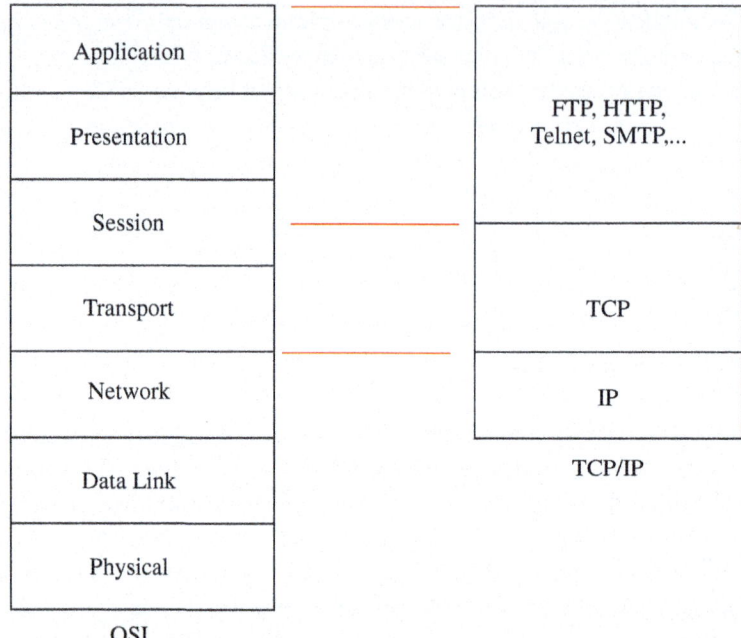

a) comparison of the OSI protocol stack model to the TCP/IP protocol stack model.

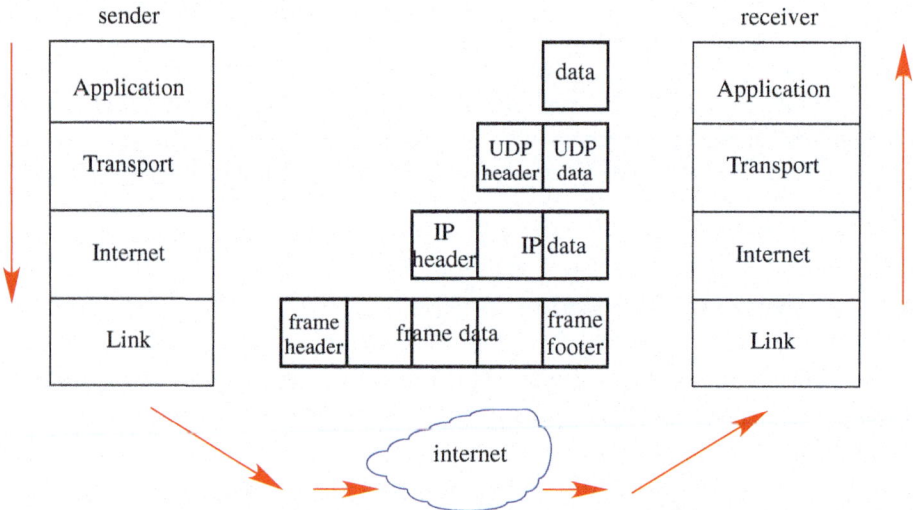

b) TCP/IP protocol for sender and receiver interaction

Fig. 3.17 OSI versus TCP/IP (Leiden, Lowe, Null)

3.8 Internet Concepts

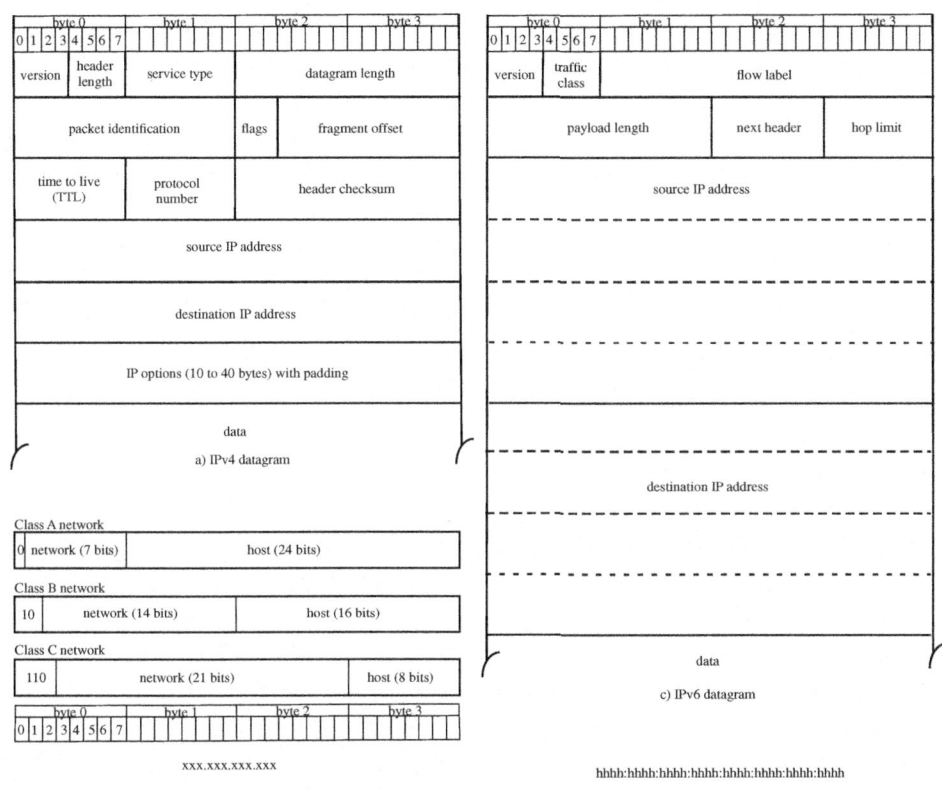

Fig. 3.18 IPv4 and IPv6 (Hanes, Null)

- version: IP protocol version. For IPv4 this field is set to $(0100)_2$.
- header length: specified as 32–bit words.
- type of service: specifies priority from low (000) to critical (101).
- total length: specifies total length of the datagram packet in bytes.
- packet ID: the packet is assigned a unique serial number.
- flags: indicates whether a larger packet can be broken into smaller packets.
- fragment offset: provides fragment location within packet.
- time to live: determines number of internet hops are allowed from source to destination.
- protocol number: indicates the type of protocol associated with the data: 0–reserved, 1–internet control message protocol (ICMP), 6–transmission control protocol (TCP), or 17–user datagram protocol (UDP).
- header checksum: holds the calculated checksum of the header.
- provides source and destination address of packet. Note each address is 32 bits in length. A unique address or source ID is provided to each source computer on the internet. The

allocation of addresses is coordinated by the Internet Corporation for Assigned Names and Numbers or ICANN (www.ICANN.org).
- IP options: provides additional control information.

The IPv4 protocol uses the 32–bit IP address configuration as shown in Fig. 3.18b. The first several bits of the address indicates the network class: Class A for large networks (0), Class B for medium sized networks (10), and Class C for smaller networks (110). The remaining IP address bits are partitioned to select a network and a specific host. The IP address is expressed as a 32–bit dotted decimal notation value (xxx.xxx.xxx.xxx). Each byte in the address is specified by its decimal equivalent (xxx) and has a value ranging from 0 to 255 (Hanes, Lowe, Null).

3.8.4.2 CIDR Addressing

To provide additional addressing flexibility a Classless Inter–Domain Routing (CIDR) addressing scheme was developed. The CIDR scheme provides for flexible subnet addressing within the IPv4 protocol. The format of the CIDR address is provided in Fig. 3.19a. A subnet is a smaller network within one of the network classes: A, B, or C.

Recall the IPv4 protocol provides for a 32–bit IP address. The CIDR protocol allows the address to be partitioned into a number of address bits allocated to identify the network and the remaining bits allocated to identify a specific host (computer) within the network. A slash character (/) follows the address with a decimal number. The decimal number indicates the number of logic ones in the subnet mask.

In the example provided in 3.19b left an IP address has been partitioned such that 24 bits will be used to identify the specific network while the remaining bits eight bits will be used to specify a specific computer within the network. This partition allocation results in a subnet mask of 255.255.255.0 or expressed in separated binary as $(11111111.11111111.11111111.00000000)_2$. Since there are 24 bits in the subnet mask containing leading ones, a /24 is appended to the IP address to communicate the partition information. The eight bits allocated for host or computer addressing will provide for $2^{host\ bits}$ or 256 unique addresses.

In Fig. 3.19b right, the IP address (190.166.2.2./24) has been partitioned into 24 bits for the subnet address and the other eight bits for specific computer addressing. This partition results in the subnet mask 255.255.255.0. When this mask is logically ANDed with the IP address the resulting subnet address 190.166.2.0 results. The specific computers within the subnet will be addressed beginning at 190.166.2.1. and ending at 190.166.2.255 (www.IETF.org).

3.8 Internet Concepts

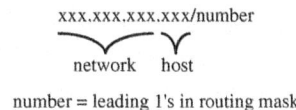

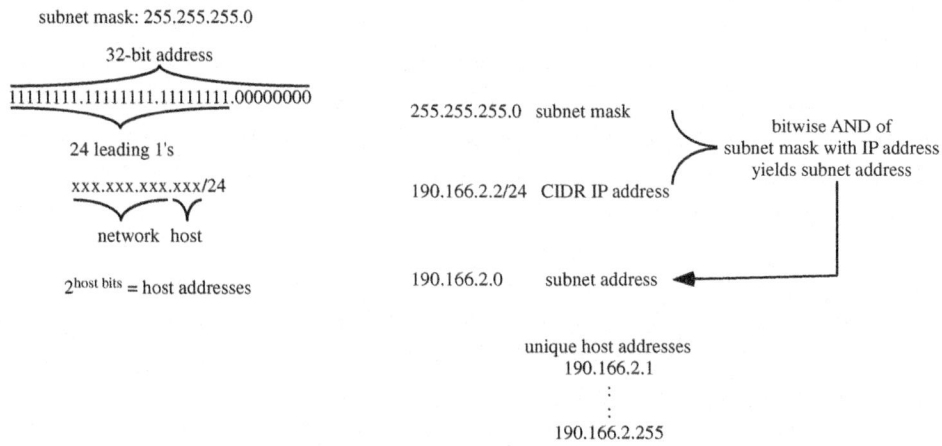

Fig. 3.19 CIDR addressing (www.IETF.org)

3.8.4.3 IPv6 Header

In the mid–1990s the IPv6 was released by the Internet Engineering Task Force (IETF). The IPv6 protocol provides for a longer 128–bit IP address space. The different fields within the IPv6 header specify:

- version: IP protocol version. For IPv6 this field is set to $(0110)_2$.
- traffic class: will specify different priority.
- flow label: will specify the type of communication in progress.
- payload length: expressed in bytes.
- next header: specifies if additional header information is provided in the payload.
- hop limit: will allow up to 256 hops from source to destination.
- source and destination addresses: 128–bits each.

The IETF developed a logical, methodical method of assigning IPv6 addresses, the Aggregatable Global Unicast Address Format, as shown in Fig. 3.20. The 128–bit address is partitioned into different fields to specify top, next, and site–level aggregation representing for an example a country, a company within the country, and networks within the

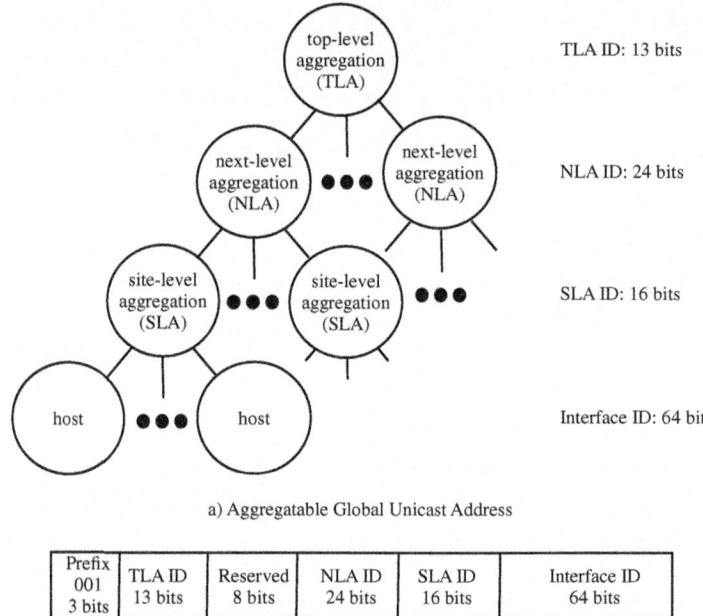

Fig. 3.20 IPv6 addressing (Hanes, Lowe, Null)

company, respectively. The 64–bit interface ID is a combination of the host device MAC address and information from a nearby router. The 128–bit IPv6 address is specified as eight 16—bit values expressed in hexadecimal and separated by colons (hhhh:hhhh:hhhh:hhhh: hhhh:hhhh:hhhh:hhhh) (Hanes, Lowe, Null).

3.8.4.4 MAC Address
The host device's Medium Access Control or MAC address is a 48–bit device specific address as shown in Fig. 3.21. The address is partitioned into six different bytes. The address specifies the Organizational Unique Identifier (OUI) and the Network Interface Controller (NIC) identifier. The NIC provides the interface between the host computer or device and the internet. The MAC addressing scheme allows each NIC to have a unique address (Hanes, Lowe, Null).

3.8.4.5 DNS and URL Addressing
Rather than memorize the IP address for individual networks and computers, descriptive, user–friendly names may be assigned using Domain Name System (DNS) techniques. The DNS serves as a distributed directory of named network assets. The directory is stored on

3.8 Internet Concepts

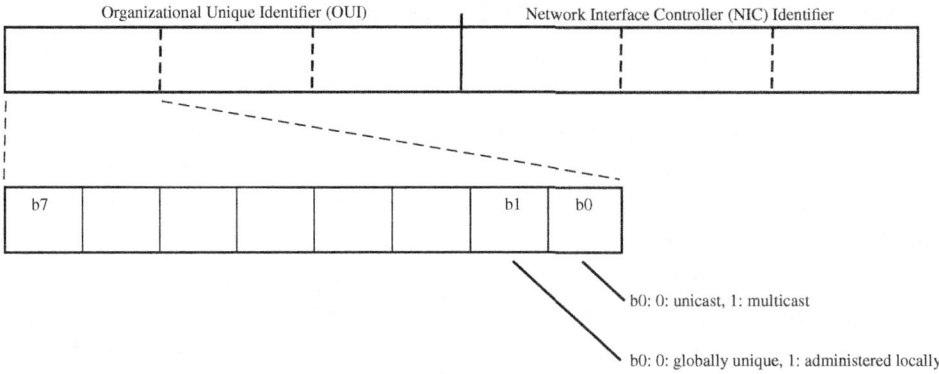

Fig. 3.21 MAC address (Hanes, Lowe, Null)

a number of DNS servers throughout the internet. You may apply for a DNS name from a DNS provider. The provider will determine if the name is available for use. The Internet Corporation for Assigned Names and Numbers (ICANN) coordinates the use of DNS names across the globe (Shuler).

To completely specify the location of a computer on the internet, a Uniform Resource Locator (URL) address is used. The URL address consists of three parts: the protocol identifier, the DNS name, and the domain name.

A common protocol is "http." The protocol type is followed by ": $//www..$" The next portion of the URL address is the DNS name followed by the domain (e.g. .edu, .org, etc.). URL domains are shown in Fig. 3.22. As an example, the website address for the main Arduino site is: http://www.arduino.cc. The ".cc" is a variant of the ".com." domain.

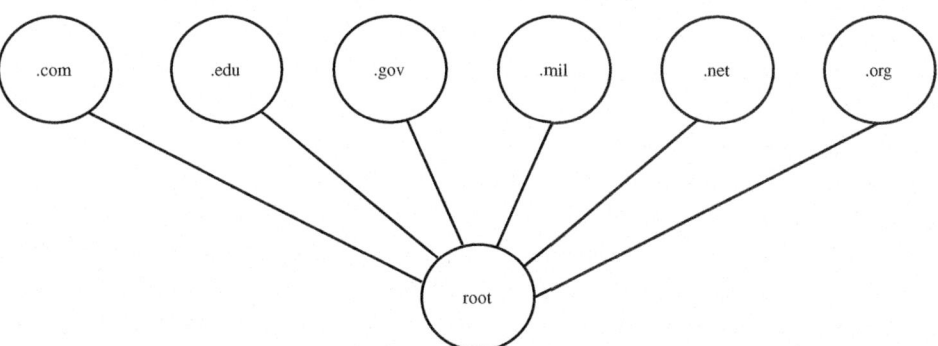

Fig. 3.22 Common URL domains (Shuler)

3.9 PMC Ethernet 10/100BASE T Port

The Arduino PMC, equipped with the Portenta H7 processor, provides access to Ethernet features via a standard RJ45 connector. The ethernet is a method of connecting computers within a Local Area Network (LAN). The 10BASE–T designator indicates a maximum transmission rate of 10 Mbps per second (10) of baseband signal transmission (BASE) with twisted pair cable (T). The 100BASE–T's maximum transmission rate is 100 Mbps per second.

Example: Ethernet 1. The following sketch is provided in the Arduino PMC Library. It provides testing for the PMC Ethernet port. The results from the sketch are provided in Figs. 3.23 and 3.24.

```
//*********************************************************************
//Portenta Machine Control - Ethernet HTTP Request Example
//
//This sketch provides a demonstration of testing the Ethernet port on
//the Portenta Machine Control.  The sketch shows how to create an HTTP
//request as a client, sending it to a remote server and receiving the
//response.
//
//Library: Arduino Portenta Machine Control
//Sketch: Ethernet
//Author: Riccardo Rizzo @Rocketct
//This code example is in the public domain.
//*********************************************************************

#include <PortentaEthernet.h>
#include <Ethernet.h>
#include <SPI.h>

EthernetClient client;
char server[] = "www.ifconfig.io";      //name address (using DNS)

unsigned long beginMicros, endMicros;
unsigned long byteCount = 0;
bool printWebData = true;               //set false for better
                                        //speed measurement
void setup()
{
Serial.begin(9600);
while(!Serial)
  {
   ;
  }

Serial.println("Ethernet example for H7 + PMC");

if(Ethernet.begin() == 0)
  {
  Serial.println("Failed to configure Ethernet using DHCP");
  while(1);
  }
 else
   {
```

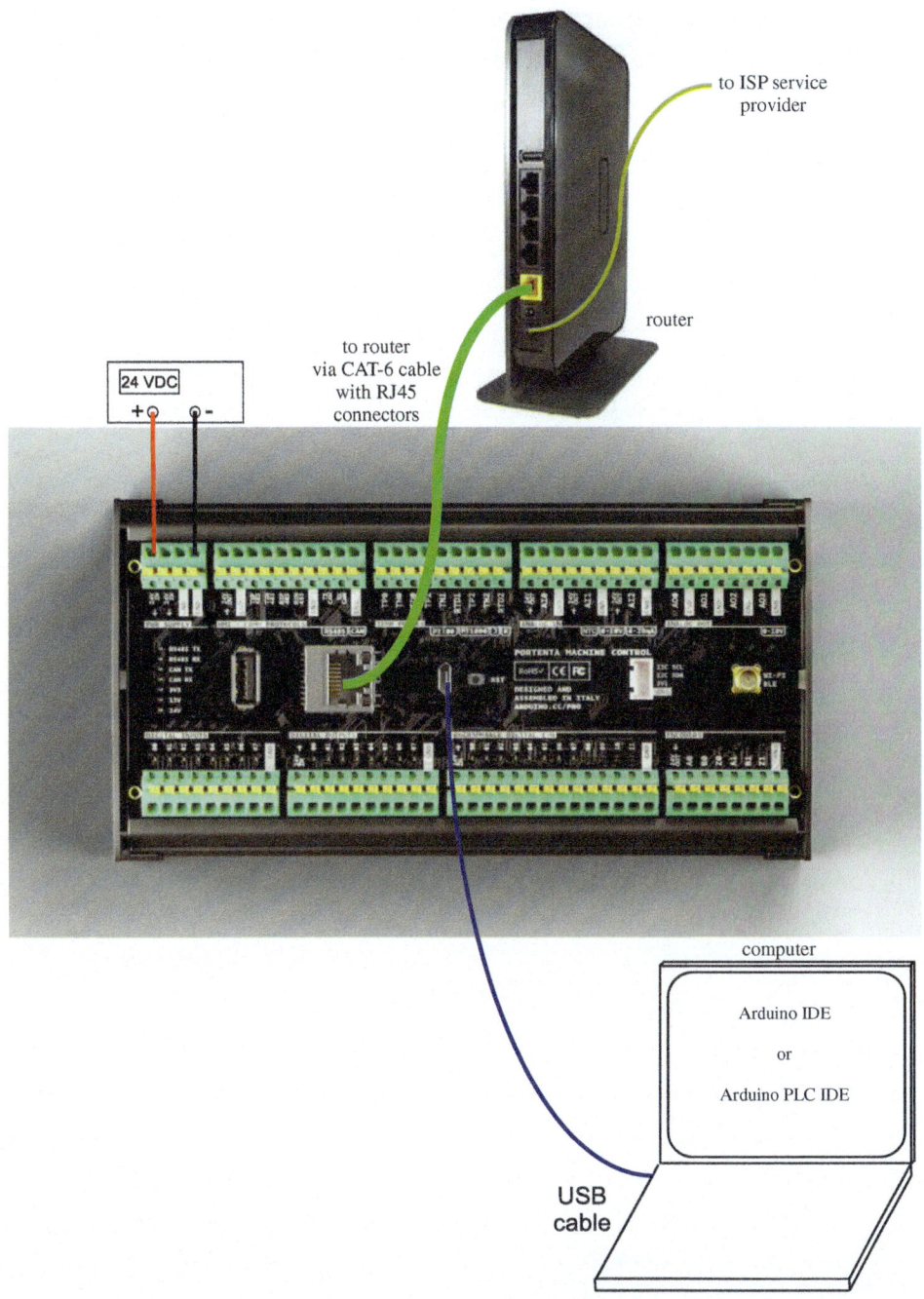

Fig. 3.23 PMC ethernet configuration. Images used courtesy of the Arduino Team (CC BY–NC–SA) (www.arduino.cc)

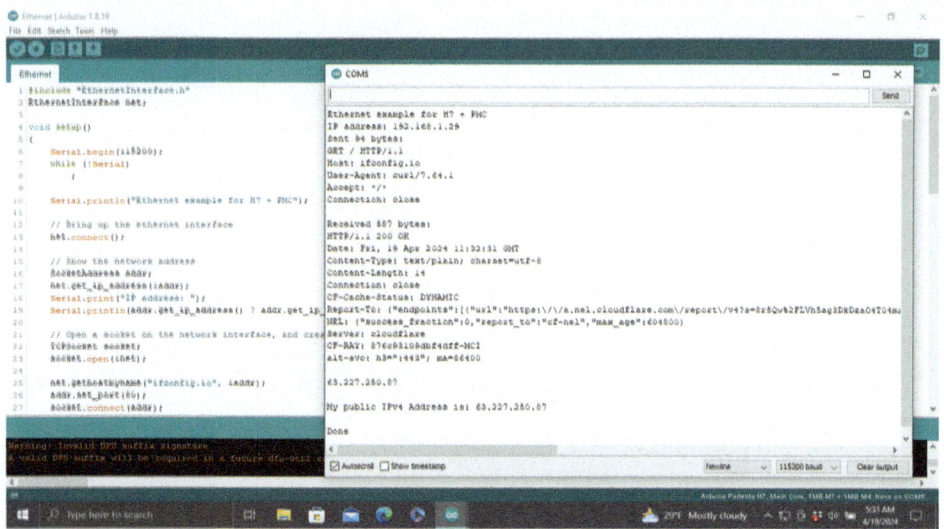

Fig. 3.24 Ethernet sketch result

```
    Serial.print("DHCP assigned IP ");
    Serial.println(Ethernet.localIP());
    }

//give the Ethernet interface a second to initialize:
delay(1000);
Serial.print("connecting to ");
Serial.print(server);
Serial.println("...");

//if you get a connection, report back via serial:
if(client.connect(server, 80))
    {
    Serial.print("connected to ");
    Serial.println(client.remoteIP());

    //Make a HTTP request:
    client.println("GET / HTTP/1.1");
    client.println("Host: ifconfig.io");
    client.println("User-Agent: curl/7.64.1");
    client.println("Connection: close");
    client.println("Accept: */*");
    client.println();
    }
else
    {
    //if you didn't get a connection to the server:
    Serial.println("connection failed");
    }

beginMicros = micros();
}
```

3.9 PMC Ethernet 10/100BASE T Port

```
void loop()
{
//if there are incoming bytes available
//from the server, read them and print them:
int len = client.available();
if(len > 0)
   {
   byte buffer[80];
   if(len > 80) len = 80;
   client.read(buffer, len);
   if(printWebData)
      {
      Serial.write(buffer, len);          //show in the serial monitor
      }
   byteCount = byteCount + len;
   }

//if the server's disconnected, stop the client:
if(!client.connected())
   {
   endMicros = micros();
   Serial.println();
   Serial.println("disconnecting.");
   client.stop();
   Serial.print("Received ");
   Serial.print(byteCount);
   Serial.print(" bytes in ");
   float seconds = (float)(endMicros - beginMicros) / 1000000.0;
   Serial.print(seconds, 4);
   float rate = (float)byteCount / seconds / 1000.0;
   Serial.print(", rate = ");
   Serial.print(rate);
   Serial.print(" kbytes/second");
   Serial.println();

   //do nothing forever
   while(true)
      {
      delay(1);
      }
   }
}
//*********************************************************************
```

The next two examples were originally written for the Arduino Opta. Since both the Opta and the PMC host the Portenta H7, the sketches operate for both devices.

Example: Ethernet 2. For this demonstration, an Arduino PMC is connected to a home router via an RJ45 cable as shown in 3.23. In the sketch the PMC is configured as a client. It will connect to a server configured computer via the internet and request the service provided. In this specific example the server is located at www.ip-api.com. This website hosts an IP Geolocation Application Programming Interface (API). When queried the API provides the geolocation of the IP address (www.ip-api.com) and provides the result on the Serial Monitor.

Note the sketch's use of the Arduino Ethernet and JSON header files. These are available for download via the Library Manager within the Arduino IDE. JSON or Java Script Object Notation is a standard for data interchange in a human readable format (www.json.org).

```
//**********************************************************************
//Web Client (Ethernet version)
//Name: opta_ethernet_web_client.ino
//Purpose: This sketch connects a PMC device to ip-api.com via Ethernet
//and fetches IP details for the device.
//
//@author Arduino PRO Content Team
//@version 2.0 15/08/23
//This code example is in the public domain.
//**********************************************************************

#include <Ethernet.h>                    //Include libraries
#include <Arduino_JSON.h>

const char* server = "ip-api.com";       //Server addr ip-api.com

String path = "/json/";                  //API path for IP details

IPAddress ip(10, 130, 22, 84);           //Static IP config for Opta

//Ethernet client instance for the communication
EthernetClient client;

//JSON variable to store and process the fetched data
JSONVar doc;

//Variable to ensure we fetch data only once
bool dataFetched = false;

void setup()
{
Serial.begin(115200);                    //serial comm at 115200 Baud
while (!Serial);                         //wait for serial port connection

//Attempt to start Ethernet connection via DHCP.
//If DHCP fails, print diagnostic message.
  if(Ethernet.begin() == 0)
    {
    Serial.println("- Failed to configure Ethernet using DHCP!");

    //Try to configure Ethernet with the predefined static IP address.
    Ethernet.begin(ip);
  }
  delay(2000);
}

void loop()
{
//Ensure we haven't fetched data already, ensure the Ethernet link is
//active, establish a connection to the server, and compose and send the
// HTTP GET request.
if(!dataFetched)
  {
  if(Ethernet.linkStatus() == LinkON)
```

3.9 PMC Ethernet 10/100BASE T Port

```
     {
  if(client.connect(server, 80))
     {
     client.print("GET ");
     client.print(path);
     client.println(" HTTP/1.1");
     client.print("Host: ");
     client.println(server);
     client.println("Connection: close");
     client.println();

     //Wait and skip the HTTP headers to get to the JSON data.
     char endOfHeaders[] = "\r\n\r\n";
     client.find(endOfHeaders);

     //Read and parse the JSON response.
     String payload = client.readString();
     doc = JSON.parse(payload);

     //Check if the parsing was successful.
     if(JSON.typeof(doc) == "undefined")
        {
        Serial.println("- Parsing failed!");
        return;
        }

     //Extract and print the IP details.
     Serial.println("*** IP Details:");
     Serial.print("- IP Address: ");
     Serial.println((const char*)doc["query"]);
     Serial.print("- City: ");
     Serial.println((const char*)doc["city"]);
     Serial.print("- Region: ");
     Serial.println((const char*)doc["regionName"]);
     Serial.print("- Country: ");
     Serial.println((const char*)doc["country"]);
     Serial.println("");

     //Mark data as fetched.
     dataFetched = true;
     }
     //Close the client connection once done.
     client.stop();
     }
  else
     {
     Serial.println("- Ethernet link disconnected!");
     }
  }
}

//*************************************************************************
```

The result is shown in Fig. 3.25.

Example: Ethernet 3. The next example, Ethernet WebClient, is available within the Arduino Mbed OS Opta Boards core. Additional instructions are provided within the "Bluetooth Low Energy, Wi–Fi and Ethernet with OPTA (https://opta.findernet.com/en/)." When

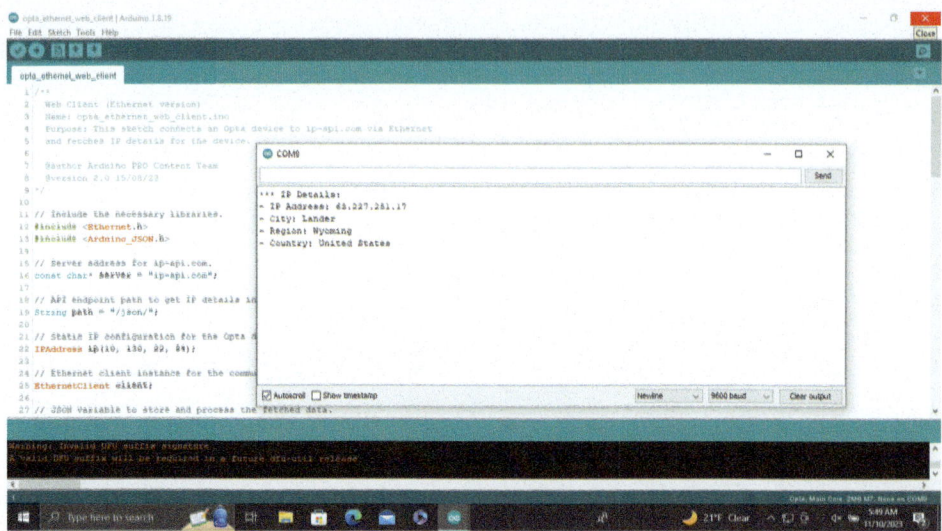

Fig. 3.25 Ethernet sketch result

compiled, uploaded, and executed; the sketch goes to the designated website and retrieves and displays the website's contents to the serial monitor.

```
//**********************************************************************
//WebClient: This sketch connects to a website (http://www.google.com)
//and displays results to the Serial Monitor.
//
//created 18 Dec 2009 by David A. Mellis
//modified 9 Apr 2012 by Tom Igoe based on work by Adrian McEwen
//This code example is in the public domain.
//**********************************************************************

#include <PortentaEthernet.h>
#include <Ethernet.h>
#include <SPI.h>

//Enter a MAC address for your controller below.
//byte mac[] = { 0xDE, 0xAD, 0xBE, 0xEF, 0xFE, 0xED };
// if you don't want to use DNS (and reduce your sketch size)
// use the numeric IP instead of the name for the server:
//IPAddress server(74,125,232,128);    //numeric IP for Google (no DNS)
char server[] = "www.google.com";      //name addr for Google (using DNS)

//Set the static IP address to use if the DHCP fails to assign
IPAddress ip(192, 168, 2, 177);
IPAddress myDns(192, 168, 2, 1);

//Initialize the Ethernet client library with the IP address and
//port of the server that you want to connect to (port 80 is default
//for HTTP):
EthernetClient client;
```

3.9 PMC Ethernet 10/100BASE T Port

```
//Variables to measure the speed
unsigned long beginMicros, endMicros;
unsigned long byteCount = 0;
bool printWebData = true;               //set false-better speed measure

void setup()
{
//Open serial communications and wait for port to open:
Serial.begin(9600);
while (!Serial)
   {
   ;// wait for serial port to connect. Needed for native USB port only
   }

//start the Ethernet connection:
Serial.println("Initialize Ethernet with DHCP:");
if(Ethernet.begin() == 0)
  {
  Serial.println("Failed to configure Ethernet using DHCP");
  //Check for Ethernet hardware present
  if(Ethernet.hardwareStatus() == EthernetNoHardware)
    {
    Serial.println("Ethernet shield was not found.Â");
    Serial.println("Sorry, can't run  without hardware. :(");
    while(true)
       {
       delay(1); //do nothing, no point running without Ethernet hardware
       }
    }
    if(Ethernet.linkStatus() == LinkOFF)
      {
      Serial.println("Ethernet cable is not connected.");
      }
    //try to congifure using IP address instead of DHCP:
    Ethernet.begin(ip, myDns);
    }
    else
      {
      Serial.print("  DHCP assigned IP ");
      Serial.println(Ethernet.localIP());
      }
//give the Ethernet shield a second to initialize:
delay(1000);
Serial.print("connecting to ");
Serial.print(server);
Serial.println("...");

//if you get a connection, report back via serial:
if(client.connect(server, 80))
   {
   Serial.print("connected to ");
   Serial.println(client.remoteIP());

   //Make a HTTP request:
   client.println("GET /search?q=arduino HTTP/1.1");
   client.println("Host: www.google.com");
   client.println("Connection: close");
   client.println()
```

```
    }
  else
    {
    //if you didn't get a connection to the server:
    Serial.println("connection failed");
    }
beginMicros = micros();
}

void loop()
{
//if there are incoming bytes available from the server,
//read them and print them:
int len = client.available();
if(len > 0)
  {
  byte buffer[80];
  if(len > 80)
     len = 80;
  client.read(buffer, len);
  if(printWebData)
    {
    Serial.write(buffer, len);            //show serial monitor
    }
        byteCount = byteCount + len;
  }
//if the server's disconnected, stop the client:
if(!client.connected())
  {
  endMicros = micros();
  Serial.println();
  Serial.println("disconnecting.");
  client.stop();
  Serial.print("Received ");
  Serial.print(byteCount);
  Serial.print(" bytes in ");
  float seconds = (float)(endMicros - beginMicros) / 1000000.0;
  Serial.print(seconds, 4);
  float rate = (float)byteCount / seconds / 1000.0;
  Serial.print(", rate = ");
  Serial.print(rate);
  Serial.print(" kbytes/second");
  Serial.println();

  //do nothing forevermore:
  while(true)
    {
    delay(1);
    }
  }
}

//*********************************************************************
```

3.10 Wi–Fi 802.11 B/G/N

The Portenta Machine Control hosts the Portenta H7 processor. The H7 is equipped with hardware to support the 802.11 b/g/n Wireless Fidelity (Wi–Fi) standard. The b/g/n suffix designates different bit rate and frequency features for 802.11 versions: b (11 Mbps, 2.4 GHz, 1999); g(6 to 54 Mbps, 2.4 GHz, 2003); and n (72 to 600 Mbps, 2.4 and 5 Ghz, 2008). The H7 Wi–Fi features also includes support for WEP, WPA, WPA2, and WPA3 security standards. For Wi–Fi operation the PMC requires a 2.4 GHz antenna (e.g. dual band Wi–Fi 2.4 GHz, 5 GHz, SMA male antenna). The antenna is connected to the PMC via the SMA connector.

The next two examples are Wi–Fi versions of the two previous ethernet examples.

Example: Wi–Fi 1. For this demonstration, an Ardunio PMC is connected to a home router via Wi–Fi. The PMC is programmed using a laptop/PC. In the sketch the PMC is configured as a client. It will connect to a server configured computer via the internet and request the service provided. In this specific example the server is located at www.ip-api.com. This website hosts an IP Geolocation Application Programming Interface (API). When queried the API provides the geolocation of the IP address (www.ip-api.com) and provides the result on the Serial Monitor.

Note the sketch's use of the Arduino Wi–Fi and JSON header files. These are available for download via the Library Manager within the Arduino IDE. JSON or Java Script Object Notation is a standard for data interchange in a human readable format (www.json.org).

```
//*********************************************************************
//Wi-Fi Web Client
//
 //Purpose: This sketch connects an PMC device to ip-api.com via Wi-Fi
//and fetches IP details.
//
//@author Arduino PRO Content Team
//@version 2.2 16/08/23
//This code example is in the public domain.
//*********************************************************************

#include <WiFi.h>
#include <Arduino_JSON.h>

//Wi-Fi network details.
const char* ssid     = "insert your network name";
const char* password = "insert your network password";

//Server address for ip-api.com.
const char* server = "ip-api.com";

//API endpoint path to get IP details in JSON format.
String path = "/json";

//Wi-Fi client instance for the communication.
WiFiClient client;

//JSON variable to store and process the fetched data.
JSONVar doc;
```

```
//Variable to ensure we fetch data only once.
bool dataFetched = false;

void setup()
{
//Begin serial communication at a baud rate of 115200.
Serial.begin(115200);

//Wait for the serial port to connect,
//This is necessary for boards that have native USB.
while (!Serial);

//Start the Wi-Fi connection using the provided SSID and password.
Serial.print("- Connecting to ");
Serial.println(ssid);
WiFi.begin(ssid, password);

while(WiFi.status() != WL_CONNECTED)
    {
    delay(1000);
    Serial.print(".");
    }

Serial.println();
Serial.println("- Wi-Fi connected!");
Serial.print("- IP address: ");
Serial.println(WiFi.localIP());
}

void loop()
{
//Check if the IP details have been fetched.
//If not, call the function to fetch IP details,
//Set the flag to true after fetching.
if(!dataFetched)
    {
    fetchIPDetails();
    dataFetched = true;
    }
}

//*********************************************************************
//Fetch IP details from defined server
//@param none
//@return IP details
//*********************************************************************

void fetchIPDetails()
    {
    if(client.connect(server, 80))
        {
        //Compose and send the HTTP GET request.
        client.print("GET ");
        client.print(path);
        client.println(" HTTP/1.1");
        client.print("Host: ");
        client.println(server);
        client.println("Connection: close");
```

3.10 Wi–Fi 802.11 B/G/N

```
        client.println();

        //Wait and skip the HTTP headers to get to the JSON data.
        char endOfHeaders[] = "\r\n\r\n";
        client.find(endOfHeaders);

        //Read and parse the JSON response.
        String payload = client.readStringUntil('\n');
        doc = JSON.parse(payload);

        //Check if the parsing was successful.
        if(JSON.typeof(doc) == "undefined")
          {
          Serial.println("- Parsing failed!");
          return;
          }

        //Extract and print the IP details.
        Serial.println("*** IP Details:");
        String query = doc["query"];
        Serial.print("- IP Address: ");
        Serial.println(query);
        String city = doc["city"];
        Serial.print("- City: ");
        Serial.println(city);
        String region = doc["regionName"];
        Serial.print("- Region: ");
        Serial.println(region);
        String country = doc["country"];
        Serial.print("- Country: ");
        Serial.println(country);
        Serial.println("");
        }
    else
        {
        Serial.println("- Failed to connect to server!");
        }
//Close the client connection once done.
client.stop();
}

//**********************************************************************
```

Example: PMC Wi–Fi 2. The next example, Wi–Fi WebClient, is available within the Arduino Mbed OS Opta Boards core. Additional instructions are provided within the "Bluetooth Low Energy, Wi–Fi and Ethernet with OPTA (https://opta.findernet.com/en/)." When compiled, uploaded, and executed; the sketch goes to the designated website and retrieves and displays the website's contents to the serial monitor.

```
//**********************************************************************
//Web client: This sketch connects to a website (http://example.com)
//using the Wi-Fi module.  This example is written for a network using
//WPA encryption. For WEP or WPA, change the Wifi.begin() call
//accordingly.
//
```

```
//created 13 July 2010 by dlf (Metodo2 srl)
//modified 31 May 2012 by Tom Igoe
//This code example is in the public domain.
//**********************************************************************

#include <WiFi.h>

#include "arduino_secrets.h"
//enter your sensitive data in the Secret tab/arduino_secrets.h
char ssid[] = SECRET_SSID;         //your network SSID (name)
char pass[] = SECRET_PASS;         //your network password (use for WPA,
                                   //or use as key for WEP)
int keyIndex = 0;                  //your network key Index number
                                   //(needed only for WEP)

int status = WL_IDLE_STATUS;
//if you don't want to use DNS (and reduce your sketch size)
//use the numeric IP instead of the name for the server:
//IPAddress server(93,184,216,34);//IP address for example.com (no DNS)
char server[] = "example.com";    //host name for example.com (using DNS)

WiFiClient client;

void setup()
{
//Initialize serial and wait for port to open:
Serial.begin(9600);
while (!Serial)
   {
   ; //wait for serial port to connect. Needed for native USB port only
   }

//check for the WiFi module:
if(WiFi.status() == WL_NO_SHIELD)
   {
   Serial.println("Communication with WiFi module failed!");
   //don't continue
   while (true);
   }

//attempt to connect to Wifi network:
while(status != WL_CONNECTED)
   {
   Serial.print("Attempting to connect to SSID: ");
   Serial.println(ssid);
   //Connect to WPA/WPA2 network. Change line if using open or WEP
   //network:
   status = WiFi.begin(ssid, pass);

   //wait 3 seconds for connection:
   delay(3000);
   }

Serial.println("Connected to wifi");
printWifiStatus();

Serial.println("\nStarting connection to server...");
//if you get a connection, report back via serial:
if(client.connect(server, 80))
```

3.10 Wi–Fi 802.11 B/G/N

```
  {
  Serial.println("connected to server");

  //Make a HTTP request:
  client.println("GET /index.html HTTP/1.1");
  client.print("Host: ");
  client.println(server);
  client.println("Connection: close");
  client.println();
  }
}

void loop()
{
//if there are incoming bytes available
//from the server, read them and print them:
while (client.available())
  {
  char c = client.read();
  Serial.write(c);
  }

//if the server's disconnected, stop the client:
if(!client.connected())
  {
  Serial.println();
  Serial.println("disconnecting from server.");
  client.stop();

  //do nothing forevermore:
  while (true);
  }
}

//***********************************************************************
void printWifiStatus(void)
{
//print the SSID of the network you're attached to:
Serial.print("SSID: ");
Serial.println(WiFi.SSID());

//print your board's IP address:
IPAddress ip = WiFi.localIP();
Serial.print("IP Address: ");
Serial.println(ip);

//print the received signal strength:
long rssi = WiFi.RSSI();
Serial.print("signal strength (RSSI):");
Serial.print(rssi);
Serial.println(" dBm");
}

//***********************************************************************
```

3.11 Controller Area Network

The Controller Area Network or CAN was developed in the early 1980s as a method of easily linking multiple electronic control units (ECUs) in a vehicle. The CAN concept was originally developed by Robert Bosch GmbH. There are several versions of the CAN standard. We concentrate on CAN 2.0 (Voss, Voss).

The ECUs or nodes serve as transceivers within the CAN network. Each transceiver can gather local data and issue control signals and then share it over a two–wire network with other ECUs for collective control. The schematic of a CAN network is provided in Fig. 3.26. Nodes within the network communicate via a two differential signal trunk line. The signal lines are designated CANL and CANH. Each ECU (node) is connected to the trunk via stubs. Processors with an onboard CAN transceiver (e.g. the PMC, Arduino UNO R4 Minima) connect directly to the trunk. A processor without an onboard CAN transceiver requires an external one (e.g. UNO R3).

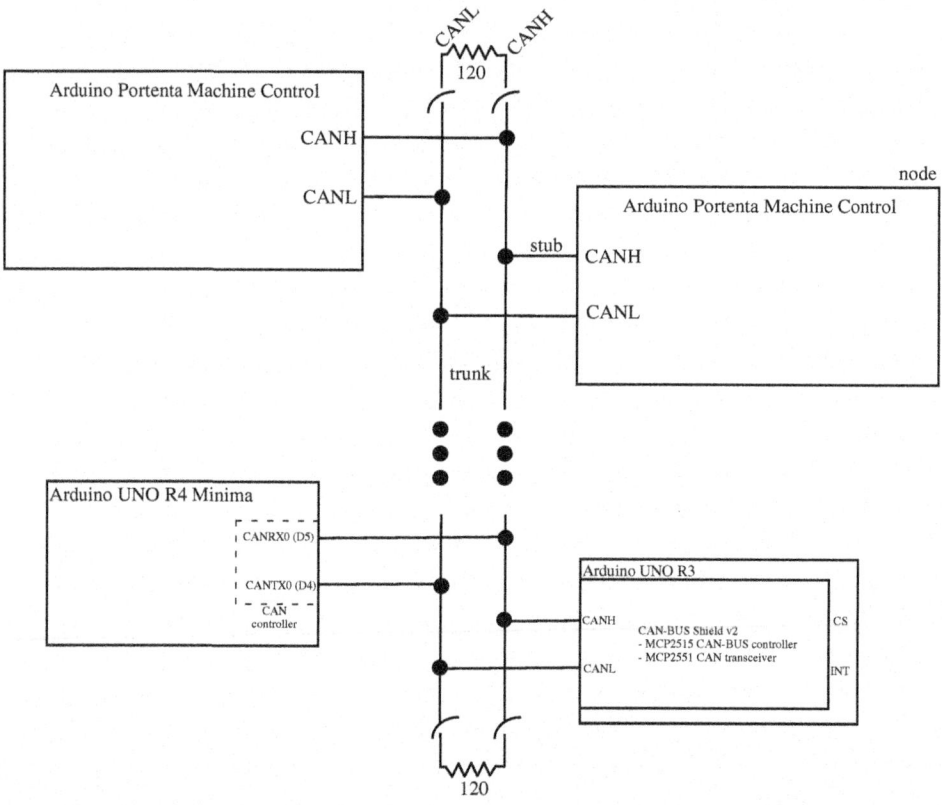

Fig. 3.26 CAN network

3.11 Controller Area Network

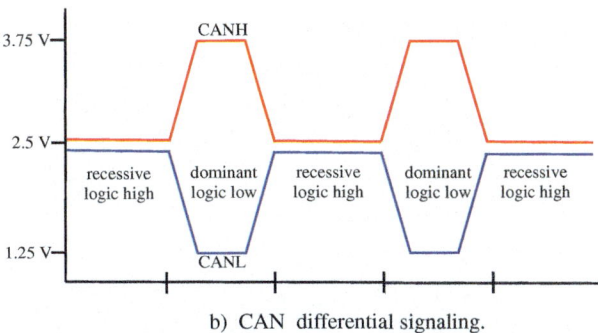

Fig. 3.27 CAN theory. **a** Can frame partition. **b** CAN differential signaling

CAN differential signaling is shown in Fig. 3.27b. A logic one is signaled as a recessive logic high. In this case the CANH and CANL signal are close to 2.5 VDC. A logic low is signaled with a dominant logic low. For this case CANH is at 3.75 VDC and CANL is at 1.25 VDC.

The CAN message packet is shown in Fig. 3.27a. Each CAN node has a separate ID placed in the arbitration portion of the frame. A lower number indicates a higher priority. The data packet may contain up to eight data bytes. The data is followed by a CRC. The beginning and end of frame is signaled with a start of frame (SOF) and end of frame (EOF) designator.

Any ECU can transmit on the CAN network. If two ECUs try to access the trunk at the same time, the ECU with the higher priority is allowed network access. All ECUs receive a transmitted message. It is up to the individual ECUs (nodes) to determine if they will respond to a specific message. As with other communications there is a tradeoff between transmission speed and trunk length. The upper bound of transmission speed is 1 Mbps with CAN 2.0.

Example. PMC to PMC CAN network. In this example two PMCs are connected via a CAN network. The CANH of each PMC is connected with one another. Also, the CANL of each PMC is connected to one another. Also, the PMCs should share a common ground. For this example, one PMC is loaded with WriteCAN from the Arduino Portenta Machine Control Library; whereas, the other PMC is configured with the ReadCAN sketch.

Example. PMC to UNO R3 CAN network. In this example a PMC is connected to an Arduino UNO R3. The UNO R3 is equipped with a seeed studio CAN–BUS Shield V2. The shield has an onboard MCP2515 CAN–BUS controller and MCP2551 CAN transceiver as shown in Fig. 3.28.

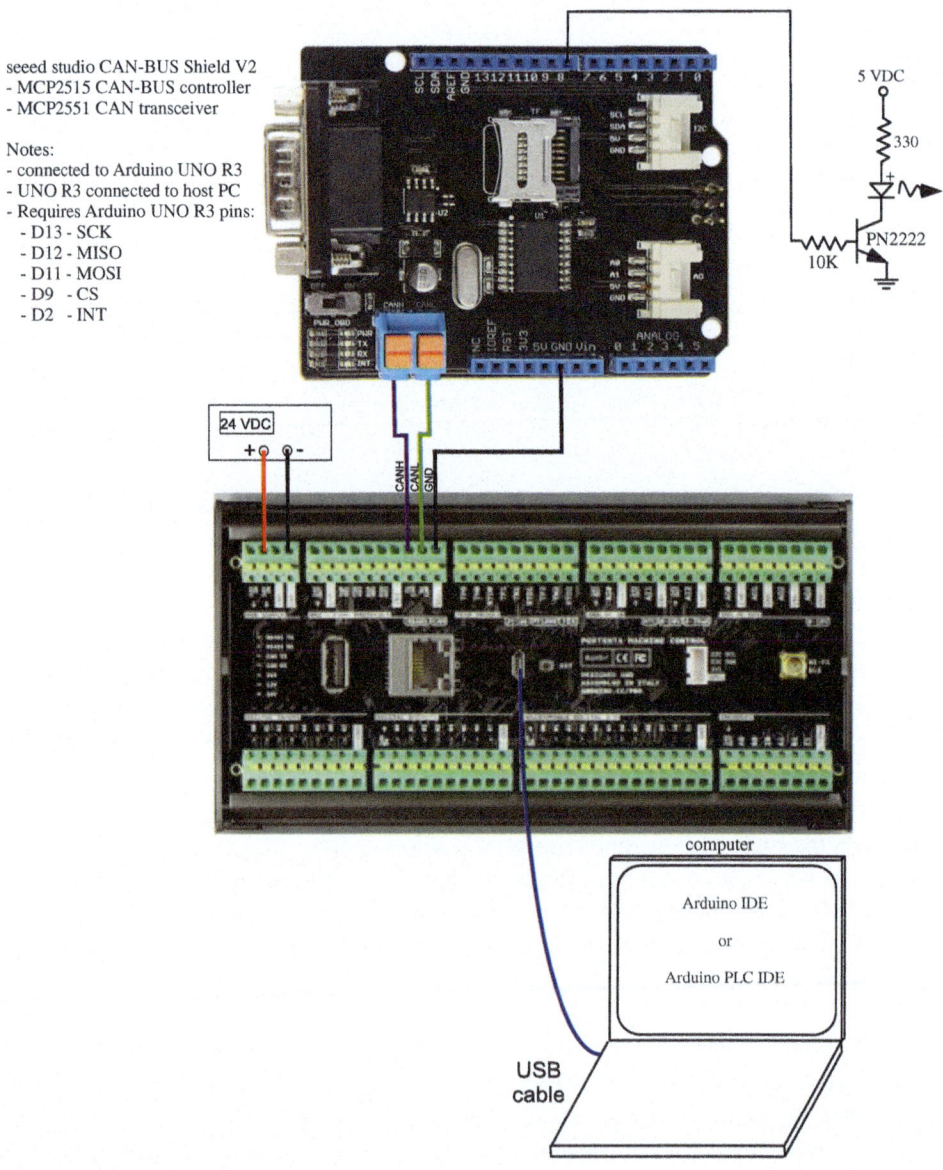

Fig. 3.28 PMC to UNO R3 CAN network. Images used courtesy of the Arduino Team (CC BY–NC–SA) (www.arduino.cc)

3.11 Controller Area Network

PMC WriteCAN sketch:

```
//***********************************************************************
//Portenta Machine Control - CAN Write Example
//This sketch shows the use of the CAN transceiver on the Portenta
//Machine Control and demonstrates how to transmit data from the TX
//CAN channel.
//Circuit:
//   - Portenta Machine Control
//
//Adapted from:
//Library: Arduino Portenta Machine Control
//Sketch: WriteCAN
//Authors: Riccardo Rizzo @Rocketct, Leonardo Cavagnis @leonardocavagnis
//Adapted:  S. Barrett
//This code example is in the public domain.
//***********************************************************************

#include <Arduino_PortentaMachineControl.h>

static uint32_t const CAN_ID = 13ul;        //CAN node identifier
static uint32_t msg_cnt = 0;
uint8_t msg_data[] = {0x01,0x02,0x03,0x04,0x05,0x06,0x07,0x08};

void setup()
{
Serial.begin(9600);
while (!Serial)
   {
   ; // wait for serial port to connect.
   }
                                         //Nodes set to same TX/RX rate
if(!MachineControl_CANComm.begin(CanBitRate::BR_500k))
   {
   Serial.println("CAN init failed.");
   while(1) ;
   }
                                         //initial data packet
//uint8_t msg_data[] = {0x01,0x02,0x03,0x04,0x05,0x06,0x07,0x08};
}

void loop()
{                                        //assemble CAN message
CanMsg msg(CAN_ID, sizeof(msg_data), msg_data);
                                         //transmit CAN message
int const rc = MachineControl_CANComm.write(msg);
if(rc <= 0)
   {
   Serial.print("CAN write failed with error code: ");
   Serial.println(rc);
   while(1) ;
   }
                                         //to serial monitor
Serial.print("CAN write message:");
for(unsigned int i = 0; i <=7; i++)
   {
   Serial.print(msg_data[i]);
   }
Serial.println(" ");
```

```
                                        //update message data
if(msg_data[7] == 0xff)
    msg_data[7] = 0x00;
msg_data[7] = msg_data[7] + 1;

Serial.print("CAN write message:");
for(unsigned int i = 0; i <=7; i++)
  {
  Serial.print(msg_data[i]);
  }
Serial.println(" ");
delay(1000);                            //message interval
}

//**********************************************************************
```

CAN sketch for UNO R3 equipped with seeed studio CAN–BUS Shield V2;

```
//**********************************************************************
//Library: Arduino seeed studio CAN_BUS_SHIELD
//Sketch: receive_blink
//This code example is in the public domain.
//**********************************************************************

#include <SPI.h>

#define CAN_2515

//Set SPI CS Pin according to your hardware
#if defined(SEEED_WIO_TERMINAL) && defined(CAN_2518FD)
const int SPI_CS_PIN  = BCM8;
const int CAN_INT_PIN = BCM25;
#else

//For Arduino MCP2515 Hat:
//the cs pin of the version after v1.1 is default to D9
//v0.9b and v1.0 is default D10
const int SPI_CS_PIN = 9;
const int CAN_INT_PIN = 2;
#endif

#ifdef CAN_2518FD
#include "mcp2518fd_can.h"
mcp2518fd CAN(SPI_CS_PIN);              //Set CS pin
#endif

#ifdef CAN_2515
#include "mcp2515_can.h"
mcp2515_can CAN(SPI_CS_PIN);            //Set CS pin
#endif

const int LED        = 8;
boolean ledON        = 1;

void setup()
{
SERIAL_PORT_MONITOR.begin(115200);
```

```
  pinMode(LED, OUTPUT);
                                          //init can bus : baudrate = 500k
  while(CAN_OK != CAN.begin(CAN_500KBPS))
    {
    SERIAL_PORT_MONITOR.println("CAN init fail, retry...");
    delay(100);
    }
    SERIAL_PORT_MONITOR.println("CAN init ok!");
}


void loop()
{
unsigned char len = 0;
unsigned char buf[8];

if(CAN_MSGAVAIL == CAN.checkReceive())
    {                                   //check if data coming
    CAN.readMsgBuf(&len, buf);          //read data, len: data length,
                                        //buf: data buf
    unsigned long canId = CAN.getCanId();

    SERIAL_PORT_MONITOR.println("-----------------------------");
    SERIAL_PORT_MONITOR.println("get data from ID: 0x");
    SERIAL_PORT_MONITOR.println(canId, HEX);

    for(int i = 0; i < len; i++)
      {                                 //print the data
      SERIAL_PORT_MONITOR.print(buf[i]);
      SERIAL_PORT_MONITOR.print("\t");
      if(ledON && i == 0)
        {
        digitalWrite(LED, buf[i]);
        ledON = 0;
        delay(500);
        }
      else if ((!(ledON)) && i == 4)
        {
        digitalWrite(LED, buf[i]);
        ledON = 1;
          }
        }
        SERIAL_PORT_MONITOR.println();
    }
}
//**********************************************************************
```

3.12 RS–485 Communication

The RS–485 (TIA/EIA–485) is an electrical standard providing for communication between devices in a noisy, industrial environment. The standard does not define a specific communication protocol but provides the electrical specifications to implement specific protocols such as MODBUS RTU.

The RS–485 standard provides for cable runs up to 1200 m ($\bar{4}000$ ft) and data rates up to 10 Mbps, and up to 32 RS–485 configured units. There is a tradeoff between maximum data rate and cable length. That is, a data rate of 10 Mbps per second is possible for a cable length of 40 ft. Whereas, at a cable length of 4000 ft data rate is limited to 100 kbps. Depending on the specific application, system requirements, and configuration; termination resistors may be required at either end of the RS–485 cable run. In general, long cable runs transmitting at high data rates require termination resistors. Typically a 120 Ω resistor is used for termination (Horowitz and Hill). The PMC RS–485 provides for either one–way (half–duplex) or two–way (full –duplex) communication between industrial devices using a differential signaling scheme. As shown in Fig. 3.29, there are four signal lines designated TXP, TXN, RXP, RXN, and a shared ground connection. A logic one (Mark) is signaled with a low signal on the TXN terminal and a high signal on the TXP terminal. A logic low (Space) is signaled with a high signal on the TXN terminal and a low signal on the TXP terminal. A twisted pair cable is used to connect RS–485 compatible devices.

Example. In this example we connect two Arduino PMCs via an RS–485 link with a 24 AWG twisted pair cable and common ground. See Fig. 3.30. We use the

We use the RS–485 half duplex sketch from the Portenta Machine Control Library. We designate the PMCs as PMC1 and PMC2 within the code and also set the sendInterval variable to 3000 for PMC1 and 5000 for PMC2. The Unified Modeling Language (UML) activity diagram for the sketch is provided in Fig. 3.31 (Fowler and Scott).

```
//**********************************************************************
//Portenta Machine Control - RS485 Half Duplex Communication Example
//
//This sketch demonstrates operation of the SP335ECR1 Multiple Protocol
//Transceiver (RS-232/422/485) onboard the Portenta Machine Control.
//This example is configured for RS-485 half duplex operation.  It
//periodically (sendInterval) sends a message on the RS-485 channel.
//
//PMC1 to PMC2 connections:
// - TXP to TXP, TXN to TXN, RXP to RXP, RXN to RXN
//
//Adapted from:
//Library: Arduino Portenta Machine Control
//Sketch: RS485_HalfDuplex
//Authors: Riccardo Rizzo @Rocketct, Leonardo Cavagnis @leonardocavagnis
//Adapted by: S. Barrett, May 5, 2024
//This code example is in the public domain.
//**********************************************************************

#include "Arduino_PortentaMachineControl.h"
                                        //message send interval (ms)
constexpr unsigned long sendInterval {3000};
unsigned long sendNow {0};
unsigned long counter {0};

void setup()
```

3.12 RS–485 Communication

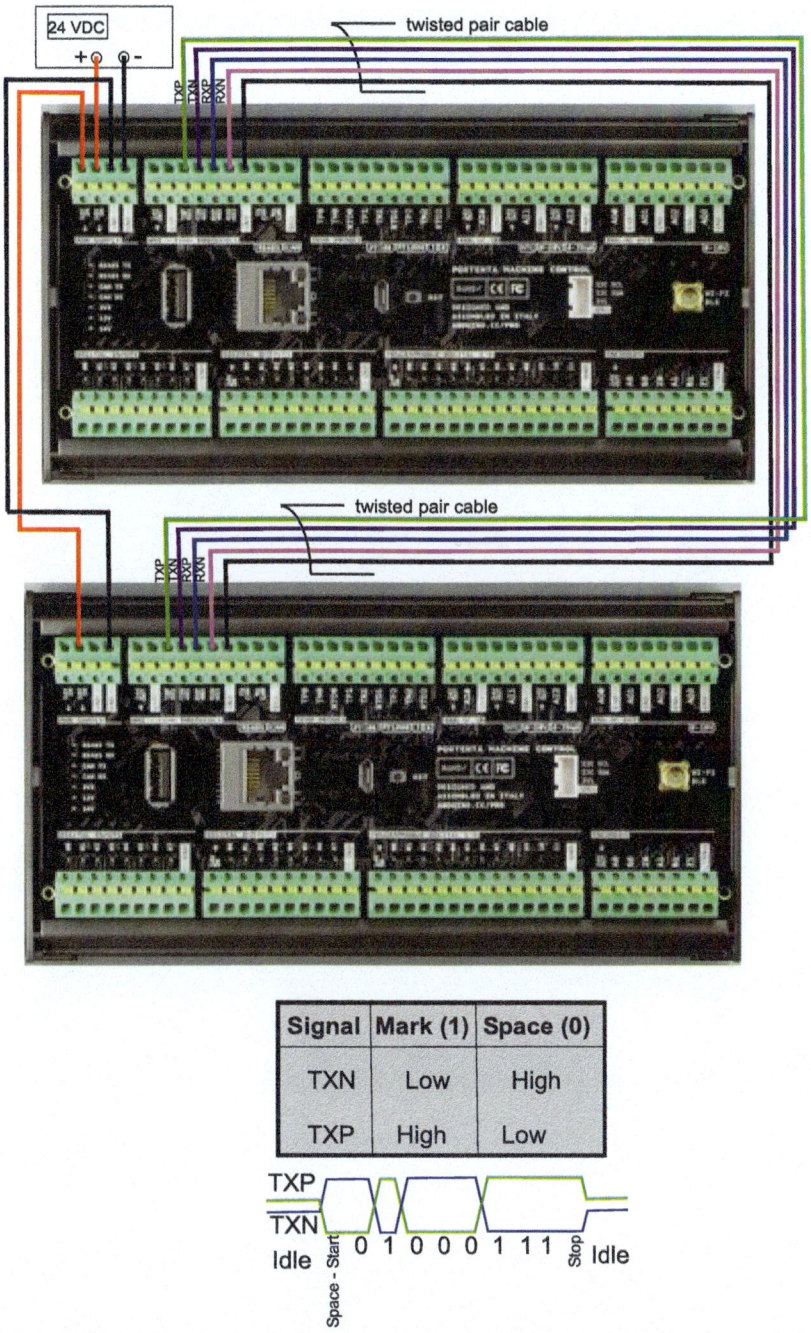

Fig. 3.29 Arduino PMC RS–485. Images used courtesy of the Arduino Team (CC BY–NC–SA) (www.arduino.cc)

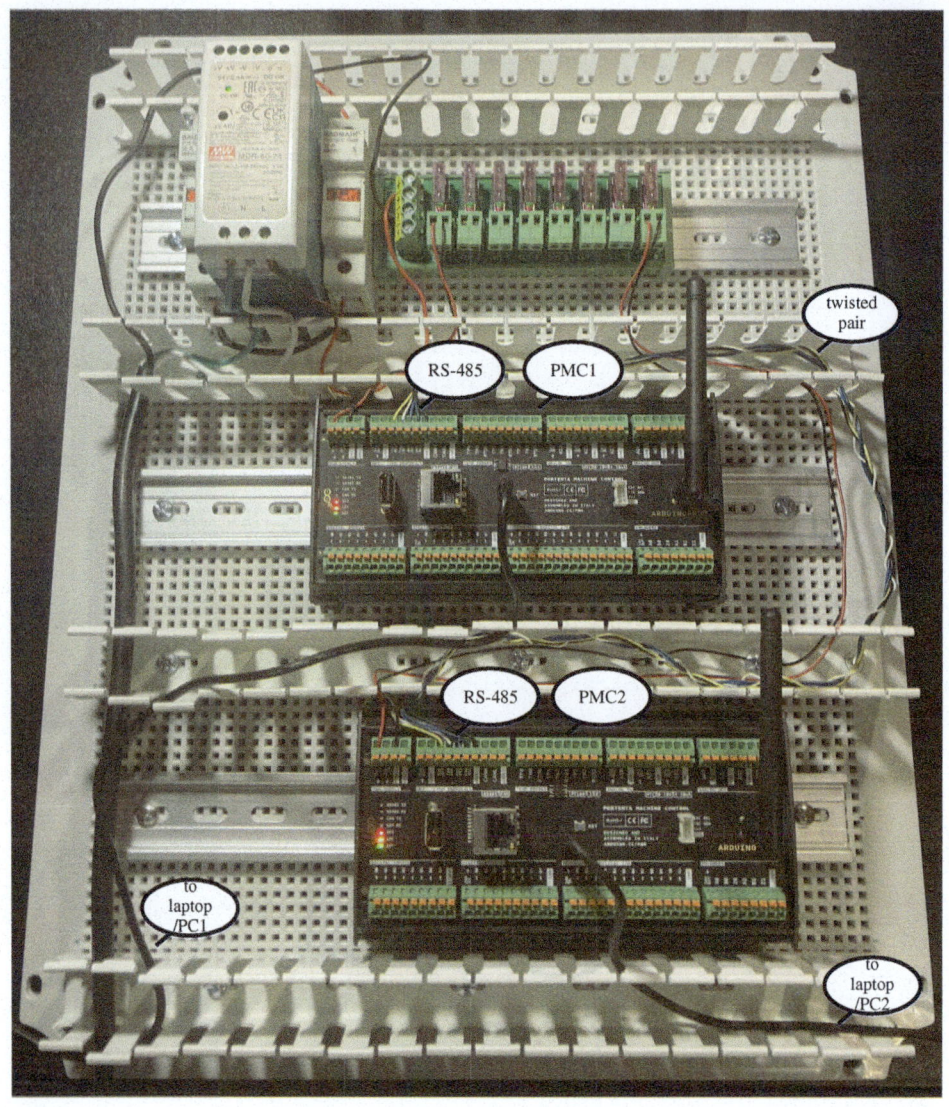

Fig. 3.30 Arduino PMC RS–485 link. Images used courtesy of the Arduino Team (CC BY–NC–SA) (www.arduino.cc)

3.12 RS–485 Communication

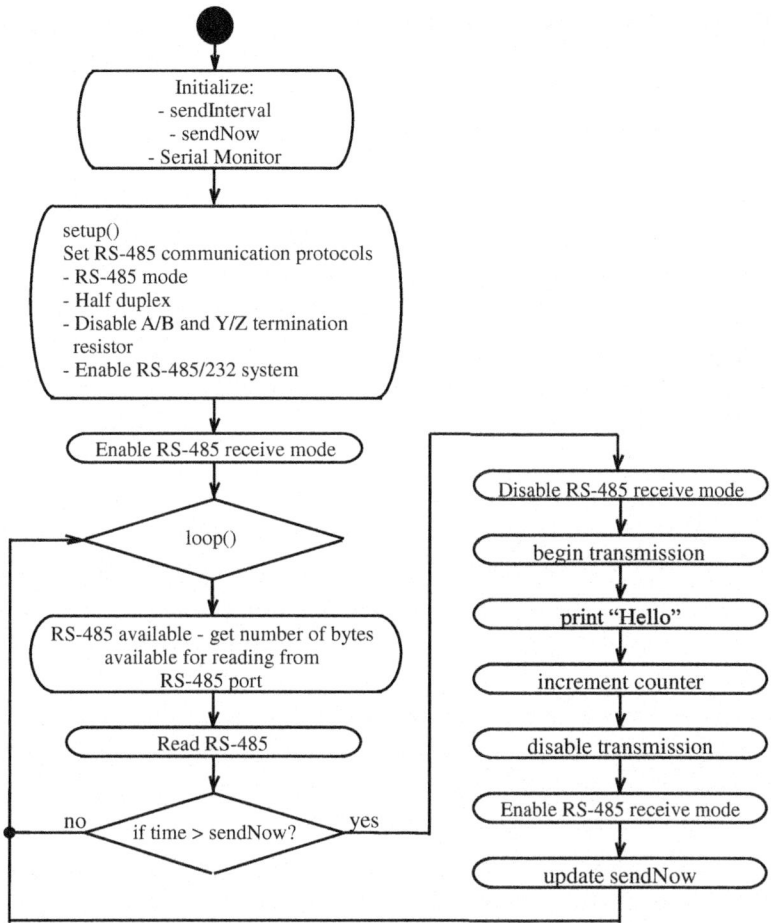

Fig. 3.31 Arduino PMC RS–485 sketch UML

```
{
Serial.begin(9600);                    //Set Baud rate for Serial Monitor
while (!Serial)
  {
  ;
  }

delay(1000);
Serial.println("Start PMC1 RS485 initialization");

//Set the PMC Communication Protocols to default config
//RS485/RS232 default config is:
// - RS485 mode
// - Half Duplex
// - No A/B and Y/Z 120 Ohm termination enabled
// - Enable the RS485/RS232 system
```

```
//Specify baudrate, and preamble and postamble times for RS485
MachineControl_RS485Comm.begin(115200, 0, 500);

//Start in receive mode
MachineControl_RS485Comm.receive();

Serial.println("PMC1 Initialization done!");
}

void loop()
{

//Get number of bytes available for reading from the RS-485 port.
//This is data that has already arrived and is stored in the serial
//receive buffer.
if(MachineControl_RS485Comm.available())
  //Reads incoming serial data.  Returns byte of incoming serial data.
  Serial.write(MachineControl_RS485Comm.read());

if(millis() > sendNow)
  {
  //Disable receive mode before transmission
   MachineControl_RS485Comm.noReceive();

  //Enables RS-485 transmission
   MachineControl_RS485Comm.beginTransmission();

  //Send message out to RS-485 link
   MachineControl_RS485Comm.print("hello from PMC1");

  //Send message count out to RS-485 link
   MachineControl_RS485Comm.println(counter++);

  //Disables RS-485 transmission
   MachineControl_RS485Comm.endTransmission();

  //Re-enable receive mode after transmission
   MachineControl_RS485Comm.receive();

  //Update timing interval
   sendNow = millis() + sendInterval;
   }
}

//***********************************************************************
```

Provided in Fig. 3.32 is a screen shot from PMC1. The first two lines are from PMC1 initialization. The next four lines (with misspellings) are received from PMC2.

The RS–485 standard provides for the interconnection of multiple (up to 32) RS–485 equipped dissimilar devices. In the next example we connect a PMC to an RS–485 equipped Arduino UNO R3. The example illustrates how two devices separated by a great distance can exchange data and control information via the RS–485 link.

Example. In this example we connect an Arduino PMC to an Arduino UNO R3 via an RS–485 link as shown in Fig. 3.33. The UNO R3 does not have a resident RS–485 transceiver. The UNO R3 is equipped with a indexTinySine RS232/RS485 transceiver Tiny-

3.12 RS–485 Communication

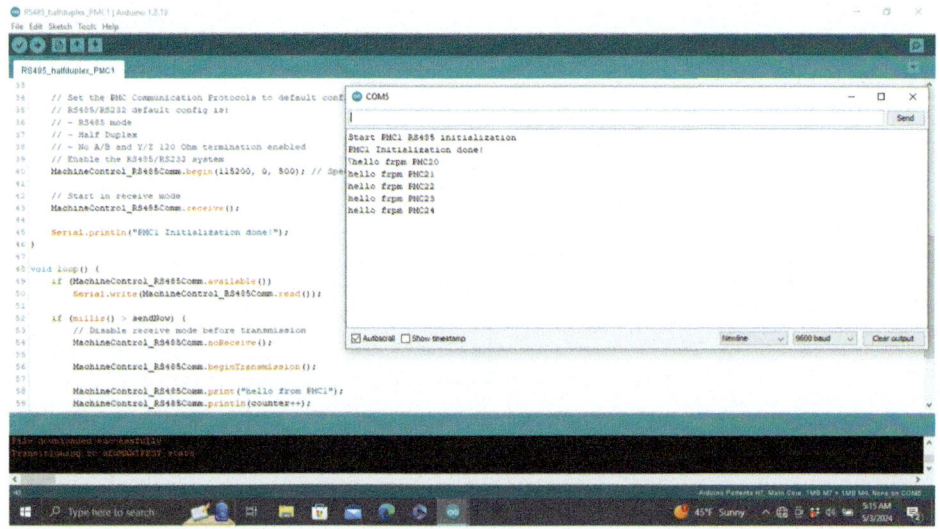

Fig. 3.32 Screen shot from PMC1 RS–485

Sine RS232/RS485 transceiver shield as shown in Fig. 3.34. The UNO R3 communicates with the shield via a Software Serial UART connection.

The PMC sketch used in the previous example is slightly modified for communication with the UNO R3. Note the BAUD rate setting must be the same for both the PMC and the UNO R3. The sketch sequentially sends a number one through four to the UNO R3. When received the UNO R3 executes a different process for the number received. When the specified process is complete the UNO R3 returns the process number to the PMC.

```
//*********************************************************************
//Portenta Machine Control - RS485 Half Duplex Communication Example
//
//This sketch demonstrates operation of the SP335ECR1 Multiple Protocol
//Transceiver (RS-232/422/485) onboard the Portenta Machine Control.
//This example is configured for RS-485 half duplex operation.  It
//periodically (sendInterval) sends a message on the RS-485 channel
//to an Arduino UNO R3.
//
//PMC1 to UNO RS connections:
// - TXP to A
// - TXN to B
//
//Adapted from:
//Library:  Arduino Portenta Machine Control
//Sketch: RS485_HalfDuplex
//Author: Riccardo Rizzo @Rocketct, Leonardo Cavagnis @leonardocavagnis
//Adapted by: S. Barrett, May 9, 2024
//This code example is in the public domain.
//*********************************************************************
```

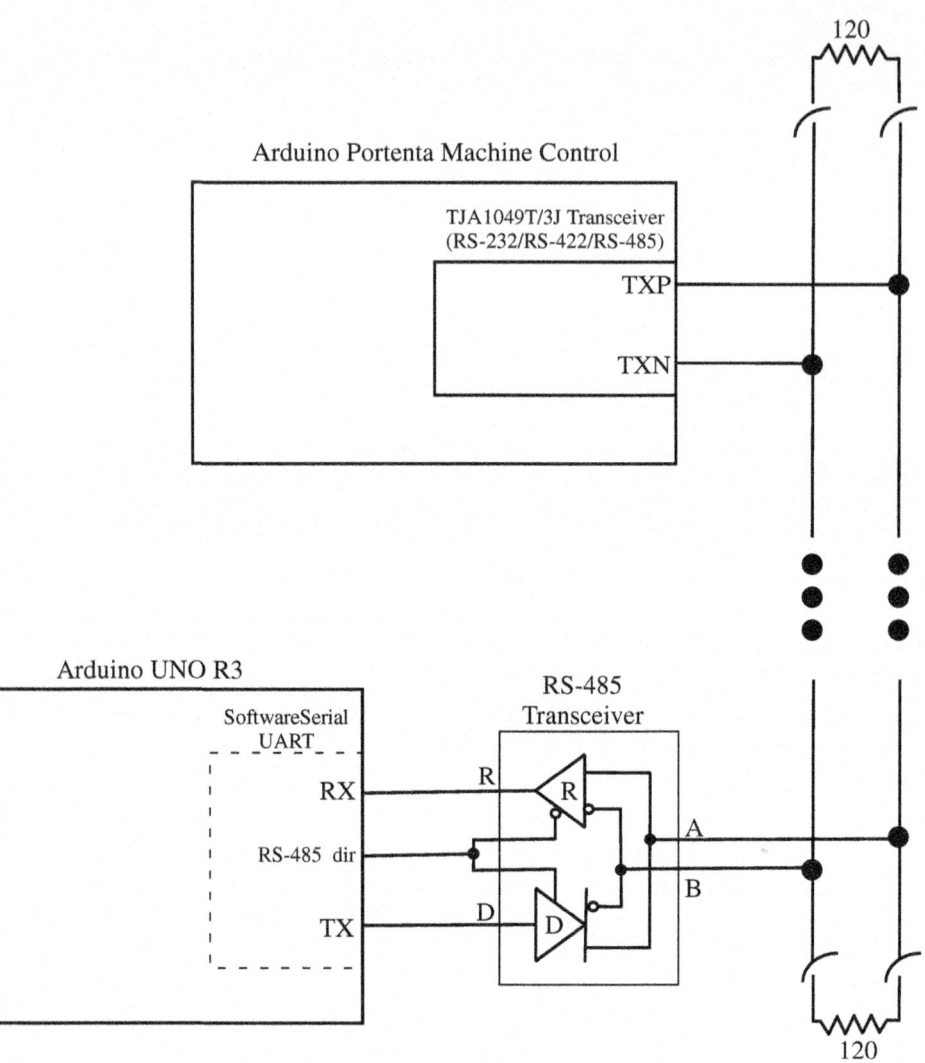

Fig. 3.33 Arduino PMC to UNO R3 RS–485 link

```
#include "Arduino_PortentaMachineControl.h"
                                    //message send interval (ms)
constexpr unsigned long sendInterval {3000};
unsigned long sendNow {0};
unsigned long counter {0};

void setup()
{
Serial.begin(9600);             //Set Baud rate for Serial Monitor
while (!Serial)
```

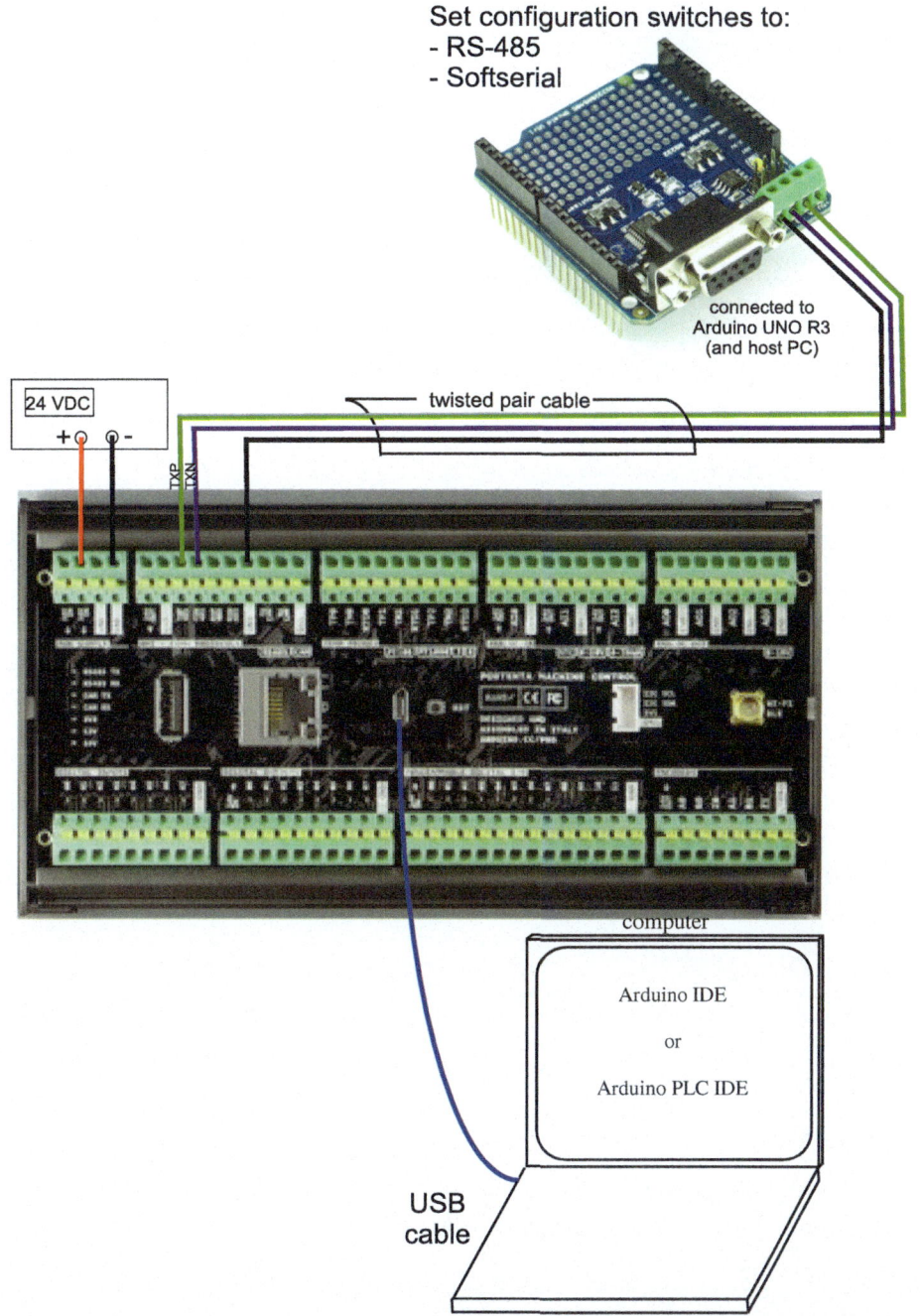

Fig. 3.34 Arduino PMC to UNO R3 transceiver. Images used courtesy of the Arduino Team (CC BY–NC–SA) (www.arduino.cc)

```
        {
          ;
        }

delay(1000);
Serial.println("Start PMC1 RS485 initialization");

//Set the PMC Communication Protocols to default config
//RS485/RS232 default config is:
// - RS485 mode
// - Half Duplex
// - No A/B and Y/Z 120 Ohm termination enabled
// - Enable the RS485/RS232 system
//Specify baudrate, and preamble and postamble times for RS485
MachineControl_RS485Comm.begin(9600, 0, 500);

//Start in receive mode
MachineControl_RS485Comm.receive();

Serial.println("PMC1 Initialization done!");
}

void loop()
{
if(MachineControl_RS485Comm.available())
  Serial.write(MachineControl_RS485Comm.read());


if(millis() > sendNow)
   {
   //Disable receive mode before transmission
   MachineControl_RS485Comm.noReceive();

   MachineControl_RS485Comm.beginTransmission();

   counter++;
   if (counter > 4) counter = 1;
   MachineControl_RS485Comm.println(counter);

   MachineControl_RS485Comm.endTransmission();

   //Re-enable receive mode after transmission
   MachineControl_RS485Comm.receive();

   sendNow = millis() + sendInterval;
   }
}

//**********************************************************************
```

The following sketch is for the Arduino UNO R3.

```
//**********************************************************************
//UNO_R3_RS485_SW_serial4
//
//Provides UNO R3 RS-485 interface via TinySine RS232/RS485 shield
//Arduino pin designations
//- 4: RS-485 direction pin (0: RX, 1: TX)
```

3.12 RS–485 Communication

```
//- 2: SoftwareSerial RX
//- 3: SoftwareSerial TX
//
//TinySine RS232/RS485 shield switch settings
//- RS232/485 Select Switch: RS-485
//- UART/Soft Serial Select Switch: Softserial
//
////Adapted by: S. Barrett, May 9, 2024
//This code example is in the public domain.
//**********************************************************************

#include <SoftwareSerial.h>

#define RS485_RX_pin    2
#define RS485_TX_pin    3
#define RS485_dir_pin   4

SoftwareSerial RS485(RS485_RX_pin, RS485_TX_pin);

void setup()
{
Serial.begin(9600);
while(!Serial)
  {
  ;
  }
pinMode(RS485_dir_pin, OUTPUT);        //RS-485 direction pin as output
digitalWrite(RS485_dir_pin, LOW);      //Set RS-485 shield to receive
delay(20);
RS485.begin(9600);                     //start RS-485 comm, BAUD rate 960
delay(100);
}

void loop()
{
digitalWrite(RS485_dir_pin, LOW);      //Set RS-485 shield to receive
delay(20);
if(RS485.available())                  //RS-485 data available?
  {
  char data = RS485.read();            //read received data character

  if ((data == '1') || (data == '2') || (data == '3') || (data == '4'))
    {
    Serial.write(data);
    switch(data)
    {
    case '1': Serial.println(" - Execute process 1");
              break;
    case '2': Serial.println(" - Execute process 2");
              break;
    case '3': Serial.println(" - Execute process 3");
              break;
    case '4': Serial.println(" - Execute process 4");
              break;
    default : Serial.println(" ");
    }
    Serial.println(" ");
    delay(100);
    digitalWrite(RS485_dir_pin, HIGH); //Set RS-485 shield to transmit
```

```
    delay(20);
    RS485.write(data);               //data back to sender
    }
  }
}

//**********************************************************************
```

Example. In this example the TinySine RS232/RS485 transceiver shield is replaced with a RS–485 MAX485 transceiver chip TTL to RS–485 module as shown in Fig. 3.35. The sketches from the prior example are used.

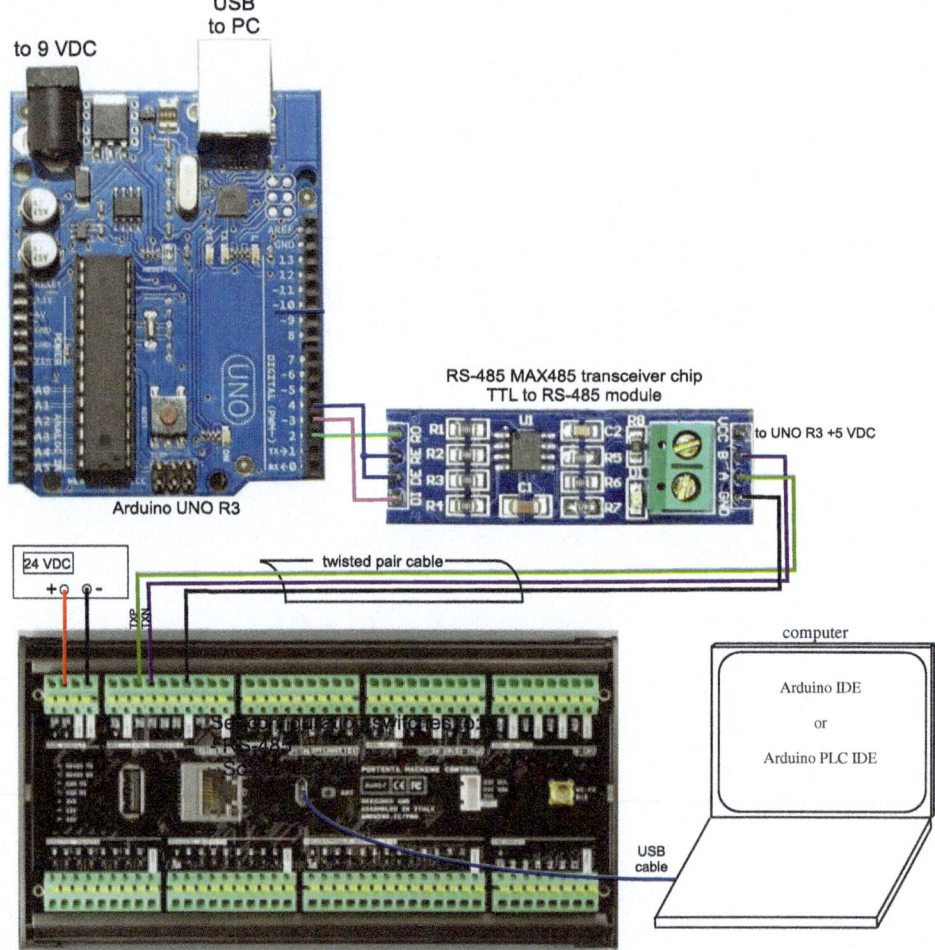

Fig. 3.35 Arduino PMC to UNO R3 transceiver

The PMC sketch used in the previous example is slightly modified for communication with the UNO R3. Note the BAUD rate setting must be the same for both the PMC and the UNO R3. The sketch sequentially sends a number one through four to the UNO R3. When received the UNO R3 executes a different process for the number received. When the specified process is complete the UNO R3 returns the process number to the PMC.

3.13 Application I: Encryption/Decryption

In Chap. 2 we discussed encryption and decryption techniques. Design and implement a PMC communication system of your choice from the chapter. Equip it with encryption and decryption features.

3.14 Application II: USB Data Logger

In this example we configure the PMC to serve as a USB data logger. This example is from the Arduino Unified Storage Library, sketch Logger.

```
//***********************************************************************
//PMC_logger - This example demonstrates the usage of the
//"Arduino_UnifiedStorage" library for logging and backing up data to
//USB storage in case a USB Mass Storage device is inserted.
//
//Notes:
//- The "logData" function simulates sensor data by reading a
//  random value every 100ms and saving it in memory.
//- Afterwards, every 1000ms "moveDataToQSPI" moves those readings to a
//  file in the Internal Storage of your board.
//- If a USB device is connected "performUpdate" function performs the
//  update process every 10 seconds by:
//  - reading the last update size from a file (number of bytes)
//  - copying the new data from the log file to a backup file
//  - and updating the last update size.
//
//Instructions:
//- Make sure you have "POSIXStorage" and "Arduino_UnifiedStorage"
//  installed
//- If you are using this sketch on an Arduino OPTA, use another board
//  or an USB adaptor to view RS485 messages
//- This sketch will log data, and check if there is any USB MSD Device
//  connected to the USB Port of your board.
//- Insert a USB Drive whenever you want
//- Every 10 seconds data is transferred from the internal storage to the
//  USB Mass storage device
//- Unplug the USB device and inspect its contents.
//
//Notes:
//- The USB device is mounted and unmounted after every update operation.
//- The first status LED is on when the USB drive is mounted. The sketch
//  has been adapted to illuminate DIGITAL OUTPUT LED 00 to indicate
//  status.
//- So as long as the status LED is off you can safely remove the drive.
//  The sketch will log to internal storage in the meantime, and wait for
//  the USB drive to be inserted again.
```

```
//
//This code example is in the public domain.
//*************************************************************************

#include <Arduino_PortentaMachineControl.h>

#define ARDUINO_UNIFIED_STORAGE_DEBUG
#include "Arduino_UnifiedStorage.h"
#include <vector>

#if defined(ARDUINO_PORTENTA_H7_M7)
#define USB_MOUNTED_LED LED_BLUE
#elif defined(ARDUINO_PORTENTA_C33)
#define USB_MOUNTED_LED LEDB
#elif defined(ARDUINO_OPTA)
#define USB_MOUNTED_LED LED_D0
#endif

InternalStorage internalStorage;
USBStorage usbStorage;

Folder backupFolder = Folder();
bool usbIntialized = false;
std::vector<String> sensorDataBuffer;

unsigned long bytesWritten = 0;
unsigned long lastLog = 0;
unsigned long lastMove = 0;
unsigned long lastBackup = 0;

volatile bool usbAvailable = false;
bool backingUP = false;



//*************************************************************************

void connectionCallback()
{
usbAvailable = true;
usbStorage.removeOnConnectCallback();
}

void disconnectionCallback()
{
usbAvailable = false;
usbStorage.onConnect(connectionCallback);
}

//*************************************************************************

//Function to run a given method periodically
void runPeriodically(void (*method)(), unsigned long interval, unsigned long* variable)
{
unsigned long currentMillis = millis();

if(currentMillis - *variable >= interval)
  {
   *variable = currentMillis;
   method();   // Call the provided method
   }
}

//*************************************************************************

// Function to log sensor data
void logDataToRAM()
```

3.14 Application II: USB Data Logger

```
{
int timeStamp = millis();
int sensorReading = analogRead(A0);
String line = String(timeStamp) + "," + String(random(9999));
sensorDataBuffer.push_back(line);
}

//*************************************************************************

void moveDataToQSPI()
{
if(!backingUP)
  {
  UFile _logFile = internalStorage.getRootFolder().createFile("log.txt",
  FileMode::APPEND);
  for(const auto& line : sensorDataBuffer)
    {
    bytesWritten += _logFile.write(line);  //Write log line to the file
    }
    _logFile.close();
    sensorDataBuffer.clear();
  }
}

//*************************************************************************

void performUpdate()
{
UFile logFile = internalStorage.getRootFolder().createFile("log.txt", FileMode::READ);
UFile backupFile = backupFolder.createFile("backup_file.txt",
                   FileMode::APPEND);        //Create or open backup file
 UFile lastUpdateFile = backupFolder.createFile("diff.txt",
 FileMode::READ);                            //Create or open last file

 backingUP = true;            //Read the last update size from the file
 unsigned lastUpdateBytes = lastUpdateFile.readAsString().toInt();

 Arduino_UnifiedStorage::debugPrint("Last update bytes: " +
 String(lastUpdateBytes));

 if(lastUpdateBytes >= bytesWritten)
   {
   Arduino_UnifiedStorage::debugPrint("No new data to copy. ");
   backupFile.close();
   lastUpdateFile.close();
   backingUP = false;
   return;
   }

logFile.seek(lastUpdateBytes);   //Move the file pointer to the last update position

unsigned long totalBytesToMove = bytesWritten - lastUpdateBytes;
Arduino_UnifiedStorage::debugPrint("New update bytes: " + String(totalBytesToMove));

uint8_t* buffer = new uint8_t[totalBytesToMove];

size_t bytesRead = logFile.read(buffer, totalBytesToMove);
                               //Only write the bytes that haven't
                               //been backed up yet
size_t bytesMoved = backupFile.write(buffer, bytesRead);
Arduino_UnifiedStorage::debugPrint("Successfully copied " + String(bytesMoved) + " new bytes. ");
                               // Open the last update file in write
                               //mode
lastUpdateFile.changeMode(FileMode::WRITE);
                               // Update the last update size
lastUpdateFile.write(String(lastUpdateBytes + bytesMoved));
backupFile.close();
```

```
logFile.close();
lastUpdateFile.close();


usbStorage.unmount();                        //Unmount the USB storage

digitalWrite(USB_MOUNTED_LED, HIGH);
MachineControl_DigitalOutputs.write(0, HIGH); //digital out 00 on
backingUP = false;
delete[] buffer;
}

//************************************************************************

// Function to backup data to USB storage
void backupToUSB()
{
if(usbAvailable && !usbIntialized)
  {
  usbStorage.begin();
  Arduino_UnifiedStorage::debugPrint("First drive insertion, creating folders... ");
  Folder usbRoot = usbStorage.getRootFolder();
  String folderName = "LoggerBackup" + String(random(9999));
  backupFolder = usbRoot.createSubfolder(folderName);
  Arduino_UnifiedStorage::debugPrint("Successfully created backup folder:
                                    " + backupFolder.getPathAsString());
  usbStorage.unmount();
  usbIntialized = true;
  }
else if(usbAvailable && usbIntialized)
  {
  Arduino_UnifiedStorage::debugPrint("USB Mass storage is available ");
  delay(100);
  if(!usbStorage.isMounted())
    {
    Arduino_UnifiedStorage::debugPrint("Mounting USB Mass Storage ");
    digitalWrite(USB_MOUNTED_LED, LOW);
     delay(2000);                         //delay 2s to allow USB removal
      MachineControl_DigitalOutputs.write(0, LOW); //digital out 00 on
      if(usbStorage.begin())
        {
        performUpdate();
        }
    }
  else if (usbStorage.isMounted())
    {
    Arduino_UnifiedStorage::debugPrint("USB Mass storage is connected, performing update ");
    performUpdate();
    }
  }
else
  {
  Arduino_UnifiedStorage::debugPrint("USB Mass storage is not available ");
  }


}
//************************************************************************

void setup()
{
MachineControl_DigitalOutputs.begin();    //init PMC digital outputs randomSeed(analogRead(A0));

#if !defined(ARDUINO_OPTA)
   Serial.begin(115200);
while(!Serial);
```

```
#else
  beginRS485(115200);
#endif

//toggle this to enable debugging output
Arduino_UnifiedStorage::debuggingModeEnabled = false;

usbStorage = USBStorage();
internalStorage = InternalStorage();

usbStorage.onConnect(connectionCallback);
usbStorage.onDisconnect(disconnectionCallback);

pinMode(USB_MOUNTED_LED, OUTPUT);
Arduino_UnifiedStorage::debugPrint("Formatting internal storage... ");
int formatted = internalStorage.format(FS_LITTLEFS);
Arduino_UnifiedStorage::debugPrint("QSPI Format status: " + String(formatted));

if(!internalStorage.begin())
  {
  Arduino_UnifiedStorage::debugPrint("Failed to initialize internal storage ");
  return;
  }
else
  {
  Arduino_UnifiedStorage::debugPrint("Initialized storage ");
  }
}

void loop()
{
runPeriodically(logDataToRAM,    100, &lastLog);
runPeriodically(moveDataToQSPI, 1000, &lastMove);
runPeriodically(backupToUSB,   10000, &lastBackup);
}
//*************************************************************************
```

Modify the sketch to log the setting of a potentiometer to the USB drive.

3.15 Summary

This chapter described methods of connecting the Portenta Machine Control (PMC) to external peripheral devices and other processors. For each range we provided the network type and also implementation technologies. The chapter began with a brief description of serial communications and related serial communication terminology. We then reviewed different connectivity technologies beginning with close range technologies, then mid–range technologies, and concluding with long range technologies.

3.16 Chapter Problems

1. Create UML activity diagrams for all chapter sketches.
2. Describe the difference in how IPv4 and IPv6 addresses are allocated.

3. What is CIDR addressing? How does it extend the IPv4 addressing space?
4. What is the relationship between a DNS and an URL address?
5. How do you choose a specific communication technology for a specific application?
6. Provides the ASCII coding for "PMC" with one even parity bit.
7. What is the relationship between ASCII and Unicode.
8. Describe the difference between a USB client and host?
9. Describe the difference between a BLE client and host?
10. Sketch the UART and RS–232 signal for the "PMC" signal above.
11. Provide a summary table for all communication technologies described in this chapter.

References

1. *Bluetooth Low Energy, Wi–Fi, and Ethernet on Opta,* https://opta.findernet.com.
2. M. Fowler with K. Scott "UML Distilled– A Brief Guide to the Standard Object Modeling Language," 2nd edition. Boston:Addison–Wesley, 2000. hero
3. *Getting Started with RS–485 on Opta,* https://opta.findernet.com.
4. Hanes D., G. Salgueiro, P. Grossetete, R. Barton, J. Henry (2017) *IoT Fundamentals–Networking Technologies, Protocols, and Use Cases for the Internet of Things,* Cisco Press.
5. Horowitz P, Hill W (2015) The Art of Electronics, third edition, Cambridge University Press.
6. Internet Corporation for Assigned Names and Numbers (ICANN), www.ICANN.org
7. Internet Engineering Task Force, www.IETF.org
8. Leiden C. and M. Wilensky (2009) TCP/IP for Dummies, 6th edition, John Wiley and Sons Publishing, Inc.
9. Levine R. and M. Levine Young (2015) *The Internet for Dummies,* John Wiley and Sons Publishing, Inc.
10. Lowe, D. (2018) Networking All–In–One for dummies, 7th edition, John Wiley and Sons Publishing, Inc.
11. Null L. and J. Lobur (2015) *Computer Organization and Architecture,* Jones and Bartlett Learning. item Shuler R., (2002) *How does the internet work?,* Rus Shuler Pomeroy IT Solutions.
12. Stenerson, J. (2004) *Fundamentals of Programmable Logic Controllers, Sensors, and Communications,* Pearson Prentice Hall.
13. Voss W. (2014), *A Comprehensible Guide to Controller Area Network,* Copperhill Media.
14. Voss W. (2008), *Controller Area Network Prototyping with Arduino,* Copperhill Media.

PMC Peripheral Interfacing 4

Objectives: After reading this chapter, the reader should be able to:

- Describe the input/output characteristics for the Portenta Machine Control digital and analog subsystems;
- Interface a wide variety of input and output devices to the Arduino Portenta Machine Control;
- Specify the voltage, current, and power requirements for a specific application;
- Design common applications for op amps including transducer interface design;
- Describe the requirement for optical isolation in certain applications; and
- Apply chapter concepts to a specific interface design.

4.1 Overview

In this chapter we explore how to interface a wide variety of input and output devices to the Arduino PMC. In Chap. 2 we discussed the rich complement of PMC Digital In, Digital Out, Programmable Digital I/O, Analog Input, and Analog output features. In this chapter we explore specific input and output devices and their interface to the Arduino PMC. Figure 4.1 provides a summary of PMC input and output features and also related peripheral devices. We discuss each type of peripheral in turn.[1]

---

[1] Portions of this chapter have been adapted and expanded for the Arduino Portenta Machine Control from "Microcontroller Fundamentals for Engineers and Scientists," Morgan and Claypool Publishers, 2006.

© The Author(s), under exclusive license to Springer Nature Switzerland AG 2026
S. F. Barrett, *Arduino VIII*, Synthesis Lectures on Digital Circuits & Systems,
https://doi.org/10.1007/978-3-031-85944-1_4

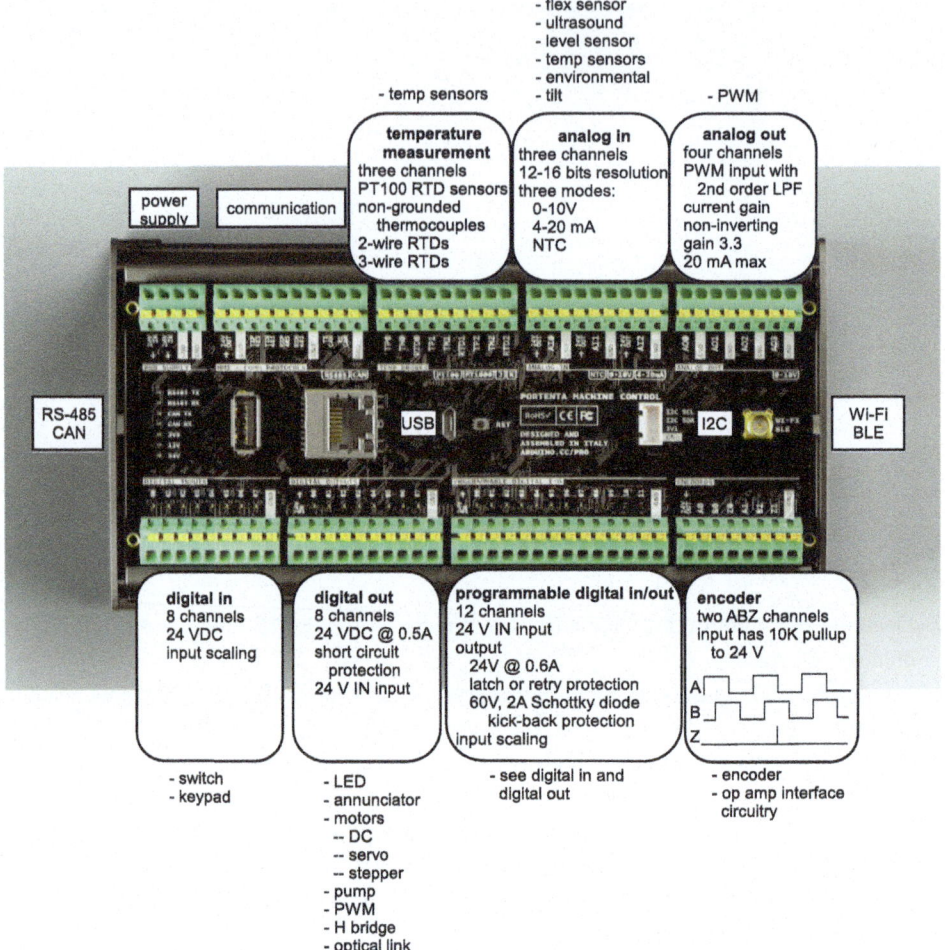

Fig. 4.1 PMC in out summary sensors. Images used courtesy of the Arduino Team (CC BY-NC-SA) (www.arduino.cc)

4.2 Power Requirements

The Arduino Portenta Machine Control is an industrial grade processor. It is at the intersection of industrial control, the Internet of Things (IoT), the Industrial Internet of Things (IIoT), and embedded systems. Due to its inherent flexibility, it may be used in a wide variety of applications within these different but related areas. We shall see that this will sometimes require a wide variety of peripheral component voltage and current requirements. In this section we explore related concepts.

4.2 Power Requirements

To provide a proper power source for a system, the following information must be determined:

- What is the voltage and current requirements of each device in the system?
- Will the system have any current surge requirements (e.g. motor starting current, motor stall current, etc.)?
- Will the system be operated where an AC source is present or will it be a remote system requiring a DC supply?
- How long must the system operate before the batteries can be replaced or recharged?
- Is an alternate power source possible (e.g. solar panel)?

Once these questions are answered, a system power supply may be assembled. In the remainder of this section, we discuss some of these power alternatives.

4.2.1 AC Operation

If a source of AC power is readily available, an AC-to-DC converter may be used. These range from a single voltage supply to a multiple DC voltage power supply with different current specifications for each voltage. For industrial applications a wide variety of DIN rail mounted supplies are available.

When selecting a power source it is important to ensure it is regulated and fused. A regulator maintains the source voltage at the same value even under different current loads. A fuse provides protection against a surge current the power supply cannot safely handle. When the current requirements for each voltage are determined, a power supply may be selected. Choose a power supply with at least double the current specification as required by the maximum demands of the project.

4.2.2 DC Operation

For a remote or portable application, a DC battery source of power may be used. To select a battery the following requirements must be known: voltage, current, polarity, capacity, and if rechargeable batteries are appropriate for the project.

- voltage: The unit for voltage is Volts. The voltage for a battery is specified for when it is new or fully charged (for a rechargeable type battery). Typical battery voltages for AAA, AA, C, and D cells are 1.5 VDC. The batteries may be placed in series to achieve higher voltages. Plastic battery packs are available for battery series stacking to increase the overall voltage rating of the power pack. Another common battery type is the 9 VDC rectangular battery with the plus and minus terminals on the same end of the battery.

- current: The unit for current is amperes or amps. The current drain of the battery is determined by the load connected to it. For many Arduino based projects the current drain may be specified in mA.
- polarity: In most projects, a positive voltage referenced to ground is required. Some circuits, for example operational amplifier-instrumentation circuits, may require both a positive and negative supply for proper operation.
- capacity: The battery capacity specification is provided in mAH or AH (amp-hours). It provides an estimate of how long a battery will last under a particular current drain. Common battery capacities are: AAA—1000 mAH, AA—2250 mAH, C—7000 mAH, D—15,000 mAH, and 9 VDC—550 mAH. These values are only estimates. The exact battery capacity is determined by battery technology and manufacturer. Capacity is typically provided within the manufacturer's specification for a battery.
- rechargeable: Rechargeable batteries are available in a wide range of voltages and capacities.

To properly match a battery to an embedded system, the battery voltage and capacity must be specified. Battery capacity is typically specified as a mAH rating. For example, a typical 9 VDC non-rechargeable alkaline battery has a capacity of 550 mAH. If the embedded system has a maximum operating current of 50 mA, it will operate for approximately eleven hours before battery replacement is required.

A battery is typically used with a voltage regulator to maintain the voltage at a prescribed level. Voltage regulators are available in a variety of voltage and maximum current specification. Figure 4.2 provides sample circuits to provide a $+5$ VDC and a ± 5 VDC portable battery source. Additional information on battery capacity and characteristics may be found in Barrett and Pack (Prentice Hall, 2005).

The Arduino PMC requires a 24 VDC supply. In the remainder of the book, we use either a DIN rail mounted power supply or a bench type power supply during development. In the remainder of the book, we explore peripheral devices with different voltage and current requirements than the PMC. For each we ensure the compatibility between the PMC and the peripheral device.

4.3 Input Sensors

An industrial controller such as the PMC is typically used in applications where data is collected by input sensors, the data is then processed by the host algorithm, and then a control decision and accompanying signals are provided by the PMC to output peripheral actuators. The input sensors may be digital or analog in nature.

4.3 Input Sensors

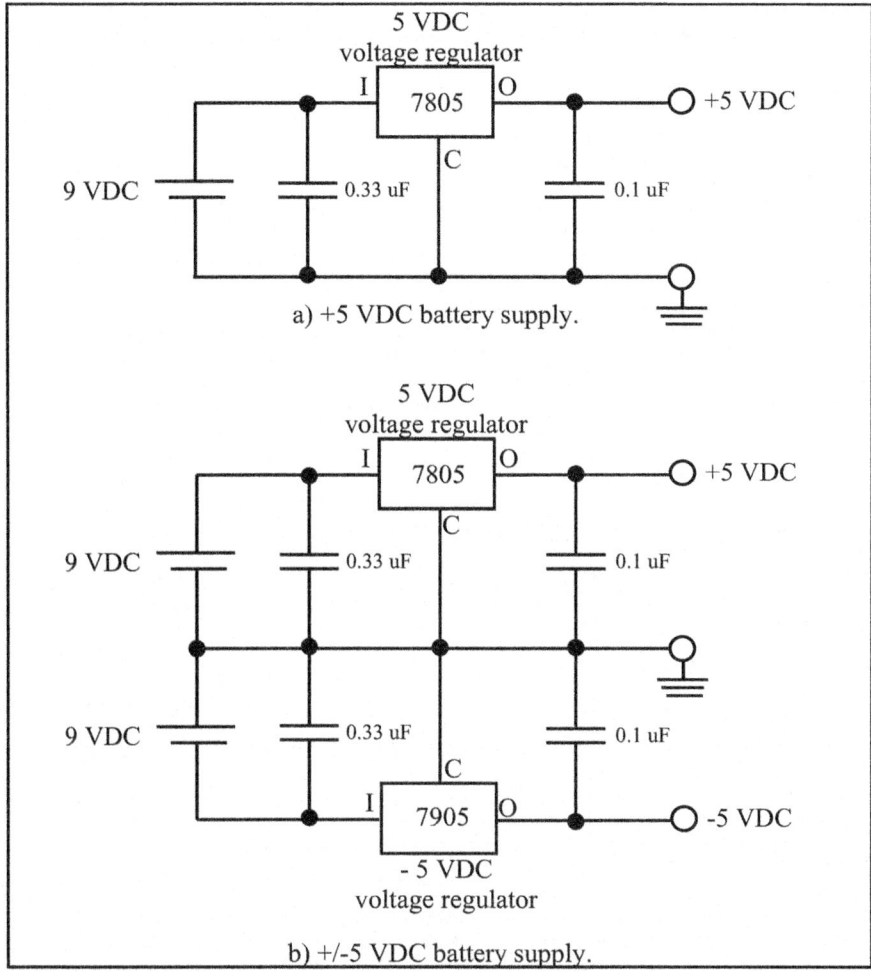

Fig. 4.2 Battery supply circuits employing a 9 VDC battery with a 5 VDC regulators

4.3.1 Digital Input Sensors

Digital sensors provide a single digital input signal or a series of digital logic pulses with sensor data encoded. The sensor data may be encoded in any of the parameters associated with a digital signal such as logic level, duty cycle, frequency, period, pulse length, or pulse rate. In the next several sections we describe how to interface a wide variety of digital input devices including switches and sensors to the Arduino PMC. We begin with switches.

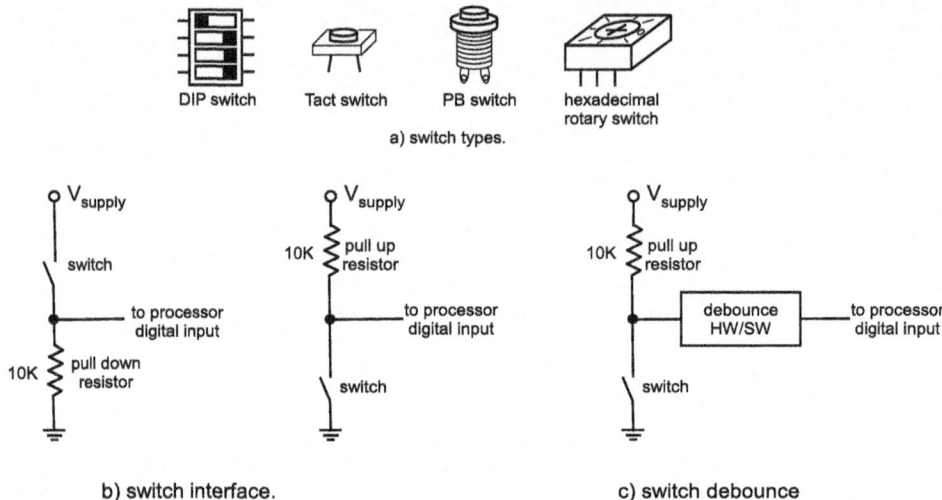

Fig. 4.3 Switch types

4.3.2 Switches

Switches come in a variety of types. A system designer chooses the appropriate switch for a specific application. Switch varieties commonly used in control applications are illustrated in Fig. 4.3a). Here is a brief summary of the different types:

- **Slide switch**: A slide switch has two different positions: on and off. The switch is manually moved to one position or the other. For control applications, slide switches are available in a panel mounted package. Also, small slide switches are available that fit in the profile of a common integrated circuit size dual inline package (DIP). A bank of four or eight DIP switches in a single package is commonly available. Slide switches may be used to select specific parameters at system startup.
- **Momentary contact pushbutton switch**: A momentary contact pushbutton switch comes in two varieties: normally opened (NO) and normally closed (NC). A normally open switch, as its name implies, does not normally provide an electrical connection between its contacts. When the pushbutton portion of the switch is depressed, the connection between the two switch contacts is made. The connection is held while the switch is depressed. When the push button is released, the connection is opened. The opposite is true for a normally closed switch. For control applications, pushbutton switches are available in a small tact type switch configuration or panel/chassis mounted configurations. Push button style switches are commonly used in ladder logic coding to start and stop a control process.

4.3 Input Sensors

- **Push on/push off switches**: This type of switch is also available in a normally open or normally closed configuration. For the normally open configuration, the switch is depressed to make connection between the two switch contacts. The pushbutton must be depressed again to release the electrical connection.
- **Hexadecimal rotary switches**: Small profile rotary switches are available for controller applications. These switches commonly have sixteen rotary switch positions. As the switch is rotated to each position, a unique four-bit binary code is provided at the switch contacts. These switches are used to select specific parameters at system startup.

Typically, a switch is interfaced to a controller via a pull down or pull up circuit as shown in Fig. 4.3b). Some processors have onboard pull up and/or pull down that may be asserted as needed.

4.3.2.1 Switch Debouncing

Mechanical switches do not make a clean transition from one position (on) to another (off). When a switch is moved from one position to another, it makes and breaks contact multiple times. This activity may go on for tens of milliseconds. A processor such as the PMC is relatively fast as compared to the action of the switch. Therefore, the processor may recognize each switch bounce as a separate and erroneous transition.

To correct the switch bounce phenomena additional external hardware components may be used or software techniques may be employed as shown in Fig. 4.3c). Software switch debouncing is accomplished by inserting a 30–50 ms lockout delay in the function responding to input changes. The delay prevents the processor from responding to the multiple switch transitions related to bouncing. The programming and interface between a switch and the PMC is described in Chap. 2.

4.3.3 Optical Encoder

An optical encoder consists of a small plastic transparent disk(s) with opaque lines etched into the disk surface. A stationary optical emitter and detector pair is placed on either side of the disks. As the disk(s) rotates with motor rotation, the opaque lines break the continuity between the optical source and detector. The signal from the optical detector is monitored to determine the rate of disk rotation as shown in Fig. 4.4.

Optical encoders are available in a variety of types depending on the information desired. There are two major types of optical encoders: incremental encoders and absolute encoders.

An absolute encoder is used when it is required to retain position information when power is lost. The absolute encoder is equipped with multiple data tracks to determine the precise location of the encoder disk (Sick Stegmann).

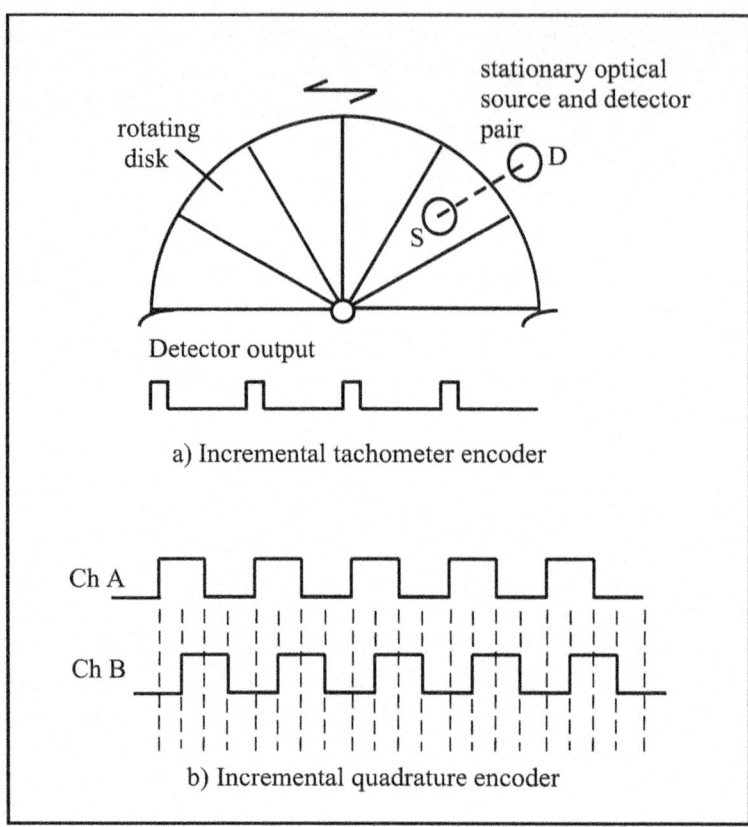

Fig. 4.4 Optical encoder

An incremental encoder is used in applications where a velocity or a velocity and direction information is required. The incremental encoder types may be further subdivided into tachometers and quadrature encoders. An incremental tachometer encoder consists of a single track of etched opaque lines as shown in Fig. 4.4a. It is used when the velocity of a rotating device is required. To calculate velocity, the number of detected pulses is counted in a fixed amount of time. Since the number of pulses per encoder revolution is known, velocity may be calculated.

The quadrature encoder contains two tracks shifted in relationship to one another by 90 °C. This allows the calculation of both velocity and direction. To determine direction, one would monitor the phase relationship between Channel A and Channel B as shown in Fig. 4.4b). Some encoders are equipped with A, B, Z channels. The Z channel provides an Index or one encoder tick per motor rotation.

The PMC is equipped with two ABZ encoder channels. The inputs for the channels are designated A0, B0, and Z0 for channel 0 and A1, B1, and Z1 for channel 1. The inputs

4.3 Input Sensors

require open collector type encoders. The encoder outputs are pulled up to 24 VDC by the PMC encoder inputs with internal 10 K Ω resistors.

Example: Three and two phase encoders. In Fig. 4.5 the PMC is connected to a three phase (ABZ) and a two phase (AB) encoder. The sketch below provides the AB state of the encoder, an incrementing/decrementing count of the encoder position, and also a count of encoder revolutions.

```
//*************************************************************************
//PMC_encoder_examples
//
//Adapted from:
//Portenta Machine Control - Encoder Read Example
//
//Sketch demonstrates use of PMC encoder features with 3 phase and 2 phase encoders
//Authors: Riccardo Rizzo @Rocketct, Leonardo Cavagnis @leonardocavagnis
//
//This code example is in the public domain.
//*************************************************************************

#include <Arduino_PortentaMachineControl.h>

void setup()
{
Serial.begin(9600);
while (!Serial)
  {
   ;                                             //wait for serial port to connect
  }
}

void loop() {
{                                                //Encoder channel 0
Serial.print("Encoder 0 State: ");               //provides current AB state
Serial.println(MachineControl_Encoders.getCurrentState(0),BIN);
                                                 //increments/decrements for
Serial.print("Encoder 0 Pulses: ");              //CW/CCW rotation
Serial.println(MachineControl_Encoders.getPulses(0));

Serial.print("Encoder 0 Revolutions: ");         //Incremented by index signal
Serial.println(MachineControl_Encoders.getRevolutions(0));
Serial.println();

                                                 //Encoder channel 1
Serial.print("Encoder 1 State: ");               //provides current AB state
Serial.println(MachineControl_Encoders.getCurrentState(1),BIN);
                                                 //increments/decrements for
Serial.print("Encoder 1 Pulses: ");              //CW/CCW rotation
Serial.println(MachineControl_Encoders.getPulses(1));

Serial.print("Encoder 1 Revolutions: ");         //Incremented by index signal
Serial.println(MachineControl_Encoders.getRevolutions(1));
Serial.println();

delay(25);
}

//*************************************************************************
```

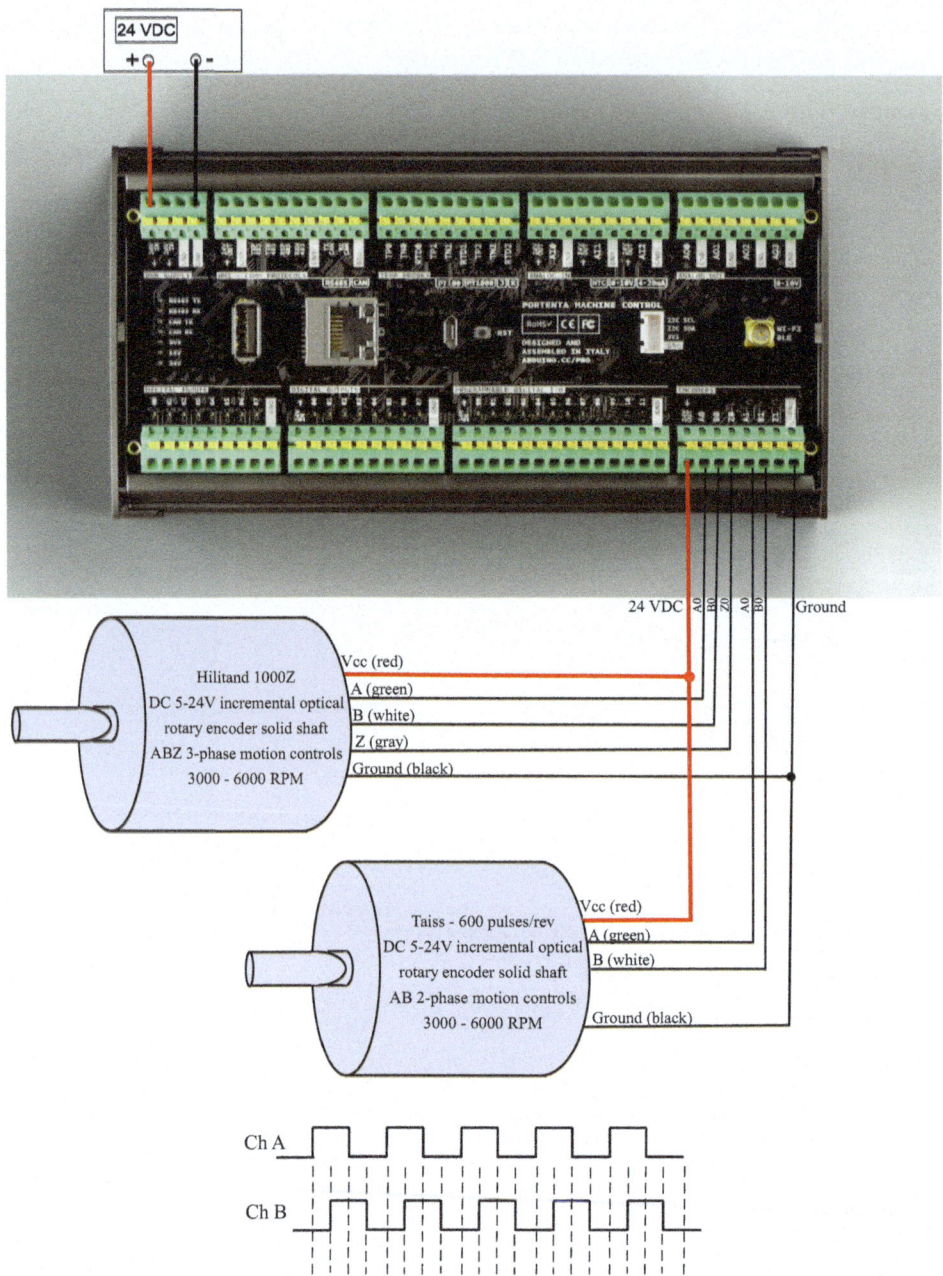

Fig. 4.5 PMC equipped with three and two phase encoders. Images used courtesy of the Arduino Team (CC BY-NC-SA) (www.arduino.cc)

The encoders may be connected to a robot wheel shaft to measure rotation. Data collected may be used to determine distance traveled, velocity, etc. The encoders may also be connected to a motor to determine motor direction, speed in revolutions per minute (RPM), etc. In the Application section we explore motor speed control using PWM techniques and a DC motor equipped with an optical tachometer.

Example: DC motor RPM measurement. In this example the RPM of a Brother DC–58–108 motor is measured using an optical tachometer as shown in Fig. 4.6. The motor is 12 VDC and is rated at 1500 RPM. The motor has A, B, and Z optical tachometer outputs. Both A and B provide 200 counts per motor revolution while Z (Index I) provides one count per revolution. We use the B channel (200 counts per revolution) and PMC input Z0 to count motor rotation and hence RPM. Note the multiple output regulator circuit to provide ± 5 VDC for the optical tachometer. Also, an LM339 op amp is configured as an open collector comparator to generate a compatible signal for PMC input Z0. The previous sketch is used to measure encoder counts. We extend this example in the Application section to measure and stabilize motor RPM.

4.4 Analog Input Sensors

Analog input sensors provide a DC voltage that is proportional to the physical parameter being measured. The analog voltage is converted to a corresponding binary representation. With analog sensors, signal preprocessing may be required to convert the sensor output to an analog DC voltage suitable for measurement by the PMC. In this section we review a series of common analog input sensors. We adapt the sketch "Analog_input_0_10V" from the Arduino Portenta Machine Control Library for each sensor.

```
//**********************************************************************
//Portenta Machine Control - Analog in 0-10 V
//
//This example provides the voltage value acquired by the Portenta
//Machine Control.
//
//Notes:
//- For each channel of the ANALOG IN connector, there is a resistor
//  divider made by a 100k and 39k, the input voltage is divided by a
//  ratio of 0.28.
//- Maximum input voltage is 10V.
//- To use the 0V-10V functionality, a 24V supply on the PWR SUPPLY
//  connector is necessary.
//
//This example code is in the public domain.
//Copyright (c) 2024 Arduino, SPDX-License-Identifier: MPL-2.0
//**********************************************************************

#include <Arduino_PortentaMachineControl.h>

const float RES_DIVIDER = 0.28057;
const float REFERENCE   = 3.0;
```

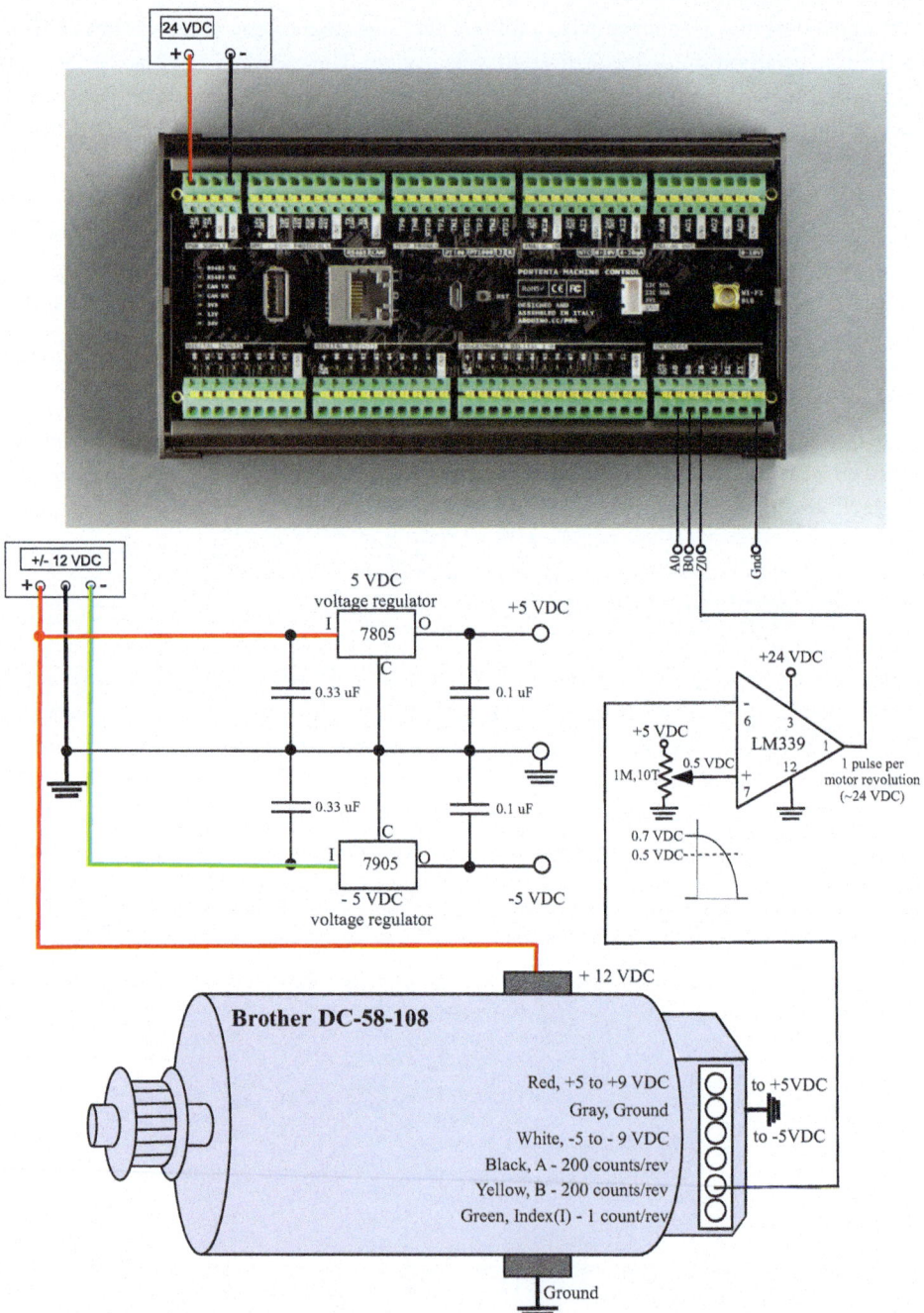

Fig. 4.6 PMC with 12 VDC motor and optical tachometer. Images used courtesy of the Arduino Team (CC BY-NC-SA) (www.arduino.cc)

4.4 Analog Input Sensors

```
void setup()
{
Serial.begin(9600);
while (!Serial)
 {
 ; // wait for serial port to connect.
 }

MachineControl_AnalogIn.begin(SensorType::V_0_10);
}

void loop()
{
float raw_voltage_ch0 = MachineControl_AnalogIn.read(0);
float voltage_ch0 = (raw_voltage_ch0 * REFERENCE) / 65535 / RES_DIVIDER;
Serial.print("Voltage CH0: ");
Serial.print(voltage_ch0, 3);
Serial.println("V");

Serial.println();
delay(250);
}
//**********************************************************************
```

4.4.1 Flex Sensor

An analog flex sensor is shown in Fig. 4.7a). The flex sensor provides a change in resistance for a change in sensor flexion. At 0 °C flex, the sensor provides 10k Ω of resistance. For 90 °C flex, the sensor provides 30–40 K Ohms of resistance. Since the PLC cannot measure resistance directly, the change in flex sensor resistance must be converted to a change in a DC voltage. This is accomplished using the voltage divider network shown in Fig. 4.7c). For increased flex, the DC voltage will increase. The voltage and hence flex can be measured using the PMC analog-to-digital converter subsystem with the "Analog_input_0_10V" sketch.

4.4.2 Ultrasound Sensor

The ultrasonic sensor pictured in Fig. 4.8 is based on the concept of ultrasound or sound waves that are at a frequency above the human range of hearing (20 Hz to 20 kHz). The ultrasonic sensor pictured in Fig. 4.8c) emits a sound wave at 42 kHz. The sound wave reflects from a solid surface and returns back to the sensor. The amount of time for the sound wave to transit to the surface and back to the sensor may be used to determine the range from the sensor to the wall.

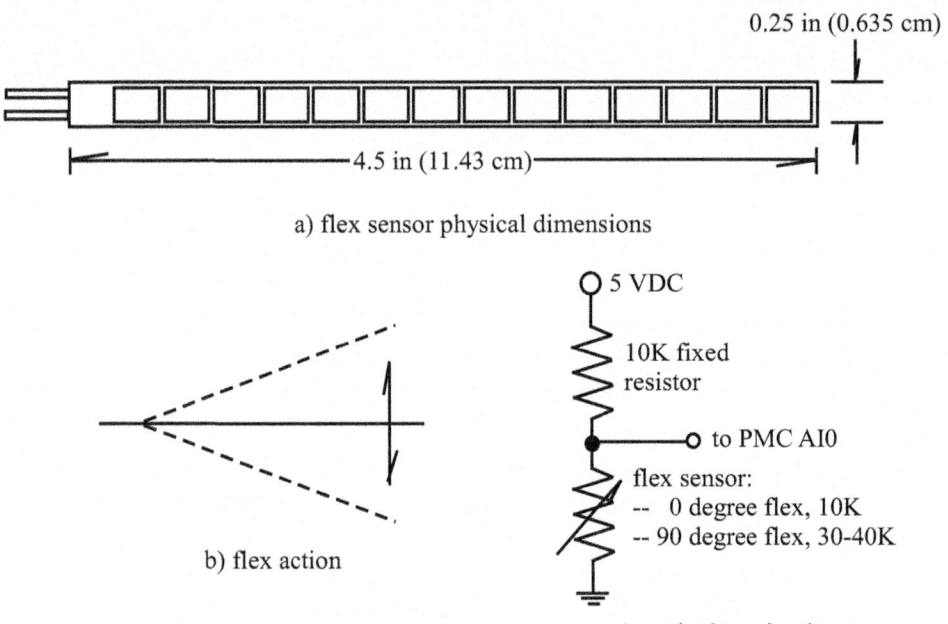

Fig. 4.7 Flex sensor

Pictured in Fig. 4.8c, d is an ultrasonic sensor manufactured by Maxbotix, the LV-Max Sonar-EZ series, MB1010–000. The sensor provides an output that is linearly related to range in three different formats: (a) a serial RS-232 compatible output at 9600 bits per second, (b) a pulse width modulated (PWM) output at a 147 us/inch duty cycle, and c) an analog output at a resolution of approximately 10 mV/inch when Vcc is 5.0 VDC. The sensor is powered from a 2.5 to 5.5 VDC source (www.sparkfun.com).

In this example we use the sensor's analog output to determine range.

```
//*********************************************************************
//PMC_sonar_analog_input_0_10V
//
//Adapted from:
//Portenta Machine Control - Analog in 0-10 V
//
//This example provides the voltage value acquired by the Portenta
//Machine Control.
//
//Notes:
//- For each channel of the ANALOG IN connector, there is a resistor
//  divider made by a 100k and 39k, the input voltage is divided by a
//  ratio of 0.28.
//- Maximum input voltage is 10V.
//- To use the 0V-10V functionality, a 24V supply on the PWR SUPPLY
//  connector is necessary.
```

4.4 Analog Input Sensors

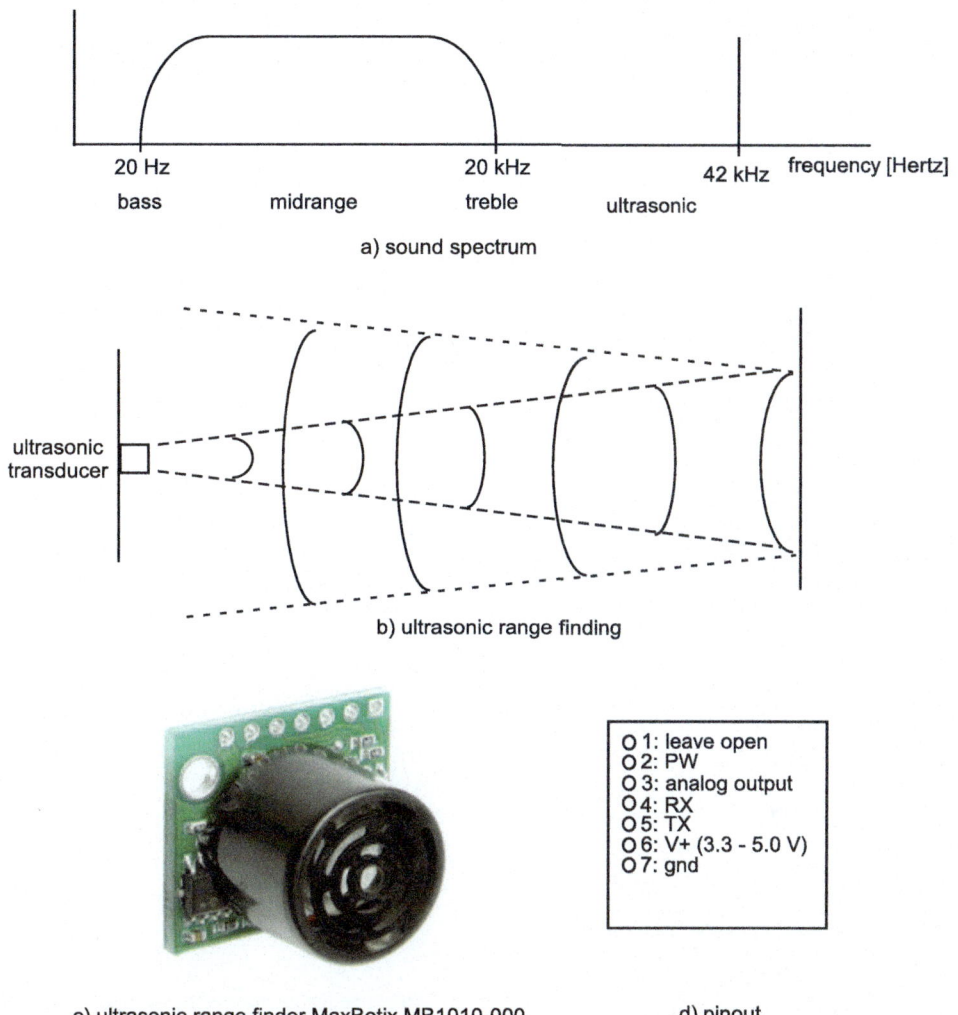

Fig. 4.8 Ultrasonic sensor. Sensor image used courtesy of SparkFun Electronics (CC BY-NC-SA) (www.sparkfun.com)

```
//- A MaxBotix MB1010-000 ultrasound sensor is used to measure range
//- Resolution is (Vcc/512)/inch
//- For 5 VDC, resolution is 9.8 mV per inch
//
//This example code is in the public domain.
//Copyright (c) 2024 Arduino, SPDX-License-Identifier: MPL-2.0
//************************************************************

#include <Arduino_PortentaMachineControl.h>
```

```
const float RES_DIVIDER = 0.28057;
const float REFERENCE   = 3.0;

void setup()
{
Serial.begin(9600);
while (!Serial)
 {
 ; // wait for serial port to connect.
 }

MachineControl_AnalogIn.begin(SensorType::V_0_10);
}

void loop()
{
float raw_voltage_ch0 = MachineControl_AnalogIn.read(0);
float voltage_ch0 = (raw_voltage_ch0 * REFERENCE) / 65535 / RES_DIVIDER;
Serial.print("Voltage CH0: ");
Serial.print(voltage_ch0, 3);
Serial.println("V");

float range = voltage_ch0 / 0.0098;
Serial.print("Range: ");
Serial.print(range, 3);
Serial.println("[inches]");


Serial.println();
delay(250);
}

//***********************************************************************
```

4.4.3 Temperature Sensors

There are several sensor types that may be used to measure temperature including integrated circuit (IC) based sensors, thermocouples, resistor temperature detectors, and thermistors. In this section we discuss each in turn. We begin with a brief review of temperature scales.

4.4.3.1 Temperature Scales

Figure 4.9 provides a summary of several common temperature scales and conversion equations (Miller).

4.4.3.2 IC Based Temperature Sensors

Due to their linearity and ease of use, we begin with the IC based sensors. Provided in Fig. 4.10 is a summary of common integrated circuit (IC) based temperature sensors. The

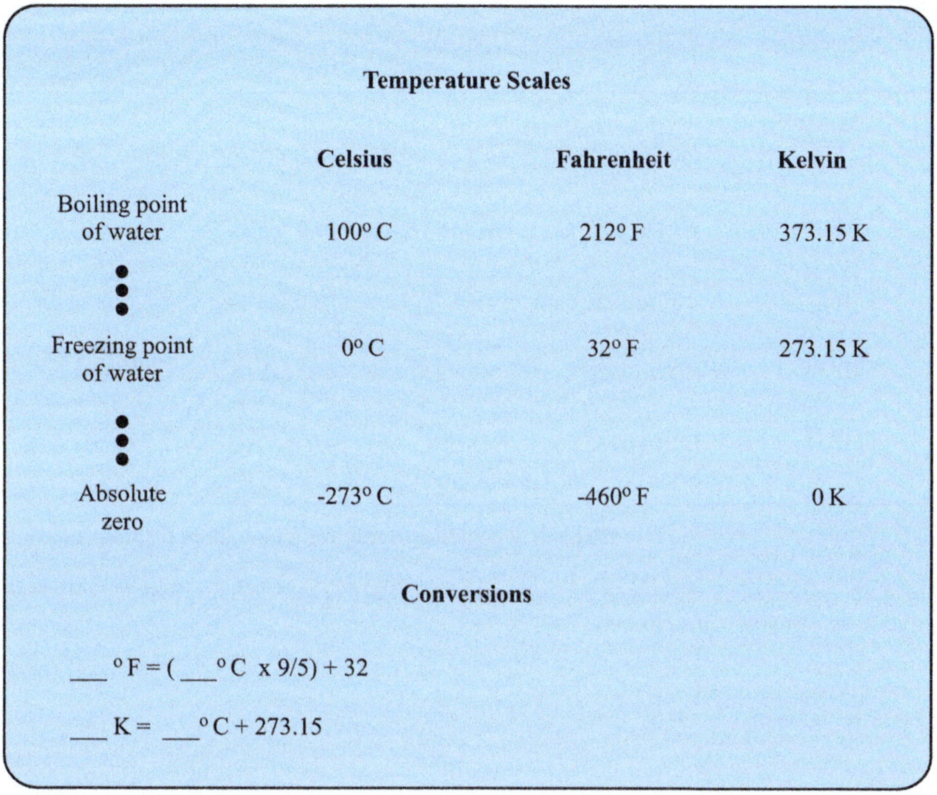

4.4 Analog Input Sensors

Temperature Scales

	Celsius	Fahrenheit	Kelvin
Boiling point of water	100° C	212° F	373.15 K
⋮			
Freezing point of water	0° C	32° F	273.15 K
⋮			
Absolute zero	-273° C	-460° F	0 K

Conversions

___ °F = (___ °C x 9/5) + 32

___ K = ___ °C + 273.15

Fig. 4.9 Temperature scales

sensors may be used over a wide temperature range and provide an output voltage as a linear function of temperature. The sensors are available in an IC type or small transistor style packaging.

Example: LM34 Temperature Sensor. In this example we use the LM34 Precision Fahrenheit Temperature Sensors manufactured by Texas Instruments to measure temperature. The LM34 provides 10 mV of output per degree Fahrenheit. The output pin of the LM34 is provided to input pin I1 (AI0) on the Arduino PMC for temperature readings. Provided below is an Arduino IDE based sketch to measure the output from the LM34, convert the LM34 output to temperature, and display the result on the Serial Monitor.

```
//************************************************************
//PMC_LM34_analog_input_0_10V
//
//Adapted from:
//Portenta Machine Control - Analog in 0-10 V
//
//This example provides the voltage value acquired by the Portenta
```

Temperature Sensor	Temperature Range	Accuracy over Range	Scale Factor	Supply Voltage
LM34	-50° to 300° F	+/- 1 to 1/2°F	10.0 mV/°F	+5 to +20 V*
LM35	-55° to 150° C	+/- 1 to 1/2°C	10.0 mV/°C	+4 to +20 V*
TMP35	10° to 125° C	+/- 2°C	10.0 mV/°C	+2.7 to +5.5V
TMP36	-40° to 125° C	+/- 2°C	10.0 mV/°C	+2.7 to +5.5V
TMP37	5° to 100° C	+/- 2°C	20.0 mV/°C	+2.7 to +5.5V

*Requires negative supply for temperature readings below 0

Fig. 4.10 Integrated circuit temperature sensors LM34, LM35 (www.ti.com), TMP35, 36, 37 temperature sensors (www.analog.com)

```
//Machine Control.
//
//Notes:
//- For each channel of the ANALOG IN connector, there is a resistor
//  divider made by a 100k and 39k, the input voltage is divided by a
//  ratio of 0.28.
//- Maximum input voltage is 10V.
//- To use the 0V-10V functionality, a 24V supply on the PWR SUPPLY
//  connector is necessary.
//- Reads analog voltage at A0 connected to LM34 IC temperature sensor.
//- The LM34 provides 10 mV of output per degree Fahrenheit.
//
//This example code is in the public domain.
//Copyright (c) 2024 Arduino, SPDX-License-Identifier: MPL-2.0
//*******************************************************************

#include <Arduino_PortentaMachineControl.h>

const float RES_DIVIDER = 0.28057;
const float REFERENCE   = 3.0;

void setup()
{
Serial.begin(9600);
while (!Serial)
 {
 ; // wait for serial port to connect.
 }

MachineControl_AnalogIn.begin(SensorType::V_0_10);
```

```
}

void loop()
{
float raw_voltage_ch0 = MachineControl_AnalogIn.read(0);
float voltage_ch0 = (raw_voltage_ch0 * REFERENCE) / 65535 / RES_DIVIDER;

//ADC output
Serial.print("ADC CH0: ");
Serial.println(raw_voltage_ch0, 3);

//ADC voltage
Serial.print("LM34 voltage: ");
Serial.print(voltage_ch0, 3);
Serial.println("V");

//ADC voltage
Serial.print("LM34 temperature [F]: ");
float LM_34_temp = voltage_ch0/.010;            //10 mv/degree F
Serial.println(LM_34_temp, 3);

Serial.println(" ");
delay(1000);
}

//*********************************************************************
```

4.4.3.3 Thermocouple

A thermocouple consists of two different metals joined at two different junctions (hot and cold) as shown in Fig. 4.11a). Due to the Seebeck Effect a voltage proportional to junction temperature is generated at each junction. For a thermocouple the hot junction is used as a probe to measure an unknown temperature. The cold junction serves as the reference. The cold junction may be placed in an ice bath or its temperature may be measured with a different sensor. The voltage difference between the hot and cold junction is proportional to the temperature (Omega).

Thermocouples come in a variety of types. We concentrate on the Type J and K as shown in Fig. 4.11b). They operate over a wide temperature range.

The PMC Temperature Measurement subsystem consists of three channels (0, 1, 2). To measure temperature the positive terminal of the thermocouple is connected to TPx and the negative terminal to TNx as shown in Fig. 4.11c).

In the sketch below a Type K thermocouple is connected to CH0 and a Type J thermocouple to CH1. The sketch is adapted from "Temp_probes_Thermocouples" from the deprecated Arduino_MachineControl library.

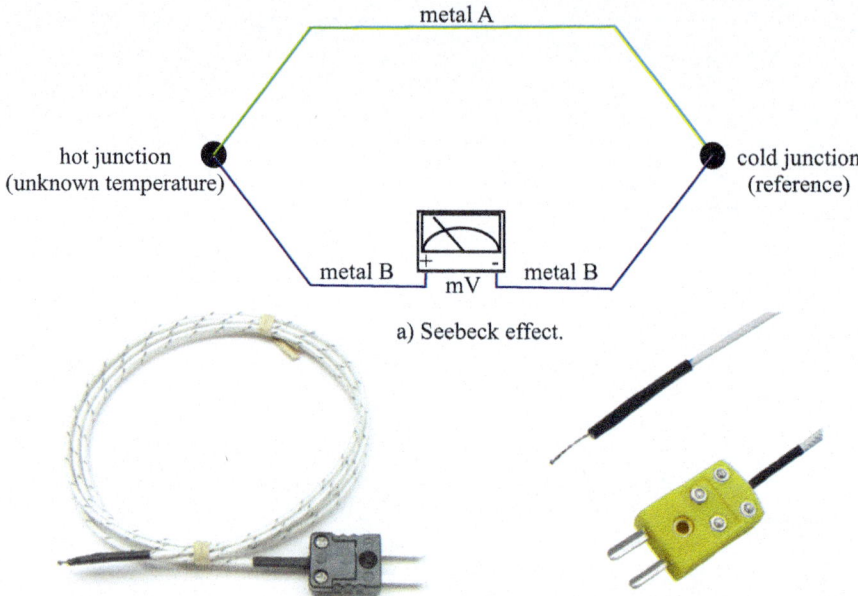

a) Seebeck effect.

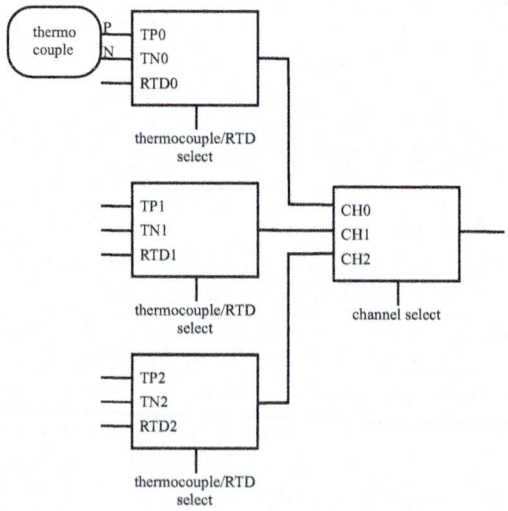

J type: black, iron/constantan, 0° C to 750 0° C (32° F to 1382° F), ~0.75% error
K type: yellow, chromel/alumel, -200° C to 1250 0° C (-328° F to 2282° F), ~0.75% error

b) Type J and Type K thermocouples.

c) PMC temperature measurement.

Fig. 4.11 Thermocouples

4.4 Analog Input Sensors

```
//*************************************************************************
//PMC_thermocouple_dep
//
//Notes:
//- Example adapted from Temp_probes_Thermocouples from deprecated
//  Arduino_MachineControl library.
//- The functions and syntax in this example are not compatible with
//  Arduino_PortentaMachineControl.
//- Example reads the temperatures measured by the thermocouples
//   -- K Type on channel 0
//   -- J Type on channel 1
//
//This example code is in the public domain.
//*************************************************************************

#include <Arduino_MachineControl.h>

using namespace machinecontrol;

void setup()
{
Serial.begin(9600);

temp_probes.tc.begin();           //initialize temperature probes
Serial.println("Temperature probes initialization done");
temp_probes.enableTC();           //enables Thermocouples chip select
Serial.println("Thermocouples enabled");
}

void loop()
{

temp_probes.selectChannel(0);    //set CH0, internal 150 ms delay
                                 //take CH0 measurement
                                 //Centigrade
float temp_ch0_C = temp_probes.tc.readTemperature();
Serial.print("Temperature CH0 [ÂºC]: ");
Serial.print(temp_ch0_C);
                                 //Fahrenheit
float temp_ch0_F = (temp_ch0_C * 9.0/5.0) + 32;
Serial.print("   Temperature CH0 [ÂºF]: ");
Serial.print(temp_ch0_F);
Serial.println();

temp_probes.selectChannel(1);    //set CH1, internal 150 ms delay
                                 //take CH1 measurement
                                 //Centigrade
float temp_ch1_C = temp_probes.tc.readTemperature();
Serial.print("Temperature CH1 [ÂºC]: ");
Serial.print(temp_ch1_C);
                                 //Fahrenheit
float temp_ch1_F = (temp_ch1_C * 9.0/5.0) + 32;
Serial.print("   Temperature CH1 [ÂºF]: ");
```

```
Serial.print(temp_ch1_F);
Serial.println();
Serial.println();
}

//*********************************************************************
```

4.4.3.4 Resistance Temperature Detectors

Resistance Temperature Detectors or RTDs provide for the precise measurement of temperature. An RTD consists of a precision trimmed piece of metal or a coil of wire wrapped around a ceramic or glass core. The RTD is calibrated to have a specific resistance at a given temperature. For example, a PT100 RTD has a resistance of $100\,\Omega$ at $0\,°C$ and a PT1000 has a resistance of $1000\,\Omega$ at $0\,°C$ (Omega).

RTDs are available in a 2, 3, and 4-wire configuration. A 3-wire version is shown in Fig. 4.12a, b. Using a three or four wire configuration, the value of R_{RTD} resistance may be isolated from the resistance of the wire leads.

The connection of an RTD to the PMC Temp Probes section is shown in Fig. 4.12c. In the following sketch, a PT100 (Adafruit #3290) is used to measure temperature. The measured temperature is provided in Centigrade and Fahrenheit.

```
//*****************************************************************
//PMC_RTD
//
//Adapted from:
//Portenta Machine Control - Temperature Probes RTD Example
//
//Notes:
//- tests 3-wire resistance temperature detectors (RTDs)
//- possible to acquire 2-wire RTDs by shorting the RTDx pin to
//  the TPx pin.
//- PMC features a precise 400 ohm 0.1% reference resistor,
//  serves as reference for the MAX31865.
//
//This example code is in the public domain.
//Copyright (c) 2024 Arduino, SPDX-License-Identifier: MPL-2.0
//*****************************************************************

#include <Arduino_PortentaMachineControl.h>

//Rref resistor - PT100: 430.0 ; PT1000: 4300.0
#define RREF      400.0

//Nominal 0-degrees-C sensor resistance
//PT100: 100.0; PT1000: 1000.0
#define RNOMINAL  100.0

void setup()
{
Serial.begin(9600);
while (!Serial)
```

4.4 Analog Input Sensors

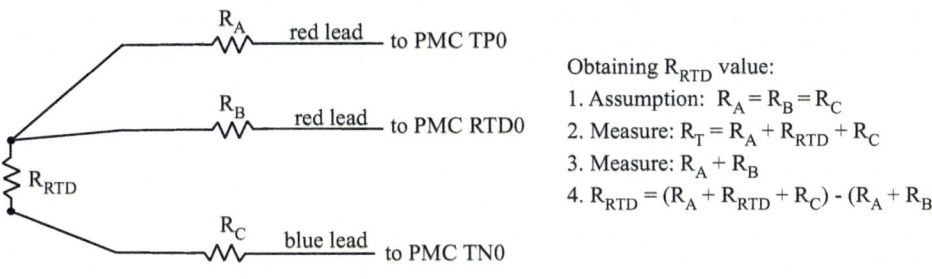

Obtaining R_{RTD} value:
1. Assumption: $R_A = R_B = R_C$
2. Measure: $R_T = R_A + R_{RTD} + R_C$
3. Measure: $R_A + R_B$
4. $R_{RTD} = (R_A + R_{RTD} + R_C) - (R_A + R_B)$

a) RTD 3-wire configuration.

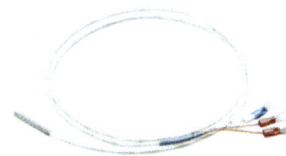

b) RTD 3-wire PT100.

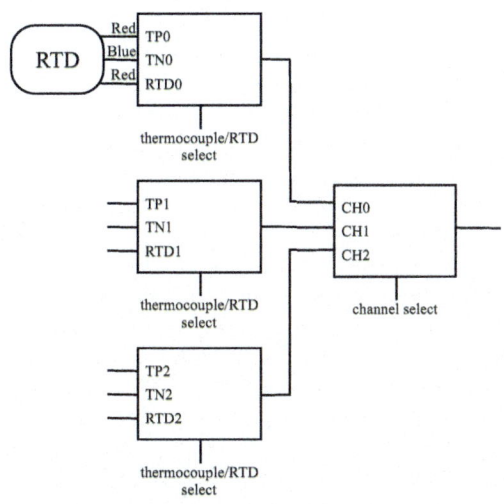

c) PMC temperature measurement.

Fig. 4.12 Resistance temperature detectors or RTDs

```
    {
      ;
    }
MachineControl_RTDTempProbe.begin(THREE_WIRE);
}

void loop()
{
MachineControl_RTDTempProbe.selectChannel(0);
Serial.println("CHANNEL 0 SELECTED");
uint16_t rtd = MachineControl_RTDTempProbe.readRTD();
float ratio = rtd;
ratio /= 32768;

//Check and report any faults via Serial Monitor
uint8_t fault = MachineControl_RTDTempProbe.readFault();
if(fault)
  {
  Serial.print("Fault 0x"); Serial.println(fault, HEX);
  if(MachineControl_RTDTempProbe.getHighThresholdFault(fault))
      {
      Serial.println("RTD High Threshold");
      }
  if(MachineControl_RTDTempProbe.getLowThresholdFault(fault))
      {
      Serial.println("RTD Low Threshold");
      }
  if(MachineControl_RTDTempProbe.getLowREFINFault(fault))
      {
      Serial.println("REFIN- > 0.85 x Bias");
      }
  if(MachineControl_RTDTempProbe.getHighREFINFault(fault))
      {
      Serial.println("REFIN- < 0.85 x Bias - FORCE- open");
      }
  if(MachineControl_RTDTempProbe.getLowRTDINFault(fault))
      {
      Serial.println("RTDIN- < 0.85 x Bias - FORCE- open");
      }
  if(MachineControl_RTDTempProbe.getVoltageFault(fault))
      {
      Serial.println("Under/Over voltage");
      }
  MachineControl_RTDTempProbe.clearFault();
  }
  else
      {
      Serial.print("RTD value: "); Serial.println(rtd);
      Serial.print("Ratio = "); Serial.println(ratio, 8);
      Serial.print("Resistance = "); Serial.println(RREF * ratio, 8);
      Serial.print("Temperature [ÂºC] = ");
      Serial.print(MachineControl_RTDTempProbe.readTemperature(RNOMINAL, RREF));

      //Centigrade
      float temp_ch0_C = MachineControl_RTDTempProbe.readTemperature(RNOMINAL, RREF);

      //Fahrenheit
      float temp_ch0_F = (temp_ch0_C * 9.0/5.0) + 32;
      Serial.print("    Temperature CH0 [ÂºF]: ");
```

4.4 Analog Input Sensors

```
    Serial.println(temp_ch0_F);
    }
Serial.println();
delay(1000);
}

//*****************************************************************
```

4.4.3.5 Thermistors

The thermal resistor or thermistor is a temperature sensing semiconductor. Thermistors enjoy repeatability, stability, and accuracy. They are available over a temperature range of −100 to +300 °C. Thermistors are available with positive temperate coefficients (PTC) or negative temperature coefficients (NTC). For NTC thermistors, resistance decreases as temperature increases (Omega).

The relationship between thermistor temperature and resistance is provided by the Steinhart—Hart equation shown in Fig. 4.13b. The simplified version of the equation is typically used. To determine the measured temperature (T), several values must be known. The value β is typically provided by the thermistor manufacturer. Also, the value of thermistor R_o and temperature T_o is also provided. All temperatures are provided in Kelvin.

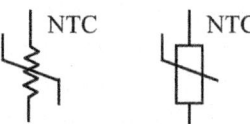

a) Thermistor schematic symbols.

$$\frac{1}{T} = a + b \ln R + c (\ln R)^3$$

T: absolute temperature in Kelvin
a,b,c: Steinhart - Hart parameters
    (for each device)
R: resistance

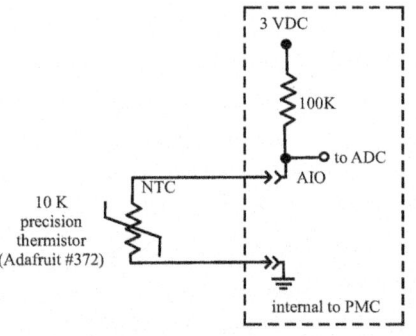

c) PMC ANALOG IN for NTC.

$$\frac{1}{T} = \frac{1}{T_o} + \frac{1}{\beta} \ln \frac{R}{R_o}$$

T: temperature in Kelvin
To: room temp 25°C = 298.15K
β: 3950 (provided for thermistor)
Ro: thermistor resistance at To

b) Steinhart - Hart equation.

Fig. 4.13 Thermistors

Example. In this example a precision 10 K NTC thermistor (Adafruit #372) is measured using the PMC. As shown in Fig. 4.13c, the thermistor is connected to ANALOG IN AI0. The values of β, R_o, and T_o are provided by Adafruit. The portion of the sketch to measure thermistor resistance is adapted from the PMC Library sketch "Analog in NTC." To convert from thermistor resistance to temperature using the Steinhart–Hart equation, a sketch from Lady Ada of Adafruit is adapted (Lady Ada).

```
//*******************************************************************************
//PMC_analog_input_NTC_thermistor
//
//Adapted from:
//Portenta Machine Control - Analog in NTC
//
//Notes - resistance calculation:
//- This example provides the resistance value acquired by the Machine Control.
//- A 3V voltage REFERENCE is connected to each channel of the ANALOG IN connector.
//- The REFERENCE has a 100k resistor in series, allowing only a low current flow.
//- The voltage sampled by the Portenta's ADC is the REFERENCE voltage divided by the
//  voltage divider composed of the input resistor and the 100k resistor in series
//  with the voltage REFERENCE.
//- To use the NTC functionality, a 24V supply on the PWR SUPPLY connector is required.
//- The resistor value is calculated by inversely applying the voltage divider formula.
//
//This example code is in the public domain.
//Copyright (c) 2024 Arduino, SPDX-License-Identifier: MPL-2.0
//
//Notes - temperature calculation:
//- Once the resistance value of the thermistor is calculated, the temperature is
//  calculated using techniques adapted from:
//    - Thermistor Example #3 from the Adafruit Learning System guide on Thermistors
//      https://learn.adafruit.com/thermistor/overview by Limor Fried, Adafruit Industries
//
//MIT License - please keep attribution and consider buying parts from Adafruit.
//SPDX-FileCopyrightText: 2011 Limor Fried/ladyada for Adafruit Industries
//SPDX-License-Identifier: MIT
//*******************************************************************************

#include <Arduino_PortentaMachineControl.h>

const float REFERENCE = 3.0;
const float LOWEST_VOLTAGE = 2.7;

#define REFERENCE_RES 100000
#define THERMISTORNOMINAL 10000      //resistance at 25 degrees C
#define TEMPERATURENOMINAL 25        //temp. for nominal resistance (almost always 25 C)
#define BCOEFFICIENT 3950            //The beta coefficient of the thermistor (usually 3000-4000)
//#define SERIESRESISTOR 10000       //the value of the 'other' resistor

void setup()
{
Serial.begin(9600);
while(!Serial)
  {
  ;    //wait for serial port to connect.
  }

MachineControl_AnalogIn.begin(SensorType::NTC);
}

void loop()
{
float raw_voltage_ch0 = MachineControl_AnalogIn.read(0);
float voltage_ch0 = (raw_voltage_ch0 * REFERENCE) / 65535;
```

4.4 Analog Input Sensors

```
float resistance_ch0;

Serial.print("Resistance CH0: ");
if(voltage_ch0 < LOWEST_VOLTAGE)
  {
  resistance_ch0 = ((-REFERENCE_RES) * voltage_ch0) / (voltage_ch0 - REFERENCE);
  Serial.print(resistance_ch0);
  Serial.println(" ohm");
  }
else
  {
  resistance_ch0 = -1;
  Serial.println("NaN");
  }

float steinhart;
steinhart =  resistance_ch0/ THERMISTORNOMINAL;      //(R/Ro)
steinhart = log(steinhart);                          //ln(R/Ro)
steinhart /= BCOEFFICIENT;                           //1/B * ln(R/Ro)
steinhart += 1.0 / (TEMPERATURENOMINAL + 273.15);    //+ (1/To)
steinhart = 1.0 / steinhart;                         //Invert
steinhart -= 273.15;                                 //convert absolute temp to C

//Centigrade
Serial.print("Temperature [C]: ");
Serial.println(steinhart);

//Fahrenheit
Serial.print("Temperature [F]: ");
steinhart = (steinhart * 9/5) + 32;
Serial.println(steinhart);
Serial.println();
delay(1000);
}
//************************************************************************
```

4.4.4 Fluid Depth Sensors

In many industrial and IoT applications it is important to measure fluid depth in a vat, storage container, well, and a rain barrel. In this section we explore several fluid depth sensors.

4.4.4.1 Milone E-Tape Fluid Sensor

Milone Technologies manufacture a line of continuous fluid level sensors. The sensor resembles a ruler and provides a near linear response. The sensor reports a change in resistance to indicate the distance from sensor top to the fluid surface. To convert the resistance change to a voltage change a voltage divider circuit is used. In this example, the Milone 0–5 VDC Resistance to Voltage Module is used. The module shown in Fig. 4.14 may be powered from 6 to 30 VDC. The output from the module ranges up to 5 VDC (www.milonetech.com).

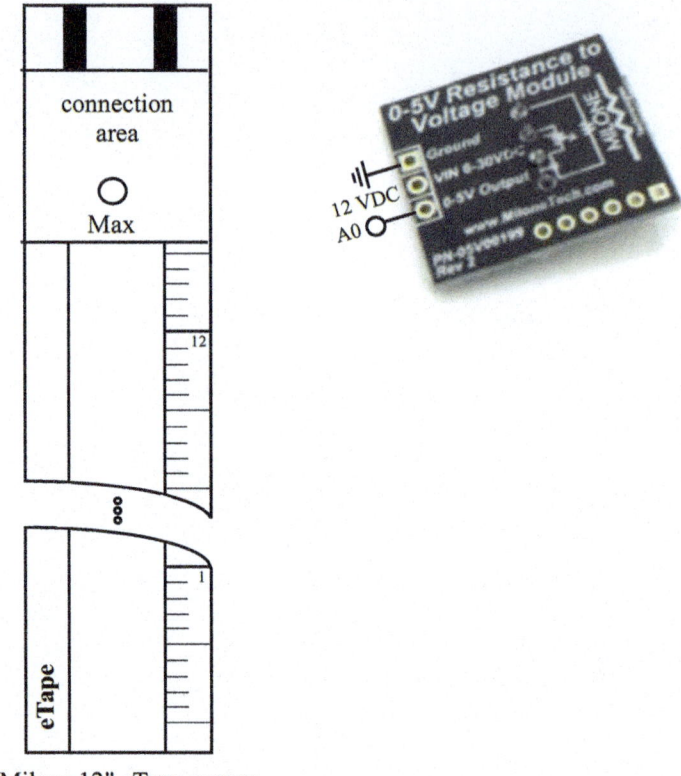

Fig. 4.14 Milone Technologies eTape liquid level sensor. Arduino illustrations used with permission of the Arduino Team (CC BY-NC-SA) (www.arduino.cc). Image of sensor used courtesy of Milone Technology (www.milonetech.com)

4.4.4.2 Piezo-Based Fluid Depth Sensor

Fluid level may also be measured using a piezo-based submersible sensor as shown in Fig. 4.15. The sensors may be used to measure fluid depths up to 500 m. In the example shown the sensor measuring range is 5 m. The sensor requires a 24 VDC power supply and provides an output from 4–20 mA (Taizhou). The relationship between fluid depth and sensor output is determined experimentally.

4.4.5 Light Sensor

There are many different types of sensors used to detect light including photoresistors, photovoltaic cells, photodiodes, and phototransistors. When choosing a light sensor it is important to match the sensor characteristics to the desired wavelength of light. Shown in

4.4 Analog Input Sensors

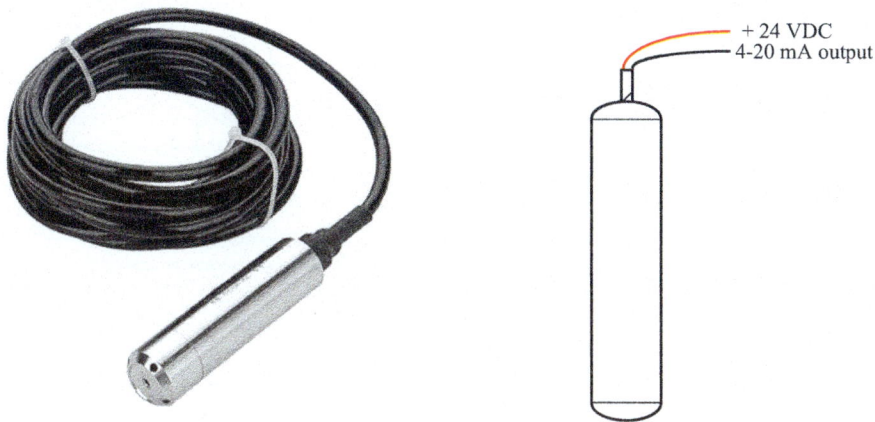

a) Submersible level sensor, 4-20 mA output.

b) Sensor configuration. Supply and measurement instrument share common ground.

Fig. 4.15 Piezo-based fluid depth sensor

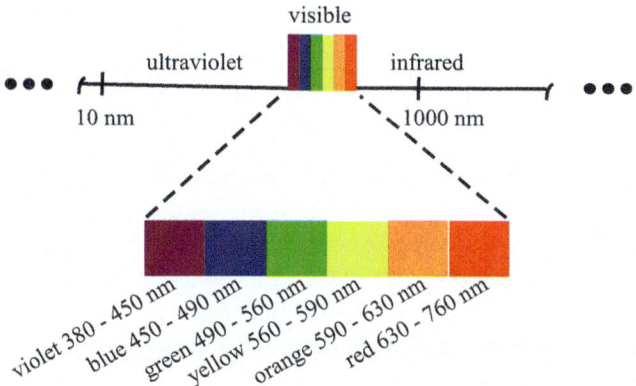

Fig. 4.16 Light spectrum (Miller)

Fig. 4.16 is a portion of the electromagnetic spectrum. The visible light spectrum ranges from approximately 380–760 nm. Each visible color has a defined range of wavelengths. The ultraviolet and infrared spectrums are adjacent to the visible spectrum.

4.4.5.1 Photoresistor

A photoresistor may be known by several different names including photoconductive cells, photo cells, and light dependent resistors. They are constructed from different types of

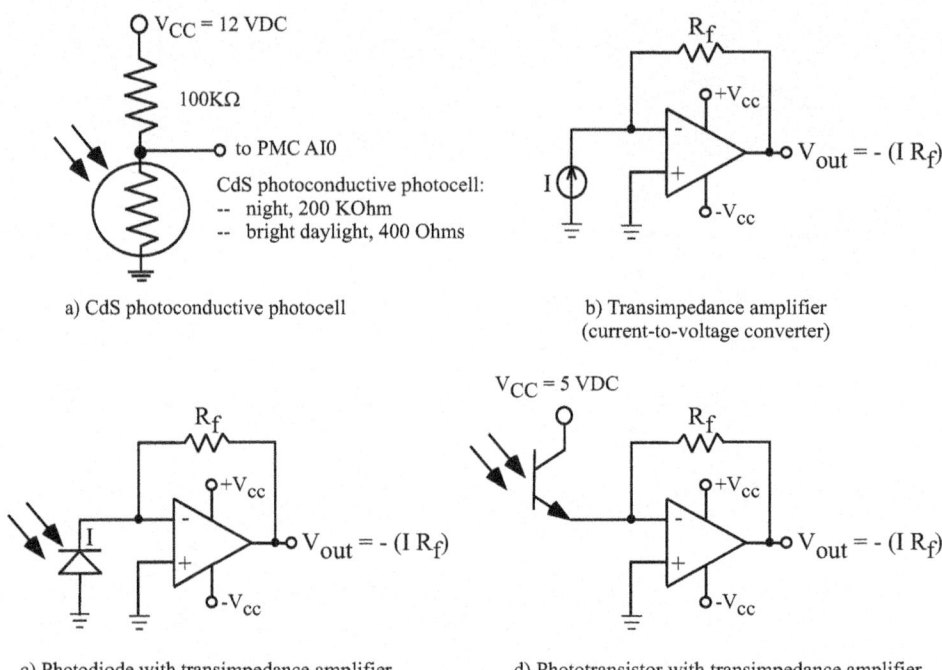

Fig. 4.17 Light sensors

semiconductor material each responsive to different light wavelengths in the infrared (IR), visible, or ultraviolet (UV) bands. When light of the appropriate wavelength impinges on the semiconductor material, current carriers are released and the resistance of the material is decreased. The photoresistor may be used in applications to detect day and night, light or dark environments, or to detect the presence of an object.

Example: PDV-P8001 Cadmium Sulfide Cell. In this example we use a PDV-P8001 Cadmium Sulfide (CdS) photoconductive photocell to determine when night has arrived. The cell is sensitive to light in the visible range (400–700 nm). The resistance of the cell ranges from 200 KΩ when in darkness to approximately 400 Ω when fully illuminated (API).

To interface to the Arduino PMC a 100 K Ω resistor is placed in series with the cell as shown in Fig. 4.17a) to implement a voltage divider circuit. The node between the photocell and the resistor is connected to Arduino PMC ADC input AI0. The threshold for determining day versus night is experimentally determined. The results are provided to the Serial Monitor for display. The example sketch was created using the Arduino IDE.

4.4 Analog Input Sensors

```
//**********************************************************************
//PMC_photocell_analog_input_0_10V
//
//Adapted from:
//Portenta Machine Control - Analog in 0-10 V
//
//This example provides the voltage value acquired by the Portenta
//Machine Control.
//
//Notes:
//- For each channel of the ANALOG IN connector, there is a resistor
//   divider made by a 100k and 39k, the input voltage is divided by a
//   ratio of 0.28.
//- Maximum input voltage is 10V.
//- To use the 0V-10V functionality, a 24V supply on the PWR SUPPLY
//   connector is necessary.
//- Reads analog voltage at A0 connected to PDV-P8001 circuit
//- Photocell in series with 100k Ohm resistor
//
//This example code is in the public domain.
//Copyright (c) 2024 Arduino, SPDX-License-Identifier: MPL-2.0
//**********************************************************************

#include <Arduino_PortentaMachineControl.h>

const float RES_DIVIDER = 0.28057;
const float REFERENCE   = 3.0;

void setup()
{
Serial.begin(9600);
while (!Serial)
 {
 ; // wait for serial port to connect.
 }

MachineControl_AnalogIn.begin(SensorType::V_0_10);
}

void loop()
{
float raw_voltage_ch0 = MachineControl_AnalogIn.read(0);
float voltage_ch0 = (raw_voltage_ch0 * REFERENCE) / 65535 / RES_DIVIDER;

//ADC output
Serial.print("ADC CH0: ");
Serial.print(raw_voltage_ch0, 3);

//ADC voltage
Serial.print("Photocell voltage: ");
Serial.print(voltage_ch0, 3);
Serial.println("V");

//determine threshold experimentally
if (raw_voltage_ch0 < 1500)
  {
  Serial.println("Day");
  }
else
```

```
    {
    Serial.println("Night");
    }
  Serial.println(" ");
  delay(1000);
  }

  //***********************************************************************
```

4.4.5.2 Photodiode and Phototransistor

The photodiode and phototransistor are semiconductor devices that generate current in the presence of light. They are typically used with an op amp based transimpedance amplifier (Fig. 4.17b) to convert the current to a voltage.[2]

The photodiode (Fig. 4.17c) is used in the reverse bias or photoconductive mode. The generated current flowing through the feedback resistor (R_f) provides the output voltage (V_{out}). The phototransistor generates more current but responds slower in response to light changes.

Optical links are used to isolate one circuit from another, provide a communication link, or provide isolation in an industrial application. An optical link may be formed by using an LED coupled with a photodiode or phototransistor responsive to the LED wavelength via an optical fiber.

Example: Optical fiber link. Industrial Fiber Optics manufacturers a series of optical emitters (LEDs) at a wide variety of wavelengths (e.g. IF-E93 (green), IF-E97 (red), IF-E92B (blue)). They also manufacture a wide variety of optical detectors including photo diodes (e.g. IF-D91B), transistors (e.g. IF-D92B), and Darlingtons (e.g. IF-D93B) (Industrial Fiber Optics).

In this application we use an optical fiber to couple a light emitting diode to a photodiode to form an optical data link. We start with some background information on optical fibers. Optical fibers are used to link two devices via light rather than an electronic signal. In a typical application an electronic signal is converted to light, transmitted down the optical fiber, and converted back to an electronic signal.

As shown in Fig. 4.18a) an optical fiber consists of several concentric layers of material including the core where light is transmitted, the cladding, the buffer, and the protective outer jacket. Light is transmitted through the fiber via the concept of total internal reflection. The core material is more optically dense than the cladding material. At shallow entry angles the light reflects from the core/cladding boundary and stays within the fiber core as shown in b). This allows for the transmission of light via fiber for long distances with limited signal degradation.

To provide an interface between an electronic signal and the fiber, an optical emitter is used as shown in c). The optical emitter contains a light emitting diode (LED) as the light

---

[2] Operational amplifiers are discussed later in the chapter.

4.4 Analog Input Sensors

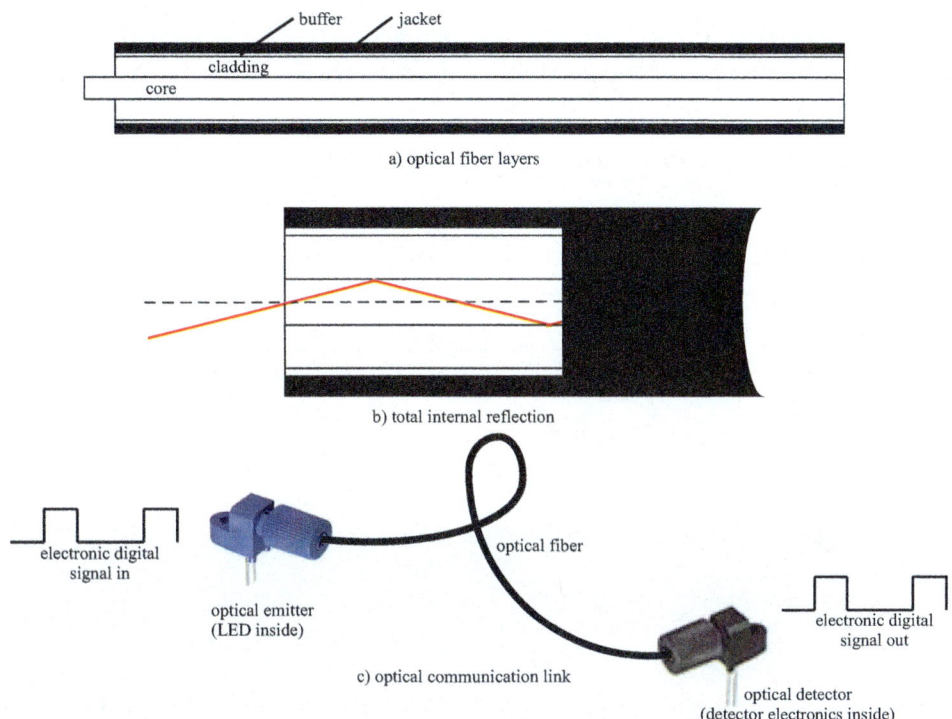

Fig. 4.18 Optical fibers

source. At the far end of the optical fiber an optical detector is used to convert the light signal back to an electronic one.

As previously mentioned, it is important to note that optical emitters, detectors, and fibers are available in a variety of wavelengths. It is essential that the emitter, detector, and fiber are capable at operating at the same optical wavelengths.

Example: Optical fiber link. In this example we use a transistor (PN2222) to switch the red LED (660 nm, IF-E97, $V_f = 1.7$, I_f 40 mA) on/off. The transistor is driven by a PMC output as shown in Fig. 4.19. On the receiving end we use a photodiode (IF-91B) with a transimpedance amplifier to convert the light signal back to an electronic signal.

We use an Arduino IDE based sketch to generate a digital signal. The resulting input and output waveforms for the optical link are shown in Fig. 4.19. Note the signal inversion from input to output. Later in the chapter we explore how to use operational amplifier building blocks to condition the signal to desired characteristics.

```
//************************************************************
//PMC_square_wave - generates 1 kHz square wave
//
//Configuration:
```

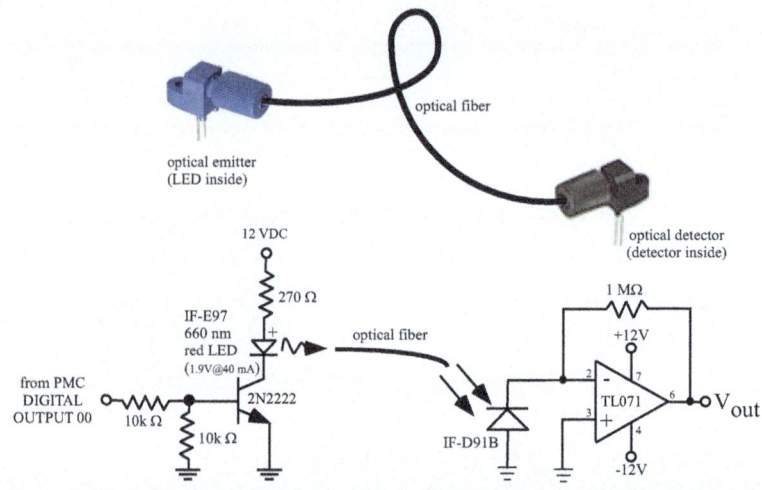

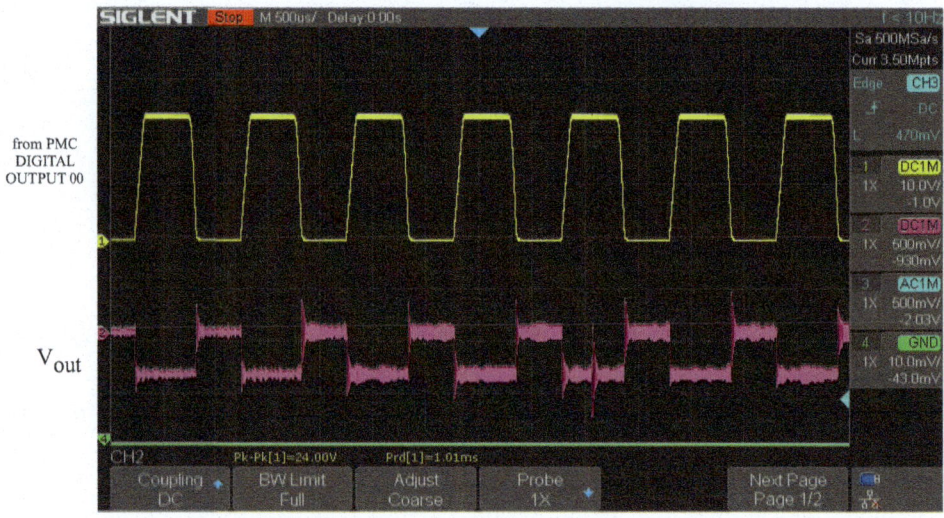

Fig. 4.19 Interface for optical fibers

```
//  - PMC connected to 24 VDC power supply
//  - Digital OUTPUTS 24V IN connected to 24 VDC power supply
//  - The 500 ms delays are used to verify circuit operation.  They
//    are commented out to generate the 1 kHz square wave.
//
//This code example is in the public domain.
//**********************************************************************

#include <Arduino_PortentaMachineControl.h>

void setup()
```

4.4 Analog Input Sensors

```
{
MachineControl_DigitalOutputs.begin();         //init PMC digital outputs
}

void loop()
{
MachineControl_DigitalOutputs.write(0, HIGH);  //digital out 00 on
//delay(500);                                  //delay in ms
delayMicroseconds(500);                        //delay in us
MachineControl_DigitalOutputs.write(0, LOW);   //digital out 00 off
//delay(500);                                  //delay in ms
delayMicroseconds(500);                        //delay in us
}

//***********************************************************************
```

4.4.6 Tilt Sensor

CTi Sensors manufactures a series of dual-axis inclinometers equipped with three-access accelerometers. The sensor outputs may either be digital or analog. We explore the TILT–15–S–90 that provides ± 90 °C inclinometer sensing for both the X and Y axis. The sensor may be powered from 4.1 to 38 VDC. We use a 5 VDC supply to power the sensor. We also use the sensor's analog output. The sensor provides 25 mV per degree over the full ± 90 °C range. The analog output voltage ranges from 0.25 V to 4.75 VDC (www.CTiSensors.com).

Example: Tilt sensor In this example we use a CTi TILT–15–X–90 sensor to measure the angle between a stationary and movable strut as shown in Fig. 4.20. We use the sensor's analog OutX output. The sensor provides 0.25V at minus 90 °C tilt and 4.75V at plus 90 °C. It has a sensitivity of 25 mV per degree at tilt measurements between the two extremes.

The OutX signal is fed to Arduino PMC input AI0. The value from the ADC converter is mapped to an angular value. At 90 °C the tilt sensor output provides an ADC reading of 2800; whereas, at minus 90 °C the tilt sensor provides an ADC reading of 27,000. We use these values with the Arduino map function to convert the ADC readings to an angular value between minus 90 and plus 90 °C.

```
//***********************************************************************
//PMC_tilt_analog_input_0_10V
//
//Adapted from:
//Portenta Machine Control - Analog in 0-10 V
//
//This example provides the voltage value acquired by the Portenta
//Machine Control.
//
//Notes:
//- For each channel of the ANALOG IN connector, there is a resistor
//  divider made by a 100k and 39k, the input voltage is divided by a
//  ratio of 0.28.
```

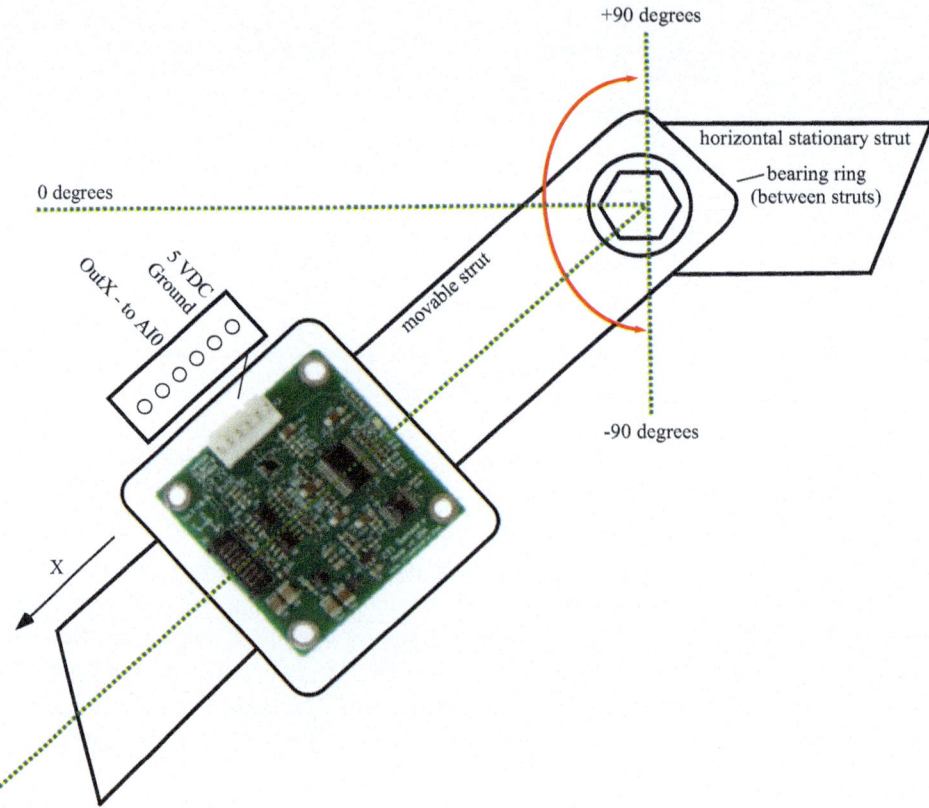

Fig. 4.20 CTi TILT–15–S–90 sensor (www.CTiSensors.com)

```
//- Maximum input voltage is 10V.
//- To use the 0V-10V functionality, a 24V supply on the PWR SUPPLY
//  connector is necessary.
//- CTi tilt sensor analog output OUTX is connected to Arduino
//  PMC pin AI0.
//- The sketch reads ADC value at AI0, maps to degree, and displays
//  value on the Serial Monitor every second.
//
//This example code is in the public domain.
//Copyright (c) 2024 Arduino, SPDX-License-Identifier: MPL-2.0
//********************************************************************

#include <Arduino_PortentaMachineControl.h>

const float RES_DIVIDER = 0.28057;
const float REFERENCE   = 3.0;
int tilt;                               //angular displacement

void setup()
{
Serial.begin(9600);
```

4.4 Analog Input Sensors

```
while (!Serial)
{
; // wait for serial port to connect.
}

MachineControl_AnalogIn.begin(SensorType::V_0_10);
}

void loop()
{
float raw_voltage_ch0 = MachineControl_AnalogIn.read(0);
float voltage_ch0 = (raw_voltage_ch0 * REFERENCE) / 65535 / RES_DIVIDER;

//ADC output
Serial.print("ADC CH0: ");
Serial.print(raw_voltage_ch0, 3);

//ADC voltage
Serial.print("Voltage CH0: ");
Serial.print(voltage_ch0, 3);
Serial.println("V");

//map to angle
unsigned int value = (unsigned int) (raw_voltage_ch0);
tilt = map(value, 27000, 2800, -90, 90);
Serial.print("Tilt reading: ");
Serial.println(tilt);              //tilt value
Serial.println(" ");
delay(1000);
}

//**********************************************************************
```

4.4.7 Environmental Sensors

In this section we concentrate on the MQ series of gas and environmental sensors. The MQ series of sensors consists of a metal oxide semiconductor active element. The electrical resistance of the active element varies in the presence of a specific gas or gases. A sample of available MQ sensors is shown in Fig. 4.21 (https://www.mysensors.org).

An MQ sensor is typically used in a voltage divider circuit as shown in Fig. 4.22a). In the presence of a specific gas or gases, the resistance of the sensing element (R_S) varies. The value of R_S is in series with a fixed load resistor (R_L) forming a voltage divider network. The output voltage is an indication of the gas concentration. Many of the sensors in the MQ series require a heater voltage as shown in Fig. 4.22b). Figure 4.22c) shows the physical configuration of the MQ series of sensors. SparkFun provides a breakout board to allow an MQ sensor to interface with a standard prototype board (https://www.mysensors.org, https://www.SparkFun.com).

To develop an interface circuit for an MQ sensor, complete the following steps:

Sensor	Description
MQ-2	liquefied petroleum gas (LPG), propane, hydrogen, methane $V_C < 24V$, $V_H = 5V$, $R_H = 31$ Ohms, $P_H < 900$ mW, preheat: 48 hours
MQ-3	alcohol, benzene, methane, hexane, LPG, carbon monoxide $V_C = 5V$, $V_H = 5V$, $R_H = 33$ Ohms, $P_H < 750$ mW, preheat: 24 hours
MQ-4	high sensitivity to methane and natural gas, lower sensitivity to alcohol and smoke $V_C = 5V$, $V_H = 5V$, $R_H = 33$ Ohms, $P_H < 750$ mW, preheat: 24 hours
MQ-6	liquefied petroleum gas (LPG) $V_C < 24V$, $V_H = 5V$, $R_H = 26$ Ohms, $P_H < 950$ mW, preheat: 24 hours
MQ-7	carbon monoxide $V_C = 5V$, $R_H = 33$ Ohms, $P_H < 350$ mW, preheat: 48 hours V_H: alternates between 5V for 60s and 1.4V for 90s
MQ-8	hydrogen $V_C = 5V$, $V_H = 5V$, $R_H = 29$ Ohms, $P_H < 900$ mW, preheat: 48 hours
MQ-9	methane, propane, carbon monoxide $V_C < 10V$, $R_H = 31$ Ohms, $P_H < 350$ mW, preheat: 48 hours V_H: alternates between 5V for 60s and 1.5V for 90s
MQ-131	ozone $V_C = 5V$, $V_H = 6V$, $R_H = 31$ Ohms, $P_H < 1,100$ mW, preheat: 24 hours
MQ-135	ammonia, nitrogen oxide, alochol, benzene, smoke, carbon dioxide $V_C = 5V$, $V_H = 5V$, $R_H = 33$ Ohms, $P_H < 800$ mW, preheat: 24 hours
MQ-138	hexane, benzene, ammonia, alcohol, smoke, carbon monoxide $V_C = 5V$, $V_H = 5V$, $R_H = 31$ Ohms, $P_H < 800$ mW, preheat: 24 hours
MQ-214	methane, LPG, butane, propane $V_C = 6V$

Fig. 4.21 MQ sensor series. Data sheets for many of the MQ sensors are provided at www.mysensors.org

- Choose a specific detectable gas of interest.
- Choose an appropriate sensor from Fig. 4.21.
- Determine key interface parameters as provided in Fig. 4.21.
- Implement the interface circuit as shown in Fig. 4.22b).
- Select an appropriate value of load resistor (R_L). A suggested value of load resistance is provided in the sensor's data sheet.[3] It is recommended to use a potentiometer that includes the value of suggested load resistance as the load resistor. This will allow the adjustment of circuit sensitivity to a specific level of gas concentration.

---

[3] Data sheets for many of the MQ sensors are provided at https://www.mysensors.org.

4.4 Analog Input Sensors

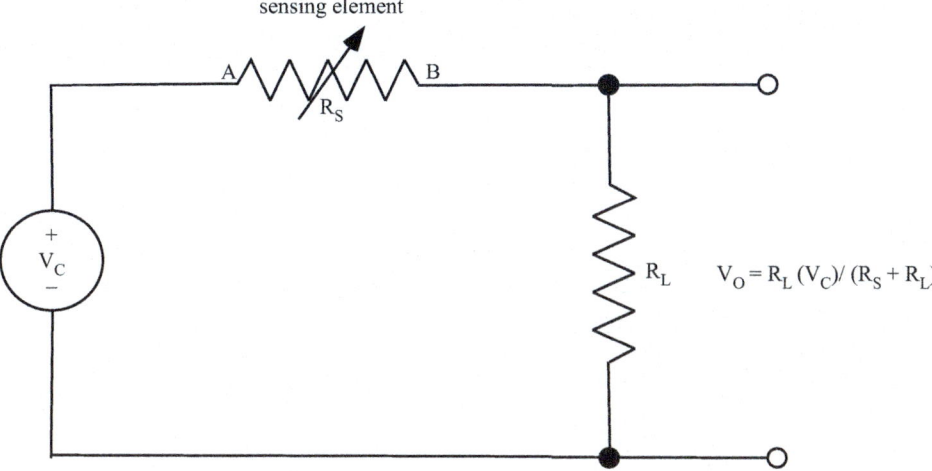

a) electrical resistance (RS) of the sensing element varies in the presence of a specific gas or gases.

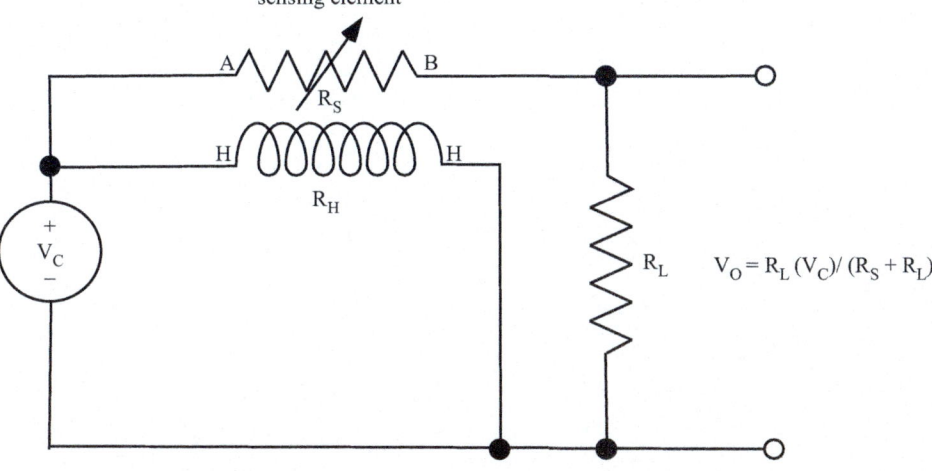

b) many of the MQ series of sensors require an applied heater voltage.

c) MQ sensors and MQ breakout board
(Sensor images courtesy of SparkFun Electronics, Inc. CC BY-NC-SA (www.sparkfun.com)).

Fig. 4.22 MQ sensor interface

- Write an Arduino sketch to read the analog output voltage from the sensor interface circuit, set a threshold for detection, and illuminate an LED and sound a buzzer when the gas is detected.

Example. MQ sensor. In this example we develop an interface circuit and an Arduino PMC sketch as a smoke detector. We use the MQ-4 sensor as the active element. The manufacturer's data sheet recommends a load resistance of 20 K Ω. A 100 K Ω potentiometer, set for 2 K Ω, serves as the load resistor as shown in Fig. 4.23 or a fixed resistor may be used. In clear air Vout is below 1.0 V. In the presence of smoke Vout approaches 2.0 V. The MQ-4 sensor draws approximately 160 mA of current.

```
//***********************************************************************
//PMC_MQ_sensor_analog_input_0_10V
//
//Adapted from:
//Portenta Machine Control - Analog in 0-10 V
//
//This example provides the voltage value acquired by the Portenta
//Machine Control.
//
//Notes:
//- For each channel of the ANALOG IN connector, there is a resistor
//  divider made by a 100k and 39k, the input voltage is divided by a
//  ratio of 0.28.
//- Maximum input voltage is 10V.
//- To use the 0V-10V functionality, a 24V supply on the PWR SUPPLY
//  connector is necessary.
//- Measures analog output from MQ-4 sensor.
//- The sketch reads ADC value at AI0, if threshold exceeded, activates
//  Sonalert on DIGITAL OUTPUT 00.
//
//This example code is in the public domain.
//Copyright (c) 2024 Arduino, SPDX-License-Identifier: MPL-2.0
//***********************************************************************

#include <Arduino_PortentaMachineControl.h>

const float RES_DIVIDER = 0.28057;
const float REFERENCE   = 3.0;
int tilt;                          //angular displacement

void setup()
{
Serial.begin(9600);
while (!Serial)
 {
 ; // wait for serial port to connect.
 }

MachineControl_AnalogIn.begin(SensorType::V_0_10);
}

void loop()
{
float raw_voltage_ch0 = MachineControl_AnalogIn.read(0);
```

4.4 Analog Input Sensors

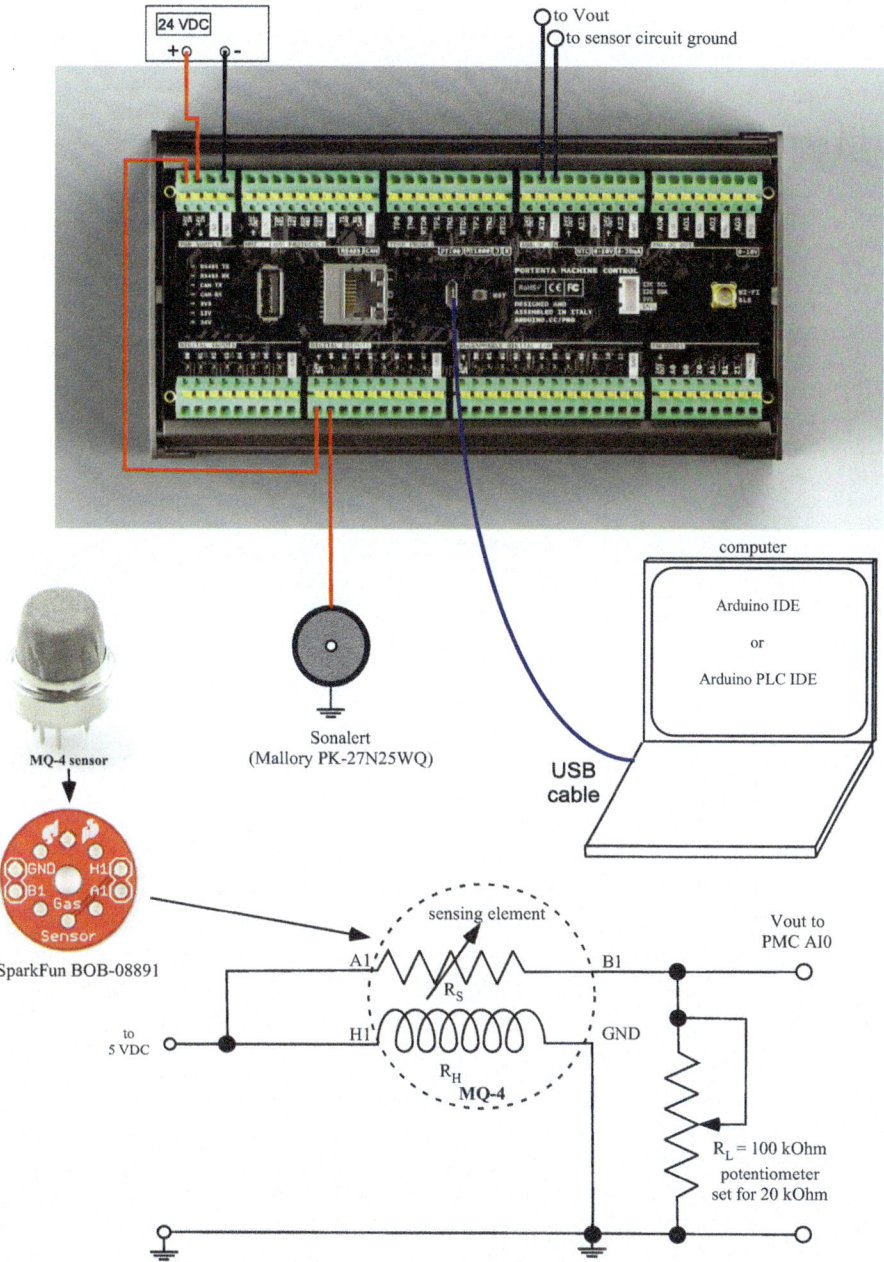

Fig. 4.23 MQ sensor test circuit. Sensor images courtesy of Sparkfun Electronics, (CC BY-NC-SA) (https://www.sparkfun.com)

```
float voltage_ch0 = (raw_voltage_ch0 * REFERENCE) / 65535 / RES_DIVIDER;

//ADC output
Serial.print("ADC CH0: ");
Serial.print(raw_voltage_ch0, 3);

//ADC voltage
Serial.print("Voltage CH0: ");
Serial.print(voltage_ch0, 3);
Serial.println("V");

//determine threshold experimentally
if (raw_voltage_ch0 > 10000)
   {
   MachineControl_DigitalOutputs.write(0, HIGH);
   Serial.println("Sonalert on");
   }
else
   {
   MachineControl_DigitalOutputs.write(0, LOW);
   Serial.println("Sonalert off");
   }
Serial.println(" ");
delay(1000);
}

//***********************************************************
```

4.4.8 Joystick

The thumb joystick is used to select a desired direction in an X–Y plane as shown in Fig. 4.24. The thumb joystick contains two built-in potentiometers (horizontal and vertical). A reference voltage of 5 VDC is applied to the VCC input of the joystick. As the joystick is moved, the horizontal (HORZ) and vertical (VERT) analog output voltages will change to indicate the joystick position. The joystick is also equipped with a digital select (SEL) button. Due to PMC logic levels, the SEL button requires a pullup resistor (10 K) to 24 VDC.

In the example, the intensity of the red and green LEDs is controlled by the joystick position. The Blue LED is illuminated when the SEL pushbutton is pressed.

```
//***********************************************************
//PMC_joystick_analog_input_0_10V
//
//Adapted from:
//Portenta Machine Control - Analog in 0-10 V
//
//This example provides the voltage value acquired by the Portenta
//Machine Control.
//
//Notes:
```

4.4 Analog Input Sensors

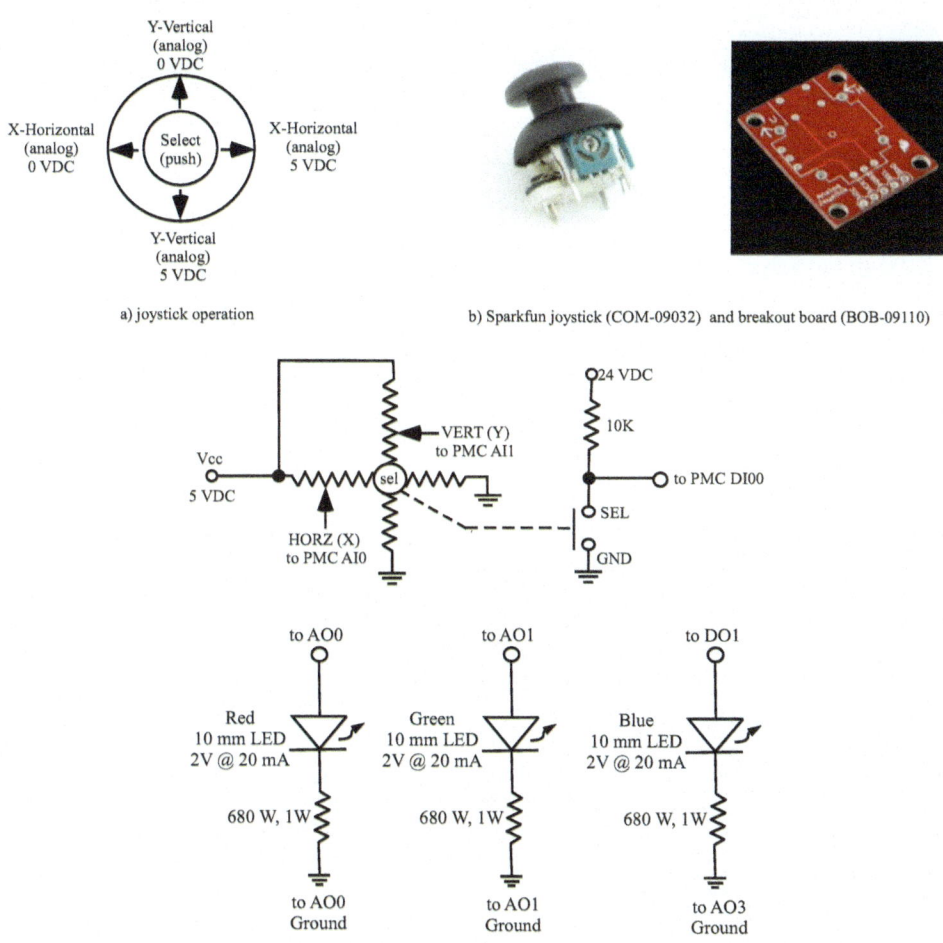

Fig. 4.24 Thumb joystick. Joystick image used courtesy of SparkFun, Electronics (CC BY-NC-SA (https://www.sparkfun.com)

```
//- For each channel of the ANALOG IN connector, there is a resistor
//  divider made by a 100k and 39k, the input voltage is divided by a
//  ratio of 0.28.
//- Maximum input voltage is 10V.
//- To use the 0V-10V functionality, a 24V supply on the PWR SUPPLY
//  connector is necessary.
//- Joystick (Sparkfun COM-09032 and BOB-09110) output X and Y
//  measured by PMC AI0 and AI1.
//- Displays value on the Serial Monitor every second.
//- Analog X and Y value controls intensity  of Red (X) and Green (Y)
```

```
//   LEDs connected A00 and A01.
//-  Blue LED (D00) controlled by Joystick SEL connected to DI0
//This example code is in the public domain.
//Copyright (c) 2024 Arduino, SPDX-License-Identifier: MPL-2.0
//*********************************************************************

#include <Arduino_PortentaMachineControl.h>

#define PERIOD_MS 4 /* 4ms - 250Hz */

const float RES_DIVIDER  = 0.28057;
const float REFERENCE    = 3.0;
uint16_t    joystick_sel = 0;

void setup()
{
Serial.begin(9600);
while (!Serial)
 {
 ; // wait for serial port to connect.
 }
Wire.begin();
MachineControl_AnalogIn.begin(SensorType::V_0_10);
MachineControl_AnalogOut.begin();
if(!MachineControl_DigitalInputs.begin())
  {
  Serial.println("Digital input GPIO expander initialization fail!!");
  }
}

void loop()
{
//Joystick X channel
float raw_voltage_x = MachineControl_AnalogIn.read(0);
float voltage_x = (raw_voltage_x * REFERENCE) / 65535 / RES_DIVIDER;
Serial.print("X: ");                       //ADC output X
Serial.print(raw_voltage_x, 3);
Serial.print(" Voltage X: ");              //X voltage
Serial.print(voltage_x, 3);
Serial.println("V");

//Red LED X - maximum output value is 10.5 V
MachineControl_AnalogOut.setPeriod(0, PERIOD_MS);
MachineControl_AnalogOut.write(0, (voltage_x * 2));
Serial.print(" Voltage (X:Red LED: ");     //voltage to Red LED
Serial.print((voltage_x * 2), 3);
Serial.println("V");

//Joystick Y channel
```

4.4 Analog Input Sensors

```
float raw_voltage_y = MachineControl_AnalogIn.read(1);
float voltage_y = (raw_voltage_y * REFERENCE) / 65535 / RES_DIVIDER;
Serial.print("Y: ");                    //ADC output Y
Serial.print(raw_voltage_y, 3);
Serial.print(" Voltage Y: ");           //Y voltage
Serial.print(voltage_y, 3);
Serial.println("V");

//Green LED Y - maximum output value is 10.5 V
MachineControl_AnalogOut.setPeriod(1, PERIOD_MS);
MachineControl_AnalogOut.write(1, (voltage_y * 2));
Serial.print(" Voltage (Y:Green LED: ");  //voltage to Green LED
Serial.print((voltage_y * 2), 3);
Serial.println("V");

//Blue LED
joystick_sel = MachineControl_DigitalInputs.read(DIN_READ_CH_PIN_00);

if(joystick_sel)
  {
  Serial.println(joystick_sel);
  MachineControl_DigitalOutputs.write(0, LOW);     //Blue LED off
  }
else
  {
  Serial.println(joystick_sel);
  MachineControl_DigitalOutputs.write(0, HIGH);    //Blue LED on
  }

Serial.println(" ");
delay(500);
}

//*********************************************************************
```

4.4.9 Greenhouse Sensors

In Application III we explore a variety of sensors to measure the temperature, humidity, soil moisture, and rain barrel water level within an instrumented greenhouse.

4.5 Output Devices and Actuators

An external device should not be connected to a controller without first performing careful interface analysis to ensure the voltage, current, and timing requirements of the controller and the external device are compatible. In this section, we describe interface considerations for a wide variety of external devices. We begin with the interface for a single light emitting diode.

4.6 Light Emitting Diodes (LEDs)

An LED is typically used as a logic or status indicator to inform the presence of a logic one or a logic zero at a specific pin of a microcontroller. An LED has two leads: the anode or positive lead and the cathode or negative lead.

To properly bias an LED, the anode lead must be biased at a level approximately 1.7 to 2.2 V higher than the cathode lead. This specification is known as the forward voltage (V_f) of the LED. The LED current must also be limited to a safe level known as the forward current (I_f). The diode voltage and current specifications are usually provided by the manufacturer.

A generic interface circuit for a controller is provided in Fig. 4.25a). A logic high control signal from a controller is applied to the transistor base. The NPN transistor acts as a switch and allows collector current flow from the voltage source (Vcc) through the load to ground. The value of the limiting resistor R_2 is calculated based on the desired load voltage and current. The resistor R_2 is calculated as:

$$R_2 = (V_{cc} - V_{LOAD})/I_{LOAD}$$

An example of an LED biasing circuit is provided in Fig. 4.25b). An NPN transistor such as a 2N2222 (PN2222 or MPQ2222) is used. A resistor value of 220 Ω is calculated for R_2 using parameters provided in b).

Figure 4.25c) provides the PLC interface circuit for a 10 mm LED with the voltage and current characteristics shown. The source voltage is set for 12 VDC.

4.7 Annunciators–Sonalerts, Beepers, Buzzers

In Fig. 4.23, we use a Sonalert (Mallory #PK–27N25WQ, 3–28 VDC) to provide an audible warning of detected smoke. The Sonalert may be directly driven by the PMC digital output. Sonalerts, beepers, and buzzers are available in a wide variety of audible frequencies.

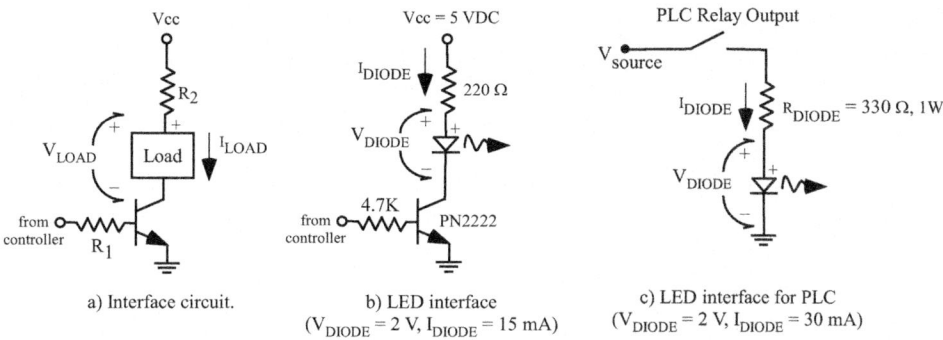

Fig. 4.25 LED interface

4.8 Electromechanical Devices

In this section we discuss the interface of a number of electromechanical devices including DC motors, linear actuators, pumps, and solenoid controlled valves.

4.9 DC Motors

Often a controller is used to control a high power motor load. To properly interface the motor to the controller, we must be familiar with the different types of motor technologies. Motor types are illustrated in Fig. 4.26.

- **DC motor**: A DC motor has a positive and negative terminal. When a DC power supply of suitable current rating is applied to the motor it will rotate. If the polarity of the supply is switched with reference to the motor terminals, the motor will rotate in the opposite direction. The speed of the motor is roughly proportional to the applied voltage up to the rated voltage of the motor.
- **Servo motor**: A servo motor provides a precision angular rotation for an applied pulse width modulation duty cycle. As the duty cycle of the applied signal is varied, the angular displacement of the motor also varies. This type of motor is used to change mechanical positions such as the steering angle of a wheel.
- **Stepper motor**: A stepper motor as its name implies provides an incremental step change in rotation (typically 2.5 °C per step) for a step change in control signal sequence. The motor is typically controlled by a two or four wire interface. For the four wire stepper motor, the microcontroller provides a four bit control sequence to rotate the motor clockwise. To turn the motor counterclockwise, the control sequence is reversed. For low power control signals, MOSFETs or power transistors are used to provide for

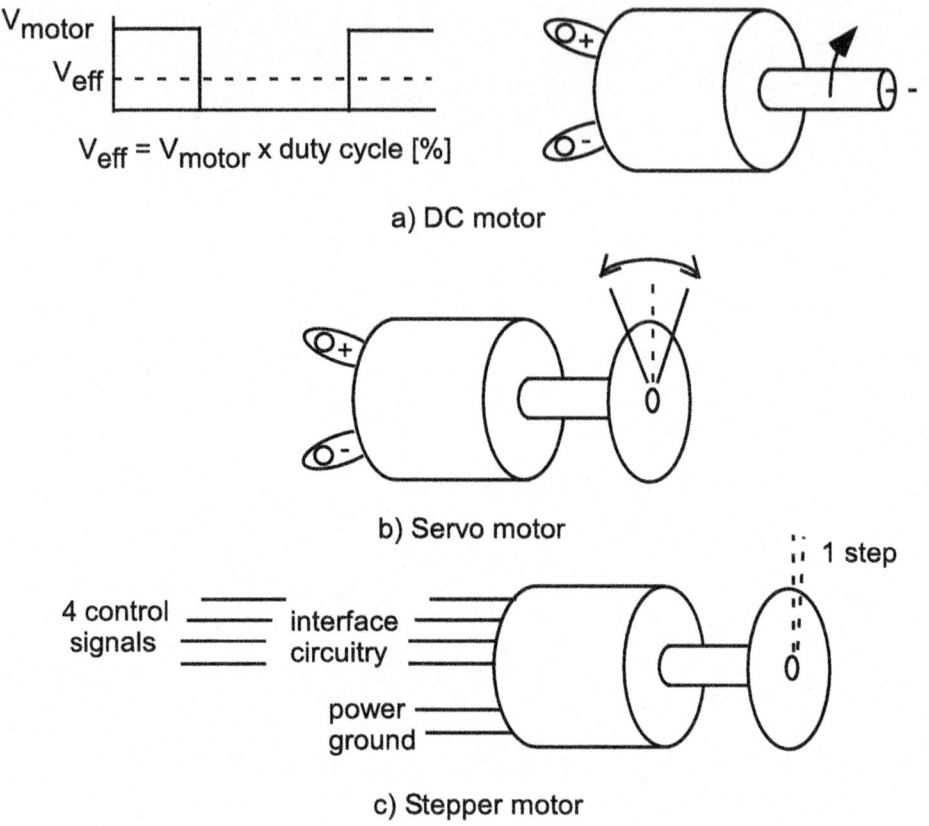

Fig. 4.26 Motor types

the proper voltage and current requirements of the pulse sequence. Many PLCs have the capability to drive the stepper motor directly.

Space does not allow a full discussion of all motor types. We will concentrate on those common in industrial control applications. Let's take a closer look at several motor types.

4.10 DC Motor Speed and Direction Control

As shown in Fig. 4.26a), the speed of a DC motor is varied by changing the effective voltage delivered. At the rated motor voltage, the motor runs at full speed. When fifty percent of the rated motor voltage is delivered to the motor, it runs at approximately 50% of rated speed, etc. The effective voltage delivered to the motor is controlled by pulse width modulation (PWM).

4.10 DC Motor Speed and Direction Control

4.10.1 Pulse Width Modulation

Motor speed may be varied by changing the applied motor voltage. PWM control signal techniques may be used to precisely control the motor speed. With PWM the duty cycle of the motor control signal is varied.

The duty cycle of the PWM signal expressed as a percentage (high time/period * 100) will also be the percentage of the motor supply voltage applied to the motor, and hence the percentage of rated full speed at which the motor will rotate. We explore DC motor control techniques in the Application section later in the chapter.

Example. In this example a Brother 12 VDC, 1500 RPM DC motor equipped with an optical tachometer is used. The hardware configuration for the example is provided in Fig. 4.27. A software generated PWM signal is generated using PMC DIGITAL OUTPUT 00. The PWC signal is routed to the gate (G) of an IRF530 power MOSFET through a resistor voltage divider network. The motor is connected between the 12 VDC motor power supply and the drain (D) connection. A reverse-biased 1N4001 diode is used for circuit protection. The MOSFET source (S) is connected to ground. Motor speed is monitored by PMC encoder Z0 connected to the optical tachometer B channel (200 counts/revolution). An LM339 open collector op amp comparator is used to condition the signal for the encoder Z0 input.

In the following sketch the PWM duty cycle delivered to the motor is varied from 100 to 30% in 10% increments. At each duty cycle the motor RPM is calculated and reported. Provided in Fig. 4.28 is the resulting motor speed at each duty cycle.

```
//*************************************************************************
//PMC_PWM_motor_speed - generates a pulse width modulated signal to vary the
//                      the speed (RPM) of a DC motor. Reports motor RPM
//
//Configuration:
//  - PMC connected to 24 VDC power supply
//  - Digital OUTPUTS 24V IN connected to 24 VDC power supply
//  - PWM signal generated in software and output on DIGITAL OUTPUTS 00
//  - Reports RPM of 12 VDC motor equipped with optical tachometer
//
//This example code is in the public domain.
//*************************************************************************

#include <Arduino_PortentaMachineControl.h>

unsigned int   PWM_on_time;                     //PWM on time
unsigned long ramp_interval = 5000;             //time at each PWM setting
unsigned long time_hack;                        //time since Arduino powered
unsigned RPM;

void setup()
{
MachineControl_DigitalOutputs.begin();          //init PMC digital out
Serial.begin(9600);
while(!Serial)
  {
  ;                                             //wait for serial monitor
  }
}
```

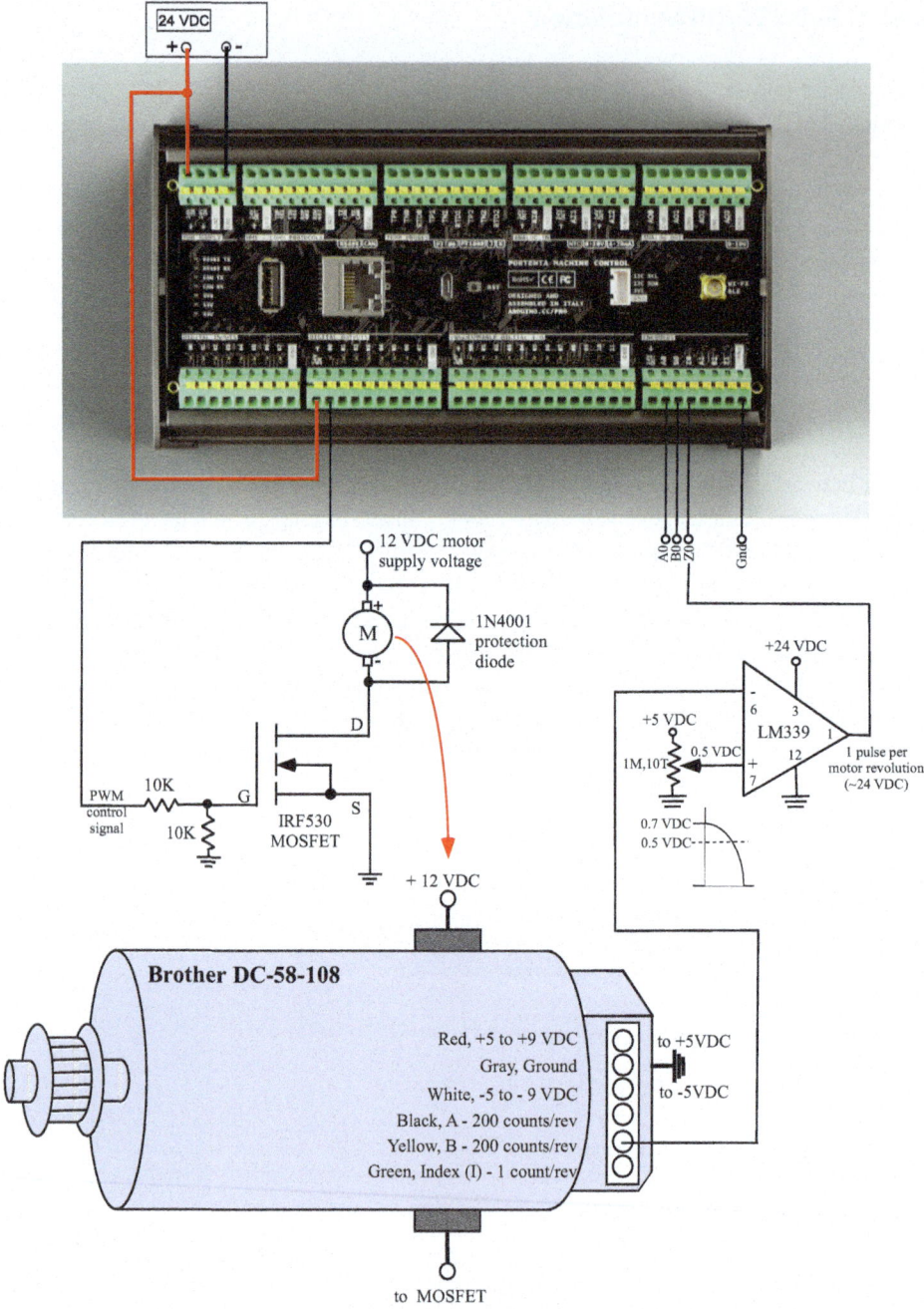

Fig. 4.27 DC motor demonstration circuit. Images used courtesy of the Arduino Team (CC BY-NC-SA) (https://www.arduino.cc)

4.10 DC Motor Speed and Direction Control

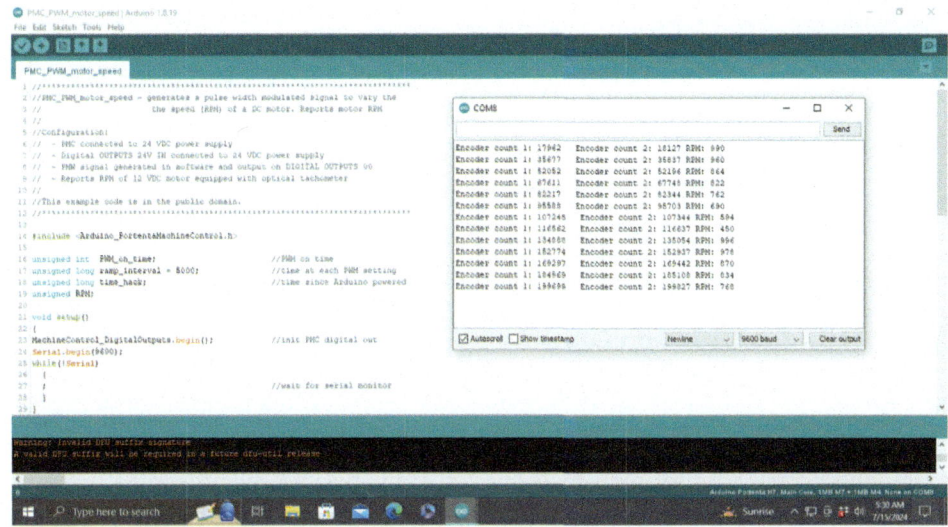

Fig. 4.28 DC motor RPM measurements. Images used courtesy of the Arduino Team (CC BY-NC-SA) (https://www.arduino.cc)

```
void loop()
{
for(PWM_on_time = 1000; PWM_on_time > 200; PWM_on_time = PWM_on_time - 100)
  {
  time_hack = millis();                              //get time hack
  while((millis()- time_hack) < ramp_interval)
    {
    MachineControl_DigitalOutputs.write(0, HIGH); //digital out 00 on
    delayMicroseconds(PWM_on_time);                //delay in us
    MachineControl_DigitalOutputs.write(0, LOW);  //digital out 00 off
    delayMicroseconds(1000 - PWM_on_time);         //PWM_off_time (delay in us)
    }
  RPM = report_RPM();
  }
}

//**************************************************************
//unsigned long report_RPM(void) -measures and reports measure RPM
//**************************************************************

unsigned long report_RPM(void)
{
                            //get 1st encoder count
unsigned long first_time = MachineControl_Encoders.getRevolutions(0);
Serial.print("Encoder count 1: ");
Serial.print(first_time);
delay(50);                  //50 ms between encoder count measurements
                            //get 2nd encoder count
unsigned long second_time = MachineControl_Encoders.getRevolutions(0);
Serial.print("   Encoder count 2: ");
Serial.print(second_time);
```

```
//RPM = ((second_time - first_time)counts/50 ms) * (60s/1 min) * (1 revolution/200 counts);
  unsigned long RPM = (second_time - first_time) * 6;
  Serial.print(" RPM: ");
  Serial.println(RPM);

return RPM;
}
//*************************************************************************
```

4.10.2 H Bridge Direction Control

For a DC motor to operate in both the clockwise and counter clockwise direction, the polarity of the DC motor supply must be changed. To operate the motor in the forward direction, the positive supply terminal must be connected to the positive motor terminal while the negative supply terminal must be provided to the negative motor terminal. To reverse the motor direction the motor supply polarity must be reversed.

An H-bridge is a circuit employed to perform this polarity switch. The H-bridge circuit consists of four electronic switches as shown in Fig. 4.29. For forward motor direction switches 2 and 3 are closed; whereas, for reverse direction switches 1 and 4 are closed.

Low power H-bridges (500 mA) come in a convenient dual in line package (e.g., 754,110). For higher power motors, an H-bridge may be constructed from discrete components as shown in Fig. 4.29. The TIP31 and TIP32 are NPN and PNP transistors with similar characteristics (40 VDC, 3 A, 40 W). The 11DQ06 are Schottky diodes. For driving higher power loads, the switching devices are sized appropriately.

If PWM signals are used to drive the base of the transistors, both motor speed and direction may be controlled by the circuit. The transistors used in the circuit must have a current rating sufficient to handle the current requirements of the motor during start and stall conditions.

4.11 Linear Actuator

A linear actuator is a specially designed motor that converts rotary to linear motion. The linear actuator is equipped with a mechanical rod that is extended when asserted in one direction and retracted when the polarity of assertion is reversed. An H-bridge may be used to control a linear actuator as shown in Fig. 4.30. In this circuit the PMC provides forward and reverse signals for the H-bridge in response to user input.

```
//*************************************************************************
//PMC_linear_actuator
//
//Reads switches at at Digital Input 00 (FWD) and Digital Input 01 (REV)
//to determine linear actuator direction.
//
```

4.11 Linear Actuator

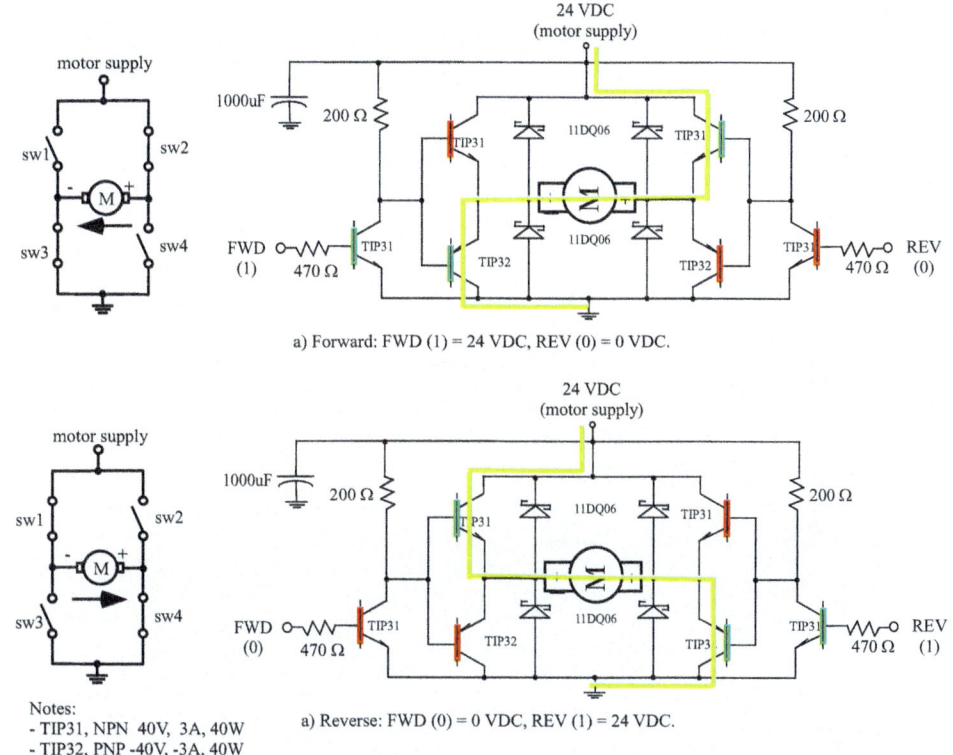

Fig. 4.29 H-bridge control circuit

```
//This code example is in the public domain.
//*******************************************************************

#include <Arduino_PortentaMachineControl.h>

void setup()
{
Serial.begin(9600);
while(!Serial)
  {
  ; //wait for serial port to connect.
  }

Wire.begin();

if(!MachineControl_DigitalInputs.begin())
  {
  Serial.println("Digital input GPIO expander initialization fail!!");
  }
}

void loop()
```

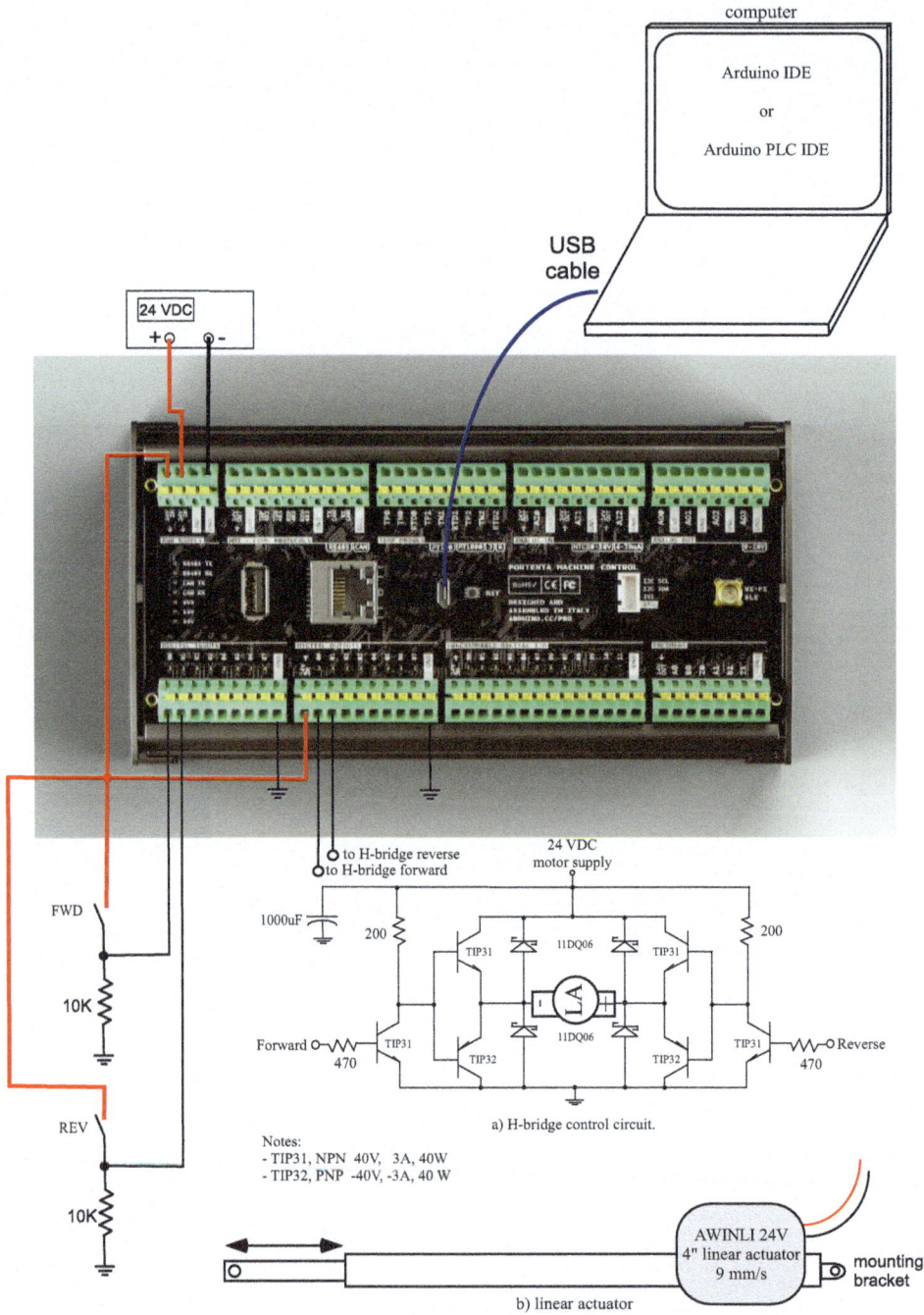

Fig. 4.30 Linear actuator control circuit (O'Berto)

```
{
if(MachineControl_DigitalInputs.read(DIN_READ_CH_PIN_00))
    {
    Serial.println("forward");                       //forward direction
    MachineControl_DigitalOutputs.write(1, LOW);
    MachineControl_DigitalOutputs.write(0, HIGH);
    delay(10);
    MachineControl_DigitalOutputs.write(0, LOW);
    delay(10);
    }
else if(MachineControl_DigitalInputs.read(DIN_READ_CH_PIN_01))
    {
    Serial.println("reverse");
    MachineControl_DigitalOutputs.write(0, LOW);
    MachineControl_DigitalOutputs.write(1, HIGH);
    delay(10);
    MachineControl_DigitalOutputs.write(1, LOW);
    delay(10);
    }
else
    {
    Serial.println("off");
    MachineControl_DigitalOutputs.write(0, LOW);
    MachineControl_DigitalOutputs.write(1, LOW);
    }
Serial.println();

delay(10);
}
//************************************************************
```

4.12 Servo Motor Control

The servo motor is used for a precise angular displacement. The displacement is related to the duty cycle of the applied control signal as shown in Fig. 4.31. A duty cycle of 3.25% equates to 0 °C rotation; whereas, a duty cycle of 6.25% equates to a rotation of 180 °C.

Example: Single channel servo. In this example a servo motor (Adafruit #1404) is controlled by the PMC via Digital Output channel 00. The output signal for a logic one from this channel is 24 VDC. A voltage divider network (47 K + 10 K) is used to scale the signal.[4] An LM324 op amp configured as a voltage follower interfaces the scaled PMC control signal to the servo motor. In the sketch below, the servo is rotated from 0 to 180 °C.

```
//***************************************************************************
//PMC_PWM_servo -        generates a pulse width modulated signal to vary the
//                       the position of a servo motor.
//
```

---

[4] Later in the chapter we discuss voltage scaling via operational amplifiers.

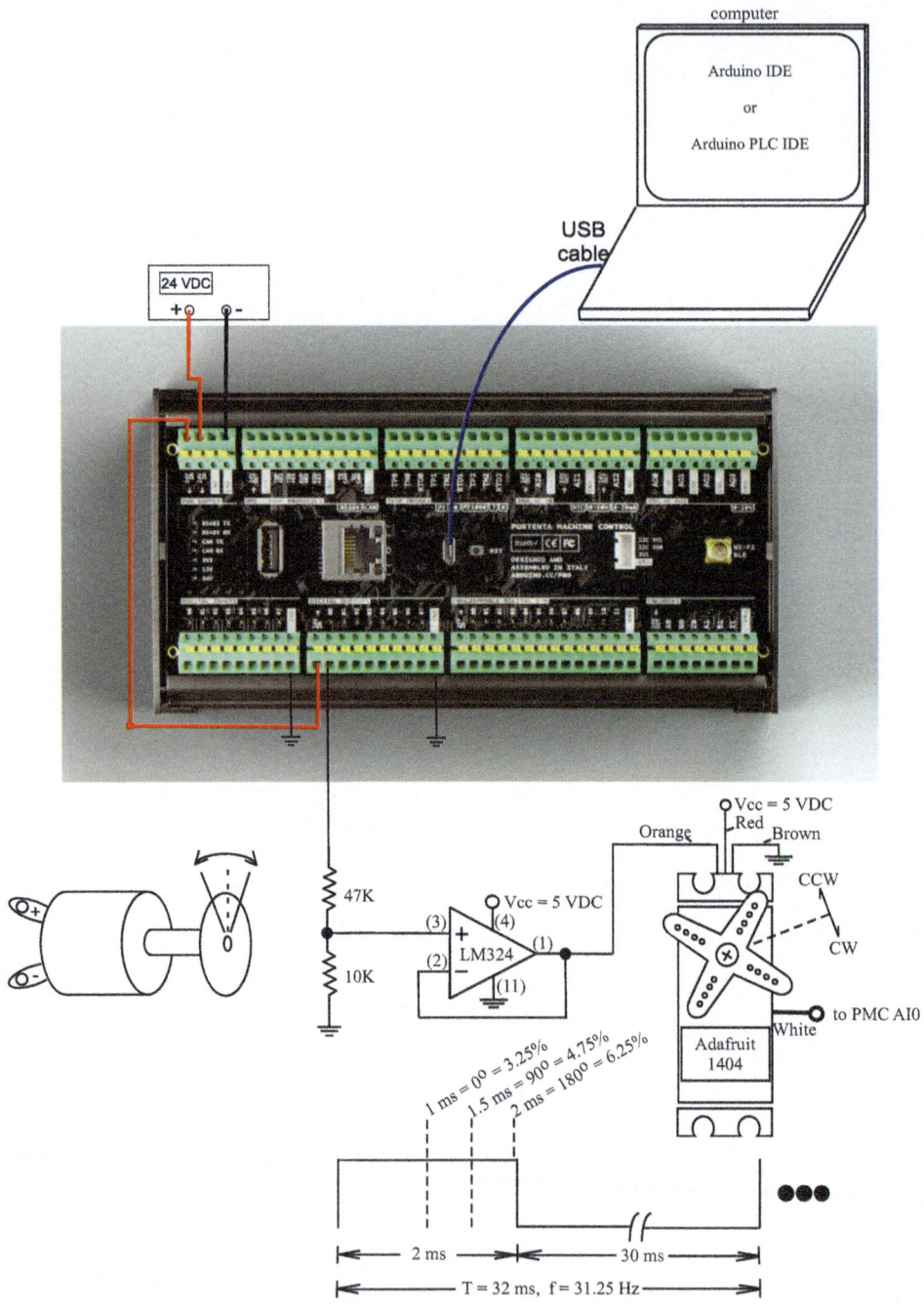

Fig. 4.31 Servo motor control

4.12 Servo Motor Control

```
//Configuration:
//   - PMC connected to 24 VDC power supply
//   - Digital OUTPUTS 24V IN connected to 24 VDC power supply
//   - PMW signal generated in software and output on DIGITAL OUTPUTS 00
//   - Servo control signal (DIGITAL OUTPUT 00) to servo via op amp
//     interface circuit.
//   - Servo wiring:
//       - Red: 5 VDC
//       - Brown: Ground
//       - Orange: control signal from PMC Digital Output 00
//       - White: analog position output to PMC Analog IN AI0
//
//This example code is in the public domain.
//*************************************************************************

#include <Arduino_PortentaMachineControl.h>

unsigned int   PWM_on_time;                        //PWM on time
unsigned long  ramp_interval = 1000;               //time at each PWM setting
unsigned long  time_hack;                          //time since Arduino powered
unsigned long  PWM_period = 20000;                 //PWM period in microceconds
unsigned RPM;

void setup()
{
MachineControl_DigitalOutputs.begin();             //init PMC digital out
Serial.begin(9600);
while(!Serial)
  {
  ;                                                //wait for serial monitor
  }
}

void loop()
{
for(PWM_on_time = 1000; PWM_on_time < 2000; PWM_on_time = PWM_on_time + 100)
  {
  Serial.print("PWM on time:  ");
  Serial.println(PWM_on_time);
  time_hack = millis();                            //get time hack
  while((millis()- time_hack) < ramp_interval)
    {
    MachineControl_DigitalOutputs.write(0, HIGH);  //digital out 00 on
    delayMicroseconds(PWM_on_time);                //delay in us
    MachineControl_DigitalOutputs.write(0, LOW);   //digital out 00 off
    delayMicroseconds(PWM_period - PWM_on_time);   //PWM_off_time (delay in us)
    }
  }
}

//*************************************************************************
```

Example: Single channel servo with position feedback. The Adafruit #1404 servo is equipped with an analog out signal to provide position feedback. In the following sketch, the servo is rotated from 0 to 180 °C and the position feedback is measured and converted to a rotation reading.

```
//***************************************************************************
//PMC_PWM_servo_analog    generates a pulse width modulated signal to vary the
//                        the position of a servo motor.
//
//Configuration:
//  - PMC connected to 24 VDC power supply
//  - Digital OUTPUTS 24V IN connected to 24 VDC power supply
//  - PMW signal generated in software and output on DIGITAL OUTPUTS 00
//  - Servo control signal (DIGITAL OUTPUT 00) to servo via op amp
//    interface circuit.
//  - Servo wiring:
//    - Red: 5 VDC
//    - Brown: Ground
//    - Orange: control signal from PMC Digital Output 00
//    - White: analog position output to PMC Analog IN AI0
//  - Measures analog feedback signal on AI0
//    - Each ANALOG IN channel is equipped with a 100k and 39k
//      resistor divider.  The input voltage is scaled by a 0.28 ratio
//    - Maximum input voltage is 10V
//
//This example code is in the public domain.
//***************************************************************************

#include <Arduino_PortentaMachineControl.h>

unsigned int  PWM_on_time;                         //PWM on time
unsigned long ramp_interval = 1000;                //time at each PWM setting
unsigned long time_hack;                           //time since Arduino powered
unsigned long PWM_period = 20000;                  //PWM period in microceconds

//for analog position feedback measurement
const float RES_DIVIDER = 0.28057;
const float REFERENCE   = 3.0;
unsigned int analog_reading;

void setup()
{
MachineControl_DigitalOutputs.begin();             //init PMC digital out
MachineControl_AnalogIn.begin(SensorType::V_0_10);//init analog in
Serial.begin(9600);
while(!Serial)
  {
  ;                                                //wait for serial monitor
  }
}

void loop()
{
for(PWM_on_time = 1000; PWM_on_time <= 2000; PWM_on_time = PWM_on_time + 100)
  {
  Serial.print("PWM on time:   ");
  Serial.println(PWM_on_time);
  time_hack = millis();                            //get time hack
  while((millis()- time_hack) < ramp_interval)
    {
    MachineControl_DigitalOutputs.write(0, HIGH); //digital out 00 on
    delayMicroseconds(PWM_on_time);                //delay in us
    MachineControl_DigitalOutputs.write(0, LOW);  //digital out 00 off
    delayMicroseconds(PWM_period - PWM_on_time);  //PWM_off_time (delay in us)
```

```
  }//end while
float raw_voltage_ch0 = MachineControl_AnalogIn.read(0);
Serial.print("Raw Voltage CH0: ");
Serial.print(raw_voltage_ch0, 3);

float voltage_ch0 = (raw_voltage_ch0 * REFERENCE) / 65535 / RES_DIVIDER;
Serial.print(" Voltage CH0: ");
Serial.print(voltage_ch0, 3);
Serial.println(" V");

//y = mx + b
//y = rotation (degrees)
//m = slope = (rotation_max - rotation_min)/(ADC_reading_max - ADC_reading_min)
//x = ADC reading
//b = determined from collected data
float rotation = (180.0/(17147.0-4454.0))* raw_voltage_ch0 - 63.0;
Serial.print(" Rotation: ");
Serial.print(rotation, 3);
Serial.println(" degrees");

  }//end for
}//end loop

//**************************************************************************
```

Example: Laser Light Show. A laser light show may be constructed from two servos. In this example we use two Futaba 180 °C range servos (Parallax 900–00005, available from Jameco #283021) mounted as shown in Fig. 4.32. The X and Y control signals are provided by the PMC. The X and Y control signals are interfaced to the servos via LM324 operational amplifiers. The laser source is provided by an inexpensive, eye safe laser pointer.

4.13 Stepper Motor Control

Stepper motors are used to provide a discrete angular displacement in response to a control signal step. There are a wide variety of stepper motors including bipolar and unipolar types with different configurations of motor coil wiring. Due to space limitations we only discuss the unipolar, five wire stepper motor. The internal coil configuration for this motor is shown in Fig. 4.33b).

Often, a wiring diagram is not available for the stepper motor. Based on the wiring configuration (reference Fig. 4.33b), one can determine the common line for both coils. It has a resistance that is one-half of all of the other coils. Once the common connection is found, one can connect the stepper motor into the interface circuit. By changing the other connections, one can determine the correct connections for the step sequence. To rotate the motor either clockwise or counterclockwise, a specific step sequence must be sent to the motor control wires as shown in Fig. 4.33c, d.

A microcontroller does not have sufficient capability to drive the motor directly. Therefore, an interface circuit is required as shown in Fig. 4.33e. The interface circuit uses power

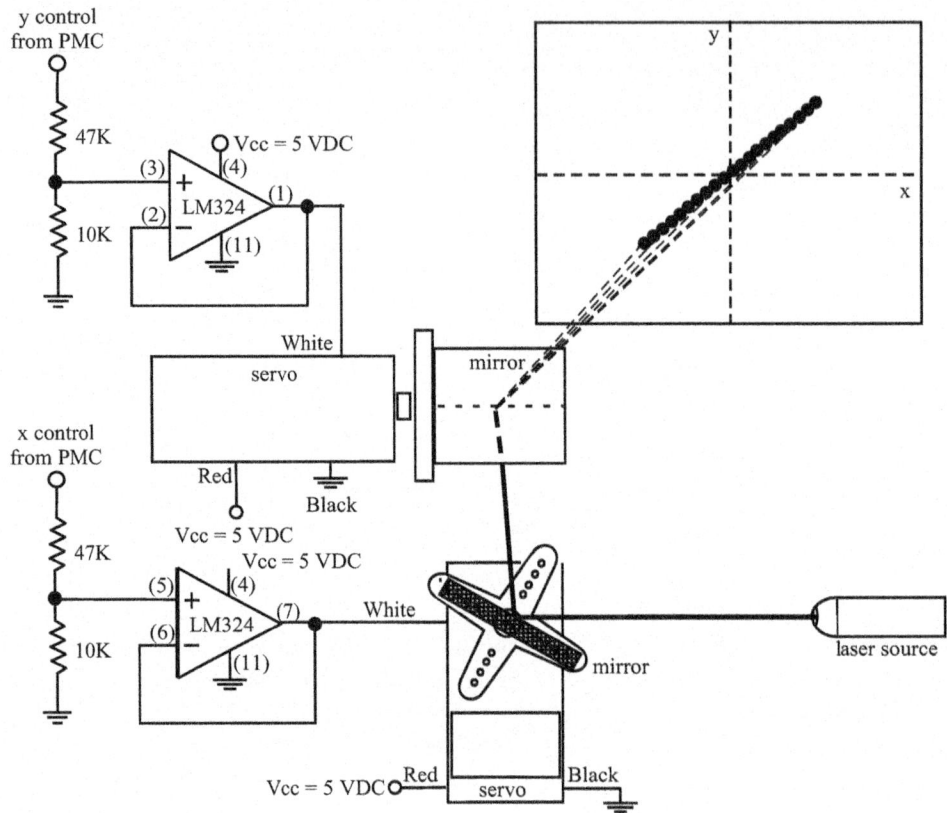

Fig. 4.32 Inexpensive laser light show

MOSFETs to activate each channel. The speed of motor rotation is determined by how fast the control sequence is completed.

Example: In this example, Arduino PMC digital outputs are connected to a stepper motor via a MOSFET based interface circuit. We use the PMC to control a five wire JRP 42BYG016 stepper motor. The motor requires 12 VDC, 160 mA per phase. As shown in Fig. 4.34 push-button switches are used to determine motor direction (forward/reverse) and a potentiometer is used to set the step delay.

```
//**************************************************************
//PMC_stepper
//
//Notes:
//- Controls stepper motor direction via two external switches (FWD, REV).
//- Motor step speed controlled by external potentiometer (step_delay_pot)
//- Generates A (Brown), B (Green), C (Red), D (Yellow) step sequence
//- Stepper motor (orange) to 12 VDC
//- Step sequence provided by PMC, through interface circuitry to stepper
```

4.13 Stepper Motor Control

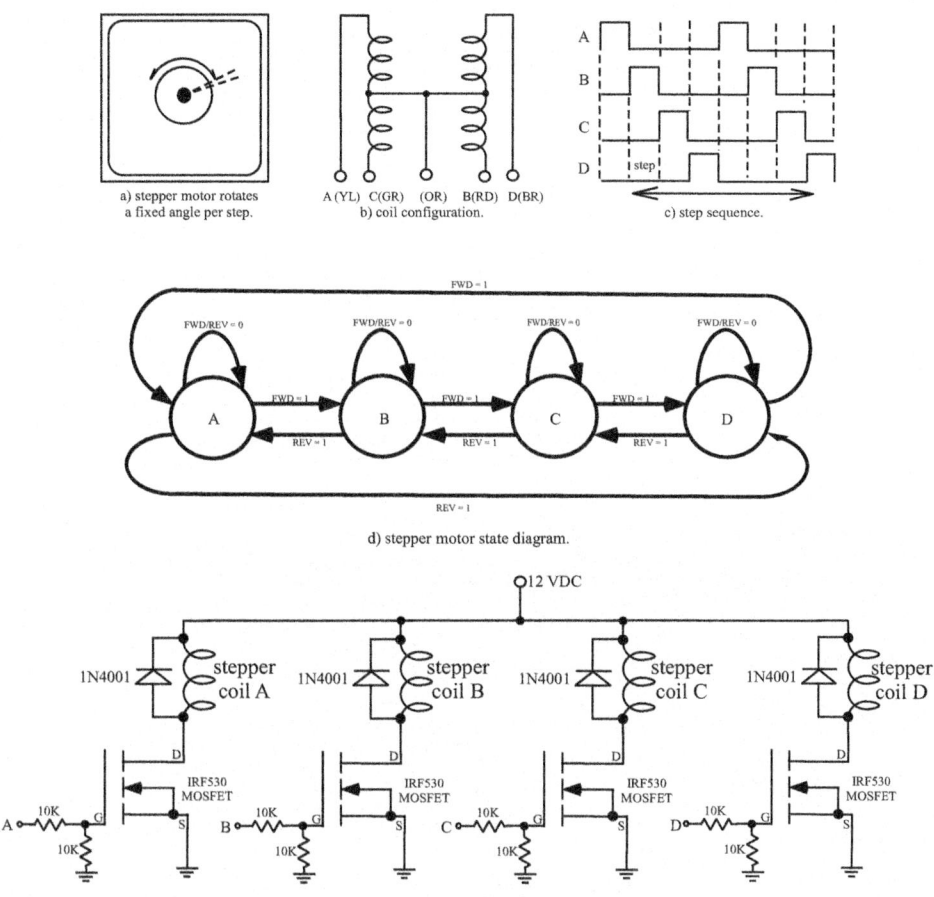

Fig. 4.33 Unipolar stepper motor control

```
//
//This code example is in the public domain.
//**********************************************************************

#include <Arduino_PortentaMachineControl.h>

//Required for analog in
const float RES_DIVIDER = 0.28057;
const float REFERENCE   = 3.0;
unsigned int analog_reading;

uint16_t stepper_direction = 0;


unsigned int  last_step = 1;
unsigned int  next_step;
```

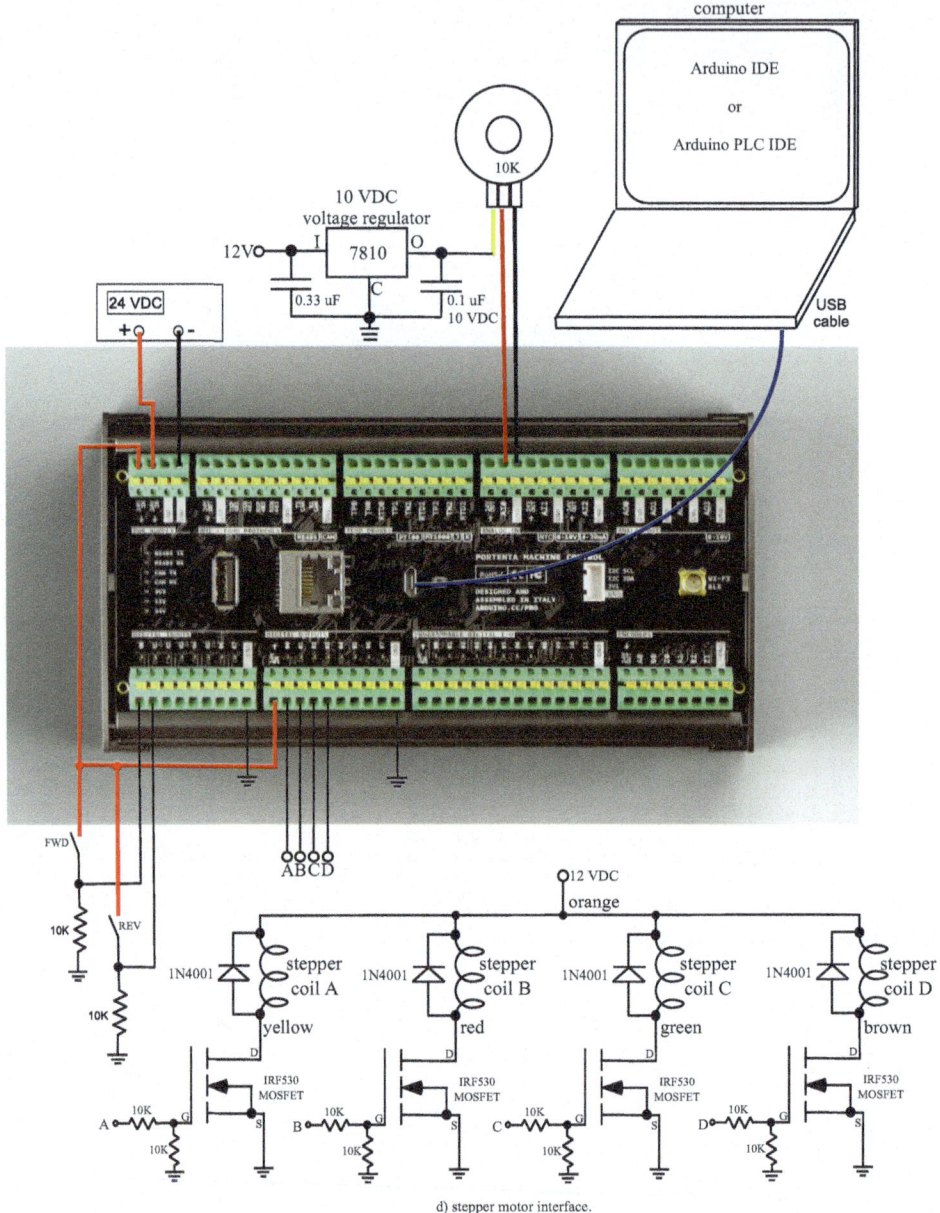

Fig. 4.34 Unipolar stepper motor control circuit. Images used courtesy of the Arduino Team (CC BY-NC-SA) (https://www.arduino.cc)

4.13 Stepper Motor Control

```
unsigned int   motor_speed;
bool           FWD_switch_value = 0;
bool           REV_switch_value = 0;

void setup()
{
Serial.begin(9600);
while(!Serial)
  {
  ; //wait for serial port to connect.
  }

Wire.begin();
MachineControl_DigitalOutputs.begin();                //init PMC digital out
if(!MachineControl_DigitalInputs.begin())             //init PMC digital in
  {
  Serial.println("Digital input GPIO expander initialization fail!!");
  }
MachineControl_AnalogIn.begin(SensorType::V_0_10);  //init PMC analog in
}

void loop()
{
float raw_voltage_ch0 = MachineControl_AnalogIn.read(0);
motor_speed = (unsigned int) (raw_voltage_ch0);
Serial.print("Raw Voltage CH0: ");
Serial.println(raw_voltage_ch0,3);
delay(1000);

Serial.print("motor speed [ms]: ");
Serial.println(motor_speed);


FWD_switch_value = MachineControl_DigitalInputs.read(DIN_READ_CH_PIN_00);
REV_switch_value = MachineControl_DigitalInputs.read(DIN_READ_CH_PIN_01);

if(FWD_switch_value == HIGH)      //FWD switch asserted
   {
   while(FWD_switch_value == HIGH)  //clockwise
      {
      if(last_step == 1)
        {
        Serial.println("FWD asserted: HIGH, step 1");
        MachineControl_DigitalOutputs.write(0, HIGH);       //A
        MachineControl_DigitalOutputs.write(1, LOW);        //B
        MachineControl_DigitalOutputs.write(2, LOW);        //C
        MachineControl_DigitalOutputs.write(3, LOW);        //D
        next_step = 2;
        }
      else if(last_step == 2)
        {
        Serial.println("FWD asserted: HIGH, step 2");
        MachineControl_DigitalOutputs.write(0, LOW);        //A
        MachineControl_DigitalOutputs.write(1, HIGH);       //B
        MachineControl_DigitalOutputs.write(2, LOW);        //C
        MachineControl_DigitalOutputs.write(3, LOW);        //D
        next_step = 3;
        }
      else if(last_step == 3)
```

```
      {
      Serial.println("FWD asserted: HIGH, step 3");
      MachineControl_DigitalOutputs.write(0, LOW);        //A
      MachineControl_DigitalOutputs.write(1, LOW);        //B
      MachineControl_DigitalOutputs.write(2, HIGH);       //C
      MachineControl_DigitalOutputs.write(3, LOW);        //D
      next_step = 4;
      }
    else if(last_step == 4)
      {
      Serial.println("FWD asserted: HIGH, step 4");
      MachineControl_DigitalOutputs.write(0, LOW);        //A
      MachineControl_DigitalOutputs.write(1, LOW);        //B
      MachineControl_DigitalOutputs.write(2, LOW);        //C
      MachineControl_DigitalOutputs.write(3, HIGH);       //D
      next_step = 1;
      }
    else
      {
      ;
      }
    last_step = next_step;
    delay(motor_speed);
    FWD_switch_value = MachineControl_DigitalInputs.read(DIN_READ_CH_PIN_00);
    }//end while
  }//end if

else if(REV_switch_value == HIGH)       //REV_switch_value asserted
  {
  while(REV_switch_value == HIGH)        //counter clockwise
    {
    if(last_step == 1)
      {
      Serial.println("REV switch: high, step 1");
      MachineControl_DigitalOutputs.write(0, HIGH);       //A
      MachineControl_DigitalOutputs.write(1, LOW);        //B
      MachineControl_DigitalOutputs.write(2, LOW);        //C
      MachineControl_DigitalOutputs.write(3, LOW);        //D
      next_step = 4;
      }
    else if(last_step == 2)
      {
      Serial.println("REV switch: high, step 2");
      MachineControl_DigitalOutputs.write(0, LOW);        //A
      MachineControl_DigitalOutputs.write(1, HIGH);       //B
      MachineControl_DigitalOutputs.write(2, LOW);        //C
      MachineControl_DigitalOutputs.write(3, LOW);        //D
      next_step = 1;
      }
    else if(last_step == 3)
      {
      Serial.println("REV switch: high, step 3");
      MachineControl_DigitalOutputs.write(0, LOW);        //A
      MachineControl_DigitalOutputs.write(1, LOW);        //B
      MachineControl_DigitalOutputs.write(2, HIGH);       //C
      MachineControl_DigitalOutputs.write(3, LOW);        //D
      next_step = 2;
      }
    else if(last_step == 4)
```

4.14 DC Solenoid Control   213

```
        {
        Serial.println("REv switch: high, step 4");
        MachineControl_DigitalOutputs.write(0, LOW);        //A
        MachineControl_DigitalOutputs.write(1, LOW);        //B
        MachineControl_DigitalOutputs.write(2, LOW);        //C
        MachineControl_DigitalOutputs.write(3, HIGH);       //D
        next_step = 3;
        }
     else
        {
        ;
        }
     last_step = next_step;
     delay(motor_speed);
     REV_switch_value = MachineControl_DigitalInputs.read(DIN_READ_CH_PIN_01);
     }//end while
  }//end if

  else
     {
     MachineControl_DigitalOutputs.write(0, LOW);           //A
     MachineControl_DigitalOutputs.write(1, LOW);           //B
     MachineControl_DigitalOutputs.write(2, LOW);           //C
     MachineControl_DigitalOutputs.write(3, HIGH);          //D
     }
}
//*************************************************************
```

4.14 DC Solenoid Control

A solenoid provides a mechanical insertion (or extraction) when asserted. Often the solenoid is coupled with valves to control fluid flow. The solenoid may be driven by the PMC via a power MOSFET based interface circuit.

Example: Water valve control: Solenoid controlled water valves are available from Adafruit (https://www.adafruit.com). There are plastic (#997) and brass (#996) valves available. The plastic valve activates from 6 VDC at 160 mA to 12 VDC at 320 mA while the brass valve activates from 6 VDC at 1.6 A to 12 VDC at 3 A. An interface circuit for the plastic water solenoid valve is provided in Fig. 4.35c).

In this first example, the sketch turns the valve on for five seconds and off for 25 s.

```
//*************************************************************
//PMC_RTC_water_valve_simple - activates water valve for prescribed time
//
//Notes:
//- Adafruit 997 water valve
//- MOSFET based interface circuit
//
//This example code is in the public domain.
//*************************************************************
```

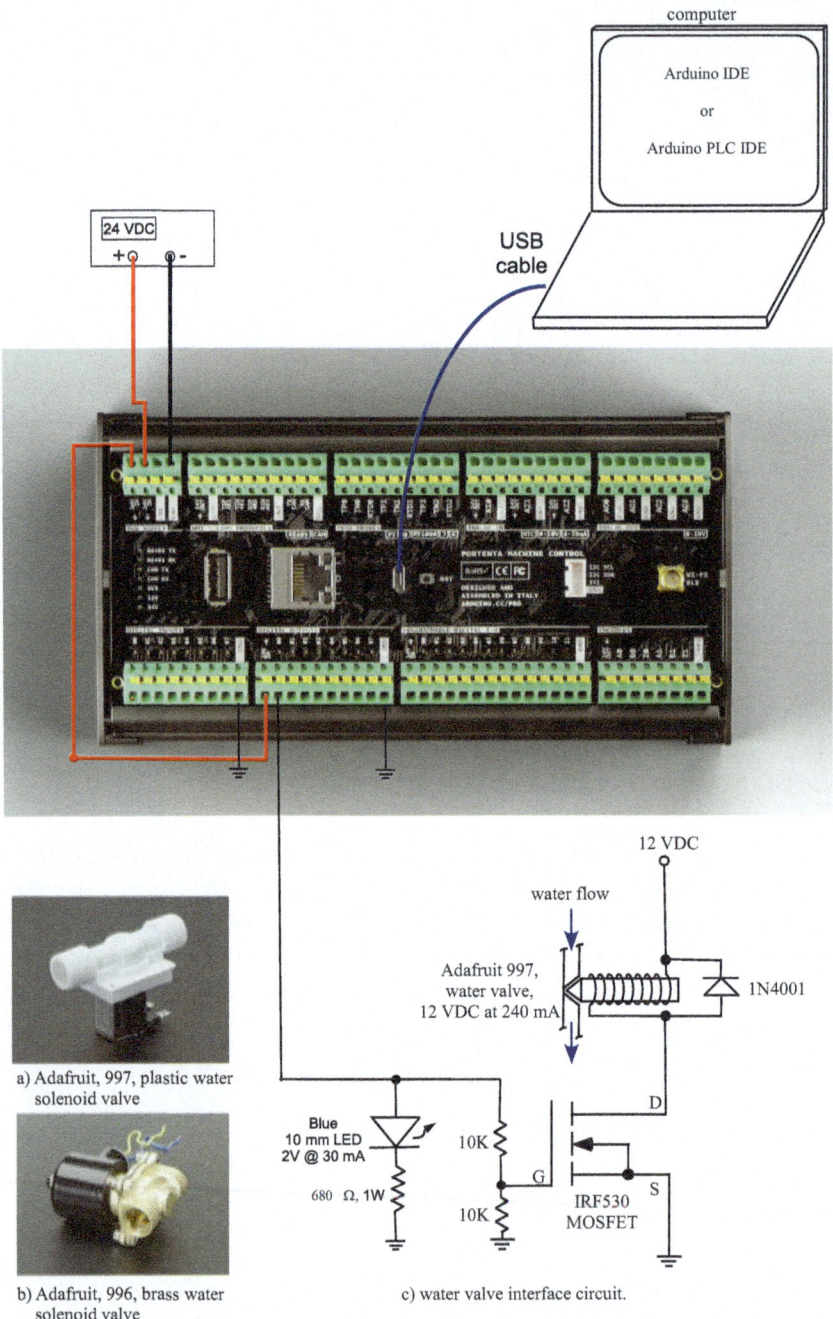

Fig. 4.35 Water valve interface circuit (https://www.adafruit.com). Illustration used with permission

4.14 DC Solenoid Control

```
#include <Arduino_PortentaMachineControl.h>

void setup()
{
Serial.begin(9600);
while (!Serial)
  {
  ;
  }
}

void loop()
{
MachineControl_DigitalOutputs.write(0, HIGH);
Serial.println("Water valve on");
delay(5000);
MachineControl_DigitalOutputs.write(0, LOW);
Serial.println("Water valve off");
delay(25000);
}

//*********************************************************************
```

In the second example the valve is asserted for a prescribed time at a desired RTC time.

```
//*********************************************************************
//PMC_RTC_water_valve - activates water valve at prescribed RTC time
//
//Adapted from:
//Portenta Machine Control - RTC Alarm Example
//
//Notes:
//- Adafruit 997 water valve
//- MOSFET based interface circuit
//
//Library: Portenta Mchine Control Library
//Copyright (c) 2024 Arduino, SPDX-License-Identifier: MPL-2.0
//This example code is in the public domain.
//*********************************************************************

#include <Arduino_PortentaMachineControl.h>

int hours    = 12;
int minutes  = 45;
int seconds  = 57;
bool alarm_flag = false;
int counter = 2;
unsigned long int i, j;

void callback_alarm();

void setup()
```

```
{
Serial.begin(9600);
while (!Serial)
  {
  ;
  }

Serial.print("RTC Initialization");
if(!MachineControl_RTCController.begin())
  {
  Serial.println(" fail!");
  }
//Serial.println(" done!");

//APIs to set date's fields: hours, minutes and seconds
MachineControl_RTCController.setHours(hours);
MachineControl_RTCController.setMinutes(minutes);
MachineControl_RTCController.setSeconds(seconds);

//Enables Alarm on PCF8563T
MachineControl_RTCController.enableAlarm();

//set the minutes at which the alarm should rise
MachineControl_RTCController.setMinuteAlarm(47);

//Attach an interrupt to the RTC interrupt pin
attachInterrupt(RTC_INT, callback_alarm, FALLING);
}

void loop()
{
if(alarm_flag)
  {
  //Serial.println("Water valve on");
  detachInterrupt(RTC_INT);
  MachineControl_RTCController.setSeconds(seconds);
  MachineControl_RTCController.setMinuteAlarm(minutes);
  MachineControl_RTCController.clearAlarm();
  attachInterrupt(RTC_INT, callback_alarm, FALLING);

  //Interrupt specific actions
                                    //valve and LED on
  MachineControl_DigitalOutputs.write(0, HIGH);
  for(j=0; j<=99; j++)              //100 x 0.1s = 10s delay
    {
    for(i=0; i<=24000000; i++)      //0.1s delay 240e5 clock cycles
      {
      asm("nop");                   //1 clock cycle
      }
```

```
        }
                                  //valve and LED off
   MachineControl_DigitalOutputs.write(0, LOW);
   //Serial.println("Water valve off");

   alarm_flag = false;

   //To disable the alarm uncomment the following line:
   //MachineControl_RTCController.disableAlarm();
   }

//APIs to get date's fields
Serial.print(MachineControl_RTCController.getHours());
Serial.print(":");
Serial.print(MachineControl_RTCController.getMinutes());
Serial.print(":");
Serial.println(MachineControl_RTCController.getSeconds());

delay(1000);
}

void callback_alarm ()
{
alarm_flag = true;
}

//*******************************************************************
```

4.15 Transducer Interface Design (TID)

A transducer is used to convert a physical variable such as temperature, pressure, or light intensity to an electrical variable (e.g. resistance, current, or voltage) for data collection and analysis by a controller. Signal conditioning circuitry is often needed to convert the electrical variable into a voltage compatible with the analog-to-digital (ADC) subsystem onboard the controller. The ADC subsystem converts the voltage to an equivalent digital value. The ADC accepts voltages between its high and low reference values for conversion (typically supply voltage and ground). It is the responsibility of the system designer to ensure transducer outputs are properly conditioned to meet these constraints.[5]

---

[5] The section on transducer interface design is adapted from "Electrical Signals and Systems," Department of Electrical Engineering, United States Air Force Academy.

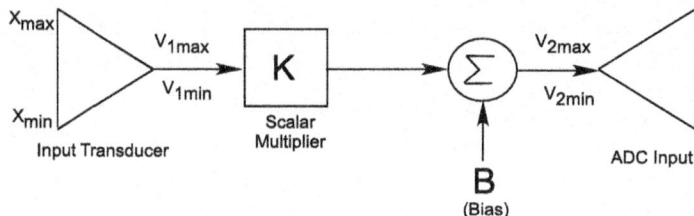

Fig. 4.36 Signal conditioning for ADC. A block diagram of the signal conditioning for an analog-to-digital converter. The range of the sensor voltage output is mapped to the analog-to-digital converter input voltage range. The scalar multiplier maps the magnitudes of the two ranges and the bias voltage is used to align two limits

The signal conditioning circuitry is called the transducer interface. The objective of the transducer interface circuit is to scale and shift the electrical signal range to efficiently map the output of the input transducer to the input range of the analog-to-digital converter which is typically 0–5 VDC or 0–3.3 VDC for microcontrollers and 0–10 VDC for the Portenta Machine Control.

Figure 4.36 shows the transducer interface circuit using an input transducer. This process assumes a linear input transducer. The output of the input transducer is first scaled by constant K. As an example, in the figure, we use a microphone as the input transducer whose output ranges from − 5 to + 5 VDC. In the example, the input to the analog-to-digital converter ranges from 0 to 5 VDC. The scalar multiplier with constant K maps the output range of the input transducer to the input range of the converter. We need to multiply all input signals by 1/2 to accommodate the mapping.

Once the range has been mapped, the signal now needs to be shifted. Note that the scale factor maps the output range of the input transducer as − 2.5 to + 2.5 VDC instead of 0 to 5 VDC. The second portion of the circuit, the bias stage, shifts the range by 2.5 VDC, thereby completing the correct mapping. Actual implementation of the circuit components is accomplished using operational amplifiers.

In general, the scaling and bias process may be described by two equations:

$$V_{2max} = (V_{1max} \times K) + B$$

$$V_{2min} = (V_{1min} \times K) + B$$

The variable V_{1max} represents the maximum output voltage from the input transducer. This voltage occurs when the maximum physical variable (X_{max}) is presented to the input transducer. This voltage must be scaled by the scalar multiplier (K) and then a DC offset bias voltage (B) is added to provide the voltage V_{2max} to the input of the ADC converter.

Similarly, The variable V_{1min} represents the minimum output voltage from the input transducer. This voltage occurs when the minimum physical variable (X_{min}) is presented to the input transducer. This voltage must be scaled by the scalar multiplier (K) and then have a DC offset bias voltage (B) added to produce voltage V_{2min} to the input of the ADC converter.

Usually, the values of V_{1max} and V_{1min} are provided with the documentation for the transducer. Also, the values of V_{2max} and V_{2min} are known. They are the high and low reference voltages for the ADC system (typically 5 VDC and 0 VDC for a microcontroller and 10 VDC and 0 VDC for the PMC). We thus have two equations and two unknowns to solve for K and B. The circuits to scale by K and add the offset B are usually implemented with operational amplifiers.

Example: A photodiode is a semiconductor device that provides an output current corresponding to the light impinging on its active surface. The photodiode is used with a transimpedance amplifier to convert the output current to an output voltage. A paired photodiode/transimpedance amplifier provides an output voltage of 0 V for maximum rated light intensity and − 2.50 VDC output voltage for the minimum rated light intensity.

Calculate the required values of K and B for this light transducer so it may be interfaced to a microcontroller's ADC system.

$$V_{2max} = (V_{1max} \times K) + B$$

$$V_{2min} = (V_{1min} \times K) + B$$

$$5.0\,V = (0\,V \times K) + B$$

$$0\,V = (-2.50\,V \times K) + B$$

The values of K and B may then be determined to be 2 and 5 VDC, respectively. The transducer interface circuit is then implemented using operational amplifiers (op amps).

Example: It was determined that the values of K and B were 2 and 5 VDC, respectively. The two-stage op amp circuitry provided in Fig. 4.37 implements these values of K and B. The first stage provides an amplification of − 2 due to the use of the inverting amplifier configuration. In the second stage, a summing amplifier is used to add the output of the first stage with a bias of − 5 VDC. Since this stage also introduces a minus sign to the result, the overall result of a gain of 2 and a bias of + 5 VDC is achieved.

4.16 Operational Amplifier Overview

The operational amplifier or op amp is used extensively in applications to interface transducers. We begin the section exploring op amp origins and development. We then describe the ideal op amp and use it as a benchmark for real world, nonideal op amps. We investi-

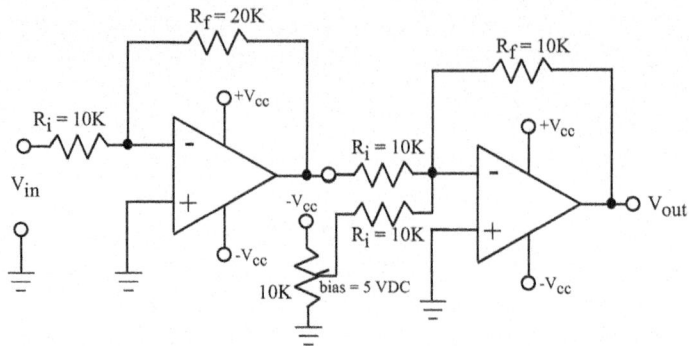

Fig. 4.37 Operational amplifier implementation of the transducer interface design (TID) example circuit

gate how to compensate for nonideal op amp parameters. Next, we review common op amp circuit configurations used extensively in instrumentation applications. We use the circuit configurations to explore transducer interface and applications.[6]

4.16.1 Operational Amplifier Origins

The operational amplifier or op amp is a two input, single output amplifier. The output is an amplified version of the difference between the two inputs. It is quite common in instrumentation applications for signal amplification, conditioning, filtering, and may be used for mathematical operations.

The first op amps were developed in the 1940s using vacuum tube technology. In 1947 John Ragazzini in his paper "Analysis of Problems in Dynamics by Electronic Circuits" was the first to use the term operational amplifier. He and is co-authors wrote "The term 'operational amplifier' is a generic term applied to amplifiers whose gain functions are such as to enable them to perform certain useful operations such as summation, integration, differentiation, or a combination of such operations (Ragazzini)."

Op amp development continued for several decades with improvement in features and implementation technology. The first monolithic, i.e. single chip, integrated circuit op amp, the uA702, was designed by Bob Widlar of Fairchild Semiconductor. It was released in 1963. An improved op amp, the uA709, also designed by Widlar, was released in 1965. Widlar also designed the LM101 for the National Semiconductor Company. The uA741 op amp was released by Fairchild Semiconductor in 1968. It quickly became quite popular and is still in production today. An op amp with a single polarity power supply, the LM324, was released in 1972. It too is still quite popular today and is used in multiple applications (Jung).

---

[6] This section on operational amplifiers is used with permission from Arduino VI: Bioinstrumentation, S. Barrett, Springer, 2024.

4.16.2 Ideal Characteristics

A generic ideal operational amplifier is illustrated in Fig. 4.38. The op amp is an active device (requires power supplies) equipped with two inputs, a single output, and several DC voltage source inputs (Sedra and Smith, Faulkenberry).

The two op amp inputs are labeled Vp, or the noninverting input, and Vn, the inverting input. The output of the op amp is determined by taking the difference between Vp and Vn and multiplying the difference by the op amp's open loop gain (A_{vol}). This gain is typically a large value much greater than 50,000.

Due to the large value of A_{vol}, it does not take much of a difference between Vp and Vn before the op amp will saturate. When an op amp saturates, it does not damage the op amp, but the output is limited to values slightly less than the supply voltages $\pm V_{cc}$. This will clip the output, and hence distort the signal, at levels slightly less than $\pm V_{cc}$. To prevent saturation, op amps are typically used in a closed loop, negative feedback configuration.

Example: Comparator level detector The comparator level detector is a common op amp building block. It is configured in an open loop configuration. One input is tied to a reference voltage threshold. The threshold setting is typically provided by a potentiometer connected between the supply voltage and ground. The input signal is provided to the other input as shown in Fig. 4.39 (Stout and Kaufman).

When the input signal is higher than the threshold voltage, the op amp saturates toward the positive supply value. When the input signal is less than threshold signal, the op amp saturates toward the negative supply value. A comparator circuit may be used to restore a degraded digital signal to its original values. In this case a single-sided op amp such as the LM324 may be used with supply voltages of 5 VDC and ground for a 5 VDC digital system.

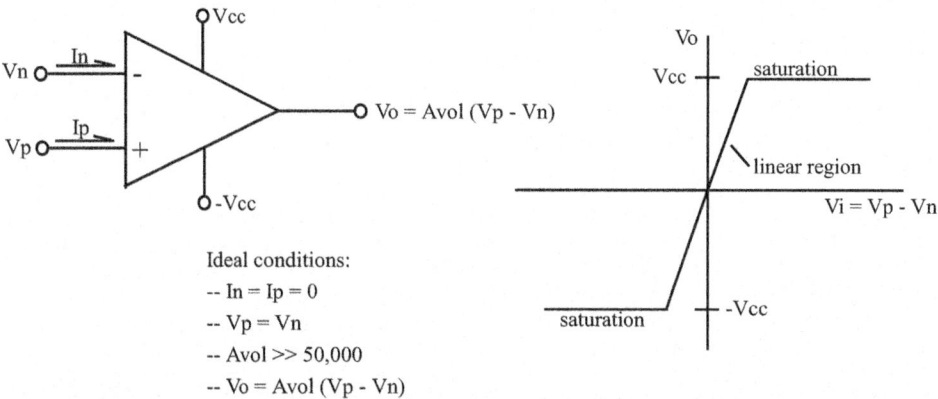

Fig. 4.38 Ideal operational amplifier characteristics

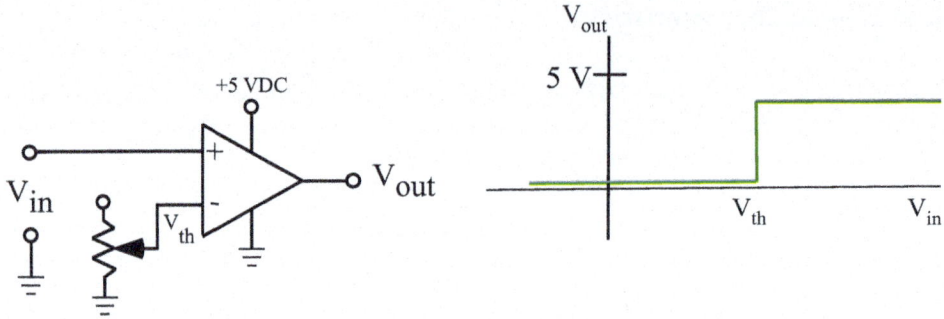

Fig. 4.39 Op amp comparator level detector circuit

An ideal operational does not exist in the real world. However, it is a good first approximation for use in developing op amp application circuits. As shown in Fig. 4.38 an op amp has the following ideal characteristics (Sedra and Smith):

- Input currents In and Ip equal to zero;
- Infinite input impedance;
- Vp and Vn input voltages equal to one another;
- Extremely high open loop gain;
- Output impedance of zero;
- Infinite bandwidth;
- High slew rate; and
- Infinite common mode rejection.

We use these ideal conditions as a close approximation when developing characteristic equations for different op amp configurations.

4.16.3 Nonideal Characteristics

Ideal op amps do not exist in the real world although some come close. To better understand nonideal op amp features, we explore their origins and potential compensation methods (Sedra and Smith, Faulkenberry, Stout and Kaufman).

- **Frequency Response**: Ideally we desire an infinite bandwidth or frequency response. That is, we want the op amp to provide the same amplification response across the frequency spectrum. Due to internal capacitance and internal configurations, the frequency response usually has a lower and upper 3 dB point. A 3 dB point is a frequency where the op amp gain is down 3 dB from its passband value. In a specific application an op amp should be chosen that operates at the desired frequencies.

4.16 Operational Amplifier Overview

- **Gain Bandwidth Product**: The Gain Bandwidth Product (GBP) is a metric that describes the tradeoff between op amp voltage gain and desired frequency of operation. The GBP is a fixed value parameter for a specific op amp. Therefore, as the desired gain is increased the corresponding bandwidth decreases and vice versa. To provide a reasonable gain and bandwidth combination, the op amp is typically used in closed loop configurations as described in the next section.
- **Offset Voltage**: Typically an op amp consists of multiple internal stages. The first stage consists of a differential transistor pair with matched characteristics for the positive and negative portion of the amplifier. Any mismatch between the two input transistor characteristics is amplified. This leads to an op amp output voltage even when both inputs are at zero volts. Some op amps are equipped with offset null compensation inputs. A potentiometer may be connected between these inputs to provide an offset compensation for the mismatched inputs.
- **Bias Current**: The differential amplifier input described above requires small bias currents to maintain the input transistors in an active state. If each of the input bias currents flow through the same equivalent resistance, they are canceled out by the differential input amplifier configuration. A compensation resistor, $R_p = R_i || R_f$, may be connected between the positive input lead V_p and ground to minimize this effect. The resistors R_i and R_f are the input and feedback resistors.
- **Input resistance**: Ideally we want the input resistance R_i to be very high. This prevents loading down a previous op amp stage or having the current stage affect another stage's operation. This effect can be minimized by the wise choice of R_i values and using a voltage follower configuration between stages. The voltage follower provides no gain but provides high input impedance. We explore the voltage follower configuration in the next section.
- **Common Mode Rejection Ratio (CMRR)**: Ideally the op amp should amplify the difference between its two inputs while cancelling any voltages common to both inputs. This allows for the cancellation of noise common to both inputs. The Common Mode Rejection Ratio or CMRR is a metric comparing an op amp's differential gain (A_d) to its common mode gain (A_{cm}). It is expressed as:

$$CMRR = 20 \log(|A_d|/|A_{CM}|)$$

- **Slew Rate**: Ideally the op amp should amplify a signal with high fidelity even for high output amplitudes rapidly changing at high frequencies. The slew rate is a parameter provided in volts per microsecond describing the op amp's capability to do this.

4.16.4 Configurations

As described in previous sections, the op amp has a very large open loop gain which minimizes bandwidth and also the differential voltage applied to the inputs. Therefore, op amps are typically used in a closed loop configuration with a controlled gain to perform a variety of functions. A sample of classic operational amplifier configurations are provided in Fig. 4.40 (Faulkenberry).

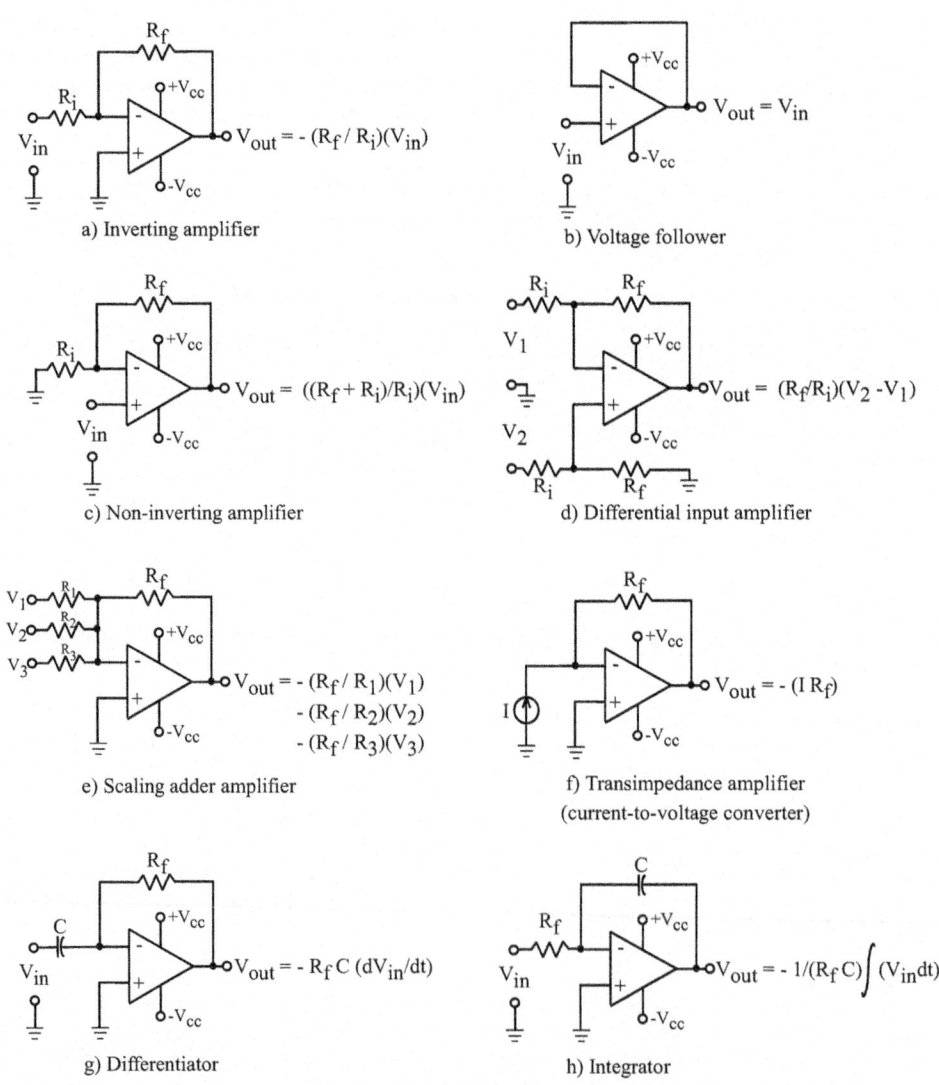

Fig. 4.40 Classic operational amplifier configurations. Adapted from [Faulkenberry]

4.16 Operational Amplifier Overview

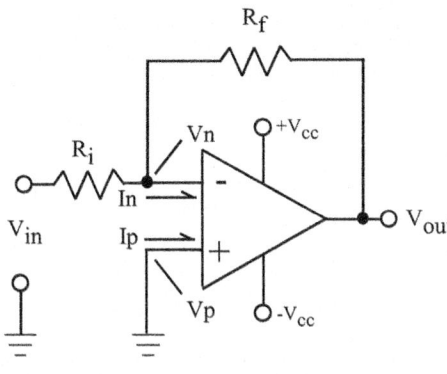

Node equation at Vn:

(Vn - Vin)/ Ri + (Vn - Vout)/Rf + In = 0

Apply ideal conditions:

In = Ip = 0

Vn = Vp = 0 (since Vp is grounded)

Solve node equation for Vout:

$V_{out} = - (R_f / R_i)(V_{in})$

Fig. 4.41 Operational amplifier analysis for the non-inverting amplifier. Adapted from (Faulkenberry)

It should be emphasized that the equations provided with each operational amplifier circuit are only valid if the circuit configurations are identical to those shown. Even a slight variation in the circuit configuration may have a dramatic effect on circuit operation.

It is important to analyze each operational amplifier circuit using the following steps:

- Write the node equation at V_n for the circuit.
- Apply ideal op amp characteristics to the node equation.
- Solve the node equation for V_o.

As an example, we provide the analysis of the noninverting amplifier circuit in Fig. 4.41. This same analysis technique may be applied to all of the circuits in Fig. 4.40 to arrive at the equations for Vout provided.

A brief description of each configurations follows.

- **Inverting amplifier**: The inverting amplifier provides a gain determined by $A_v = - R_f/R_i$. As indicated by the minus sign, the amplifier inverts the polarity of the input signal to produce the output signal. The value of R_i should be kept high to maintain a high input impedance. Also, the gain should be limited to prevent saturation of the output signal.
- **Voltage follower**: The voltage follower circuit provides a high impedance buffer for use between op amp stages in multi-stage designs. As the name implies, the output signal follows the input signal.
- **Noninverting amplifier**: The noninverting amplifier provides a noninverted gain. The value of R_i should be kept high to maintain a high input impedance. Also, the gain should be limited to prevent saturation of the output signal.

- **Difference amplifier**: The difference amplifier provides the amplified difference of its two input signals. Voltages common to both inputs are not amplified (e.g. noise).
- **Summing amplifier**: The summing amplifier provides the amplified sum of the input signals. The input values of R_i may be chosen to determine the relative proportions of each input signal within the output signal. Also note the signal inversion at the output. The individual value of the input resistances (R_1, R_2, R_3) should be kept high to maintain high input impedance.
- **Transimpedance amplifier**: The transimpedance amplifier translates a current input to a voltage output. This configuration is commonly used to convert the current output from certain transducers to a voltage suitable for conversion by a microcontroller.
- **Integrator**: The integrator performs the mathematical integration of the input signal.
- **Differentiator**: The differentiator performs the mathematical integration of the input signal.

Example: TMP36 Interface. An Analog Devices TMP36 low voltage temperature sensor is used to measure the interior temperature of a greenhouse. The TMP36 will be interfaced to the PMC ADC using a transducer interface circuit. The TMP36 is a linear sensor providing a 10 mV per degree Centigrade output and measures temperatures between −40 to +125 °C C. It provides 750 mV of output at 25 °C (https://www.analog.com).

To design the interface circuit, the minimum and maximum value of the temperature variable and corresponding output voltage must be known. A spreadsheet is provided in Fig. 4.42 to determine these values. Using two equations developed from the transducer interface design process, the values of K and B are determined to be 6.06 V/V and −0.606 V respectively. A block diagram and corresponding circuit diagram is provided in Fig. 4.43.

Example: One-bit ADC. Provided in Fig. 4.44 is a one-bit analog-to-digital converter. The circuit is used to compare a sensor output against a specific threshold. The circuit outputs a logic one or zero based on the relationship between the sensor output and the established threshold.

4.17 Application I: DC Motor Speed Control

The goal of this example is to stabilize the speed of a DC motor using several different concepts discussed in this chapter. A block diagram of a circuit to stabilize motor speed is shown in Fig. 4.45.

The control algorithm is hosted on an Arduino PMC. It takes as input the desired motor speed. The algorithm provides a pulse width modulated (PWM) signal to control motor speed. Motor speed is measured in real time using an optical tachometer. The actual motor speed is provided as another input to the PMC. The control algorithm compares desired to actual motor speed to update the PWM control signal. The PWM signal parameters (on time and off time) are varied, which adjusts the effective voltage supplied to the motor, to operate

4.17 Application I: DC Motor Speed Control

Temp [oC]	V1	V2
-40	0.100	0.00
-35	0.150	0.30
-30	0.200	0.61
-25	0.250	0.91
-20	0.300	1.21
-15	0.350	1.52
-10	0.400	1.82
-5	0.450	2.12
0	0.500	2.42
5	0.550	2.73
10	0.600	3.03
15	0.650	3.33
20	0.700	3.64
25	0.750	3.94
30	0.800	4.24
35	0.850	4.55
40	0.900	4.85
45	0.950	5.15
50	1.000	5.45
55	1.050	5.76
60	1.100	6.06
65	1.150	6.36
70	1.200	6.67
75	1.250	6.97
80	1.300	7.27
85	1.350	7.58
90	1.400	7.88
95	1.450	8.18
100	1.500	8.48
105	1.550	8.79
110	1.600	9.09
115	1.650	9.39
120	1.700	9.70
125	1.750	10.00

K	6.060
Bias (B)	-0.606

TMP 36 temperature sensor
Temp in C = [(Vout in mV) - 500]/10

V2 = (V1 * K) + B

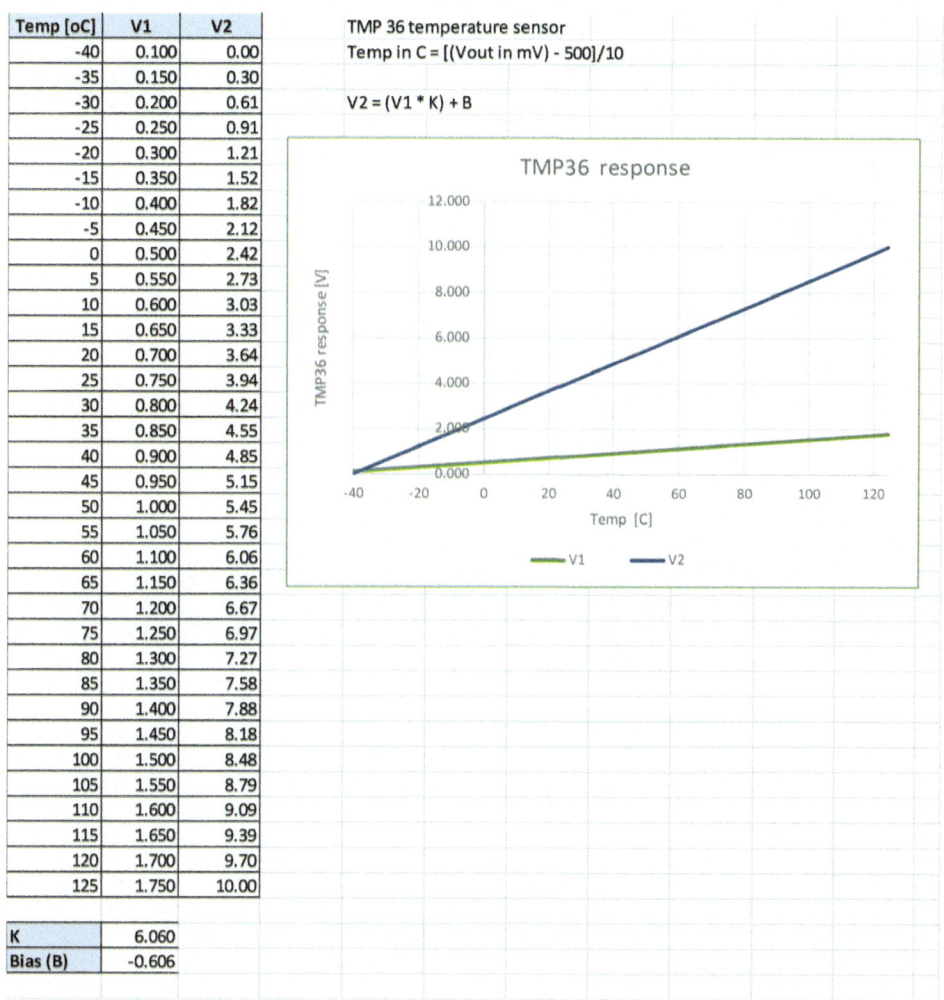

Fig. 4.42 TMP36 temperature sensor

the motor at the desired speed. The sum of the on time and off time (PWM period) should remain constant. The circuit of 4.27 is used for this exercise.

4.17.1 Motor Control Software Configuration

The UML activity diagram for the motor control algorithm is provided in Fig. 4.46. Also, the relationship between duty cycle and RPM determined earlier in the chapter are used to set initial duty cycle parameters.

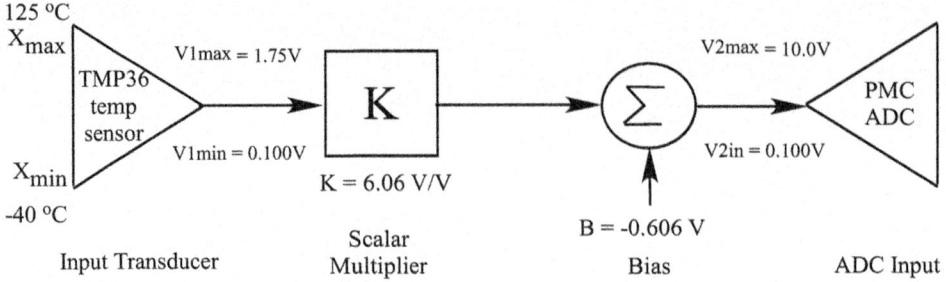

a) transducer interface block diagram.

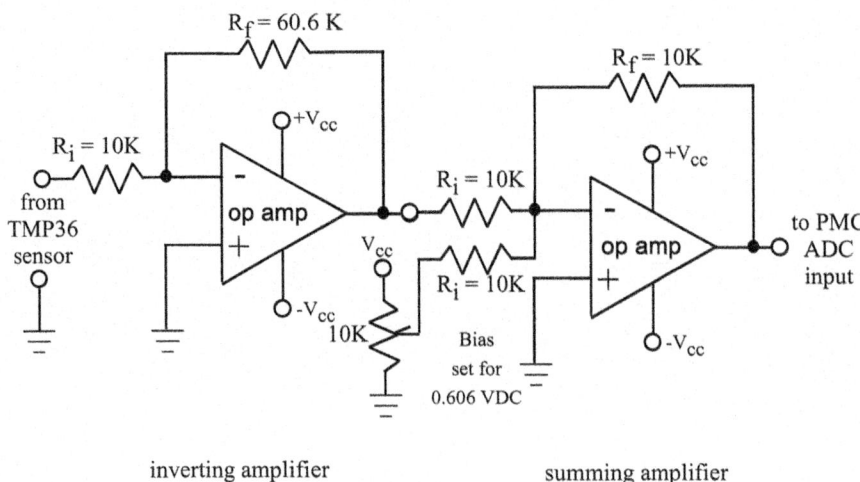

b) transducer interface circuit diagram.

Fig. 4.43 TMP36 temperature sensor interface

Activities:

1. Write a test a sketch to stabilize the DC motor at a specific motor speed. **Hint**: Adapt sketch "PMC_PWM_motor_speed" provided earlier in the chapter.
2. Use a potentiometer connected to PMC ANALOG IN 00 to set the desired speed on the DC motor.

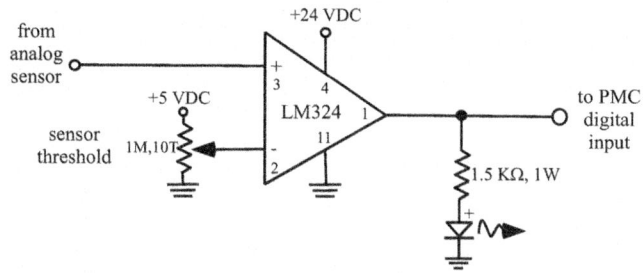

a) One-bit ADC configuration.

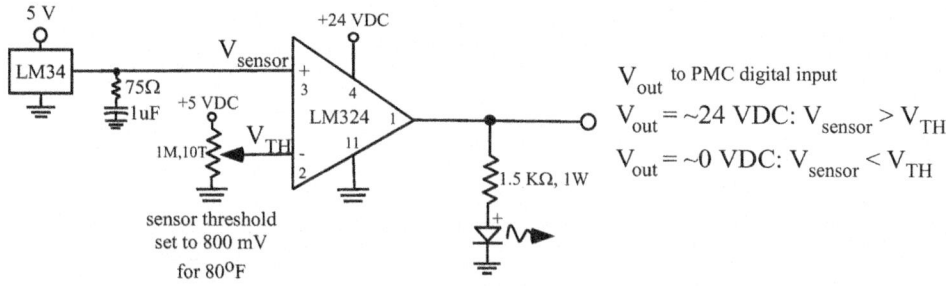

b) One-bit ADC configuration. Logic 1: $V_{sensor} > V_{TH}$, Logic 0: $V_{sensor} > V_{TH}$.

Fig. 4.44 One-bit analog-to-digital converter

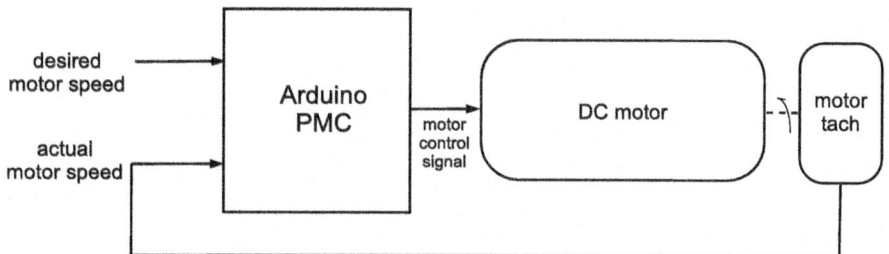

Fig. 4.45 DC motor control. Images used courtesy of the Arduino Team (CC BY-NC-SA) (https://www.arduino.cc)

4.18 Application II: PCA9685 16-Channel PWM and Servo Driver

In the previous example we stabilized motor speed by adjusting PWM parameters. Without a dedicated PWM system, a processor is tied up generating the PWM signal. In this example we provide the PMC with a 16-channel PWM and servo driver.

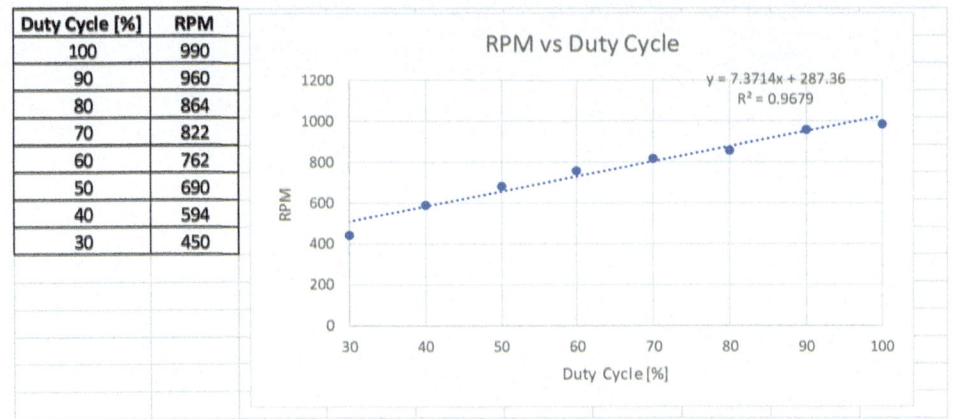

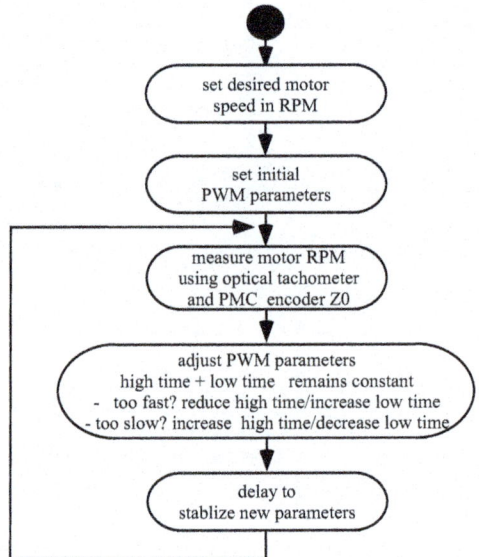

Fig. 4.46 DC motor control UML

4.18.1 Application IIa: PCA9685 16-Channel Servo Driver

The PCA9685 is a 16-channel PWM and servo driver (Adafruit #815). It is equipped with 12-bit resolution to set PWM and servo parameters. The PCA9685 is connected to the PMC via the I2C system as shown in Fig. 4.47.

In this example we configure the PCA9685 to control a servo (Adafruit #815) connected to Channel 0.

4.18 Application II: PCA9685 16-Channel PWM and Servo Driver

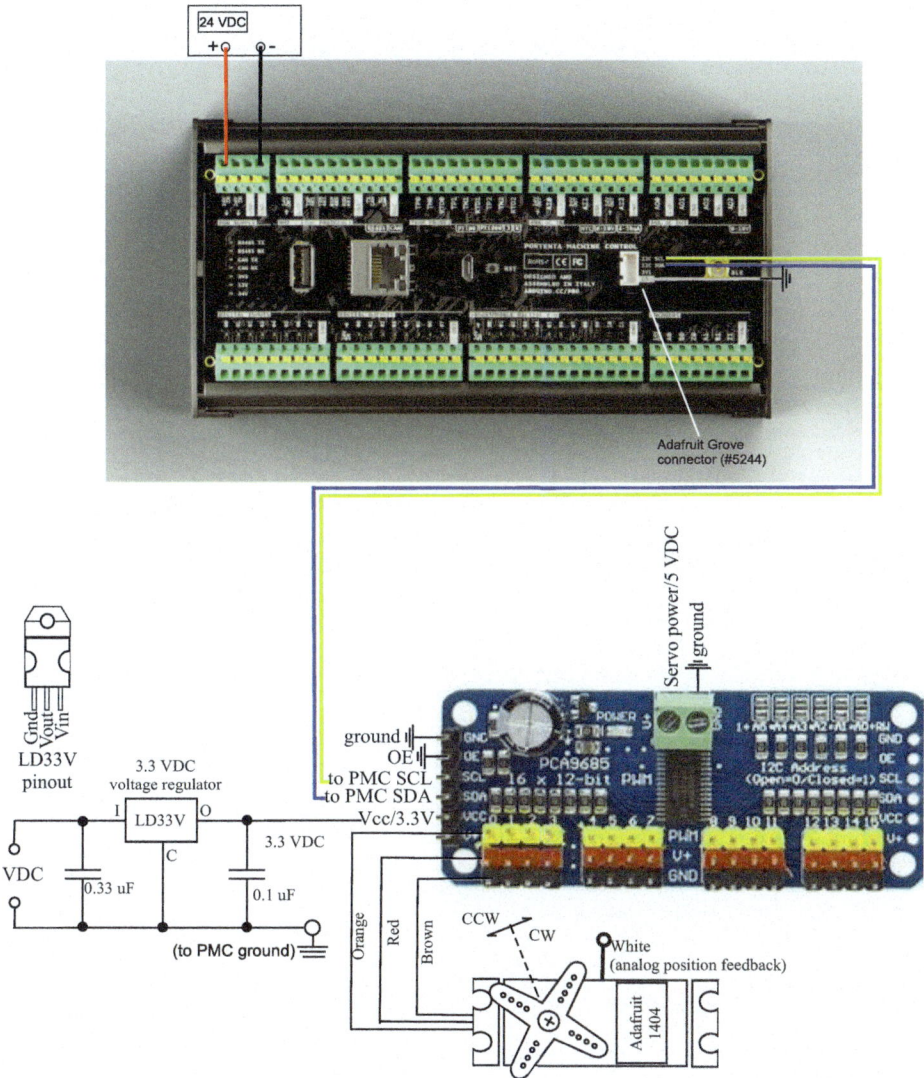

Fig. 4.47 Servo control with 16-channel PWM and servo driver. PMC image used courtesy of the Arduino Team (CC BY-NC-SA) (https://www.arduino.cc). PCA9685 image courtesy of Adafruit (https://www.adafruit.com)

```
//*********************************************************************
//PMC_PCA9685_servo
//
//Adapted from Adafruit PWM Servo Driver Library
// - Sketch: Servo
//
//Notes:
// - Example uses the Adafruit PCA9685 16-channel PWM and Servo
//   Driver (#815) to control servo (#1404) connected to CH0.
//-  The Portenta Machine Control uses I2C to communicate with
//   the PCA9685 via SCL and SDA.
//
//Pick one up today in the adafruit shop!
//------> http://www.adafruit.com/products/815
//
//Adafruit invests time and resources providing this open source code,
//please support Adafruit and open-source hardware by purchasing
//products from Adafruit!
//
//Written by Limor Fried/Ladyada for Adafruit Industries.
//BSD license, all text above must be included in any redistribution
//
//Adapted for this PMC example.
//*********************************************************************

#include <Wire.h>
#include <Adafruit_PWMServoDriver.h>

Adafruit_PWMServoDriver pwm = Adafruit_PWMServoDriver();
#define SERVOMIN  150      //'minimum' pulse length count (out of 4096)
#define SERVOMAX  600      //'maximum' pulse length count (out of 4096)
#define USMIN  600         //'minimum' microsecond length based on 150
#define USMAX  2400        //'maximum' microsecond length based on 600
#define SERVO_FREQ 50      //Analog servos run at ~50 Hz updates


uint8_t servonum = 0;      //servo #

void setup()
{
Serial.begin(9600);
Serial.println("channel 0 servo test!");

pwm.begin();                //see servo sketch for notes on freq setting
pwm.setOscillatorFrequency(27000000);
pwm.setPWMFreq(SERVO_FREQ);//analog servos run at ~50 Hz updates
delay(10);
}

void loop()
{
Serial.println(servonum);
//Drive servo from min to max using setPWM() function
for(uint16_t pulselen = SERVOMIN; pulselen < SERVOMAX; pulselen++)
  {
  pwm.setPWM(servonum, 0, pulselen);
  }

delay(500);
```

```
//Drive servo from max to min using setPWM() function
for(uint16_t pulselen = SERVOMAX; pulselen > SERVOMIN; pulselen--)
  {
  pwm.setPWM(servonum, 0, pulselen);
  }

delay(500);

//Drive servo from min to max using writeMicroseconds() function
for(uint16_t microsec = USMIN; microsec < USMAX; microsec++)
  {
  pwm.writeMicroseconds(servonum, microsec);
  }

delay(500);
//Drive servo from max to min using writeMicroseconds() function
for(uint16_t microsec = USMAX; microsec > USMIN; microsec--)
  {
  pwm.writeMicroseconds(servonum, microsec);
  }

delay(500);

}
//*********************************************************************
```

4.18.2 Application IIb: PCA9685 16-Channel PWM

The PCA9685 can also generate PWM signals for motor speed control applications. In the sketch a 1000 Hz square wave with a 50% duty cycle is generated on PCA9685 channel 0. The start point (on time) and the stop point (off time) is set using an integer from 0 to 4095. The signal can be verified by observing the signal at PCA9685 CH0 PWM pin with an oscilloscope.

```
//*********************************************************************
//PMC_pwmtest
//
//Adapted from Adafruit PWM Servo Driver Library
//- Sketch: pwmtest
//
//Notes:
//- Example for Adafruit PCA9685 16-channel PWM and Servo driver
//- Generates PWM signal on channel 0.
//- Portenta Machine Control uses I2C to communicate with PCA9685
//
//Pick one up today in the adafruit shop!
//     ------> http://www.adafruit.com/products/815
```

```
//
//Adafruit invests time and resources providing this open source code,
//please support Adafruit and open-source hardware by purchasing
//products from Adafruit!
//
//Written by Limor Fried/Ladyada for Adafruit Industries.
//BSD license, all text above must be included in any redistribution.
//Adapted for this specific example.
//*****************************************************************

#include <Wire.h>
#include <Adafruit_PWMServoDriver.h>

uint8_t pwmnum = 0;                       //PWM channel number
uint16_t on_point  = 0;                   //signal start point (0-4096)
uint16_t off_point = 2048;                //signal stop point  (0-4096)
float pwm_freq     = 1000.0;              //PWM frequency
float pwm_duty_cycle;

Adafruit_PWMServoDriver pwm = Adafruit_PWMServoDriver();

void setup()
{
Serial.begin(9600);
Serial.println("Channel 0 PWM test!");
pwm.begin();
pwm.setOscillatorFrequency(27000000);
pwm.setPWMFreq(1000);                     //max PWM frequency 1600
Wire.setClock(400000);
}

void loop()
{
//Vary PWM parameters
pwm.setPWM(pwmnum, on_point, off_point); //ch, on, off
Serial.print("PWM frequency [Hz]:  ");
Serial.println(pwm_freq);
Serial.print("PWM on point:  ");
Serial.println(on_point);
Serial.print("PWM off point:  ");
Serial.println(off_point);
pwm_duty_cycle = (float)(off_point - on_point)/4096 * 100.0;
Serial.print("PWM duty cycle [%]:  ");
Serial.println(pwm_duty_cycle);
Serial.println(" ");
delay(5000);
}

//*****************************************************************
```

Activities:

1. Adapt the sketch to vary the duty cycle of the 1000 Hz signal.
2. Adapt the circuit of Application I to control the speed of a 12 VDC motor.

4.19 Application III: PMC Weather Station

In this exercise we provide the initial design for a weather station with the following requirements:[7]

- Design a weather station to sense wind direction and speed, ambient temperature, humidity, and accumulated rainfall; and
- Collected weather data should be displayed on an LCD and logged to bulk storage.

4.19.1 Structure Chart

To begin the design process, a structure chart is used to provide an overall system picture and partition the system into definable pieces. The structure chart for the weather station is provided in Fig. 4.48. We employ a top-down design/bottom-up implementation approach. The system is partitioned until the lowest level of the structure chart contains "doable" pieces of hardware components or software functions. Data flow is shown on the structure chart as directed arrows.

The main PMC controller subsystems needed for this project are the I2C system for the LCD display; the ADC system for wind direction, temperature, and humidity sensing; the PMC Real Time Clock (RTC) module; the USB for data logging; and the encoder system for the wind speed and the rain gauge.

4.19.2 Circuit Diagram

The circuit diagram for the weather station and subsystems is provided in Fig. 4.49. We have purposely partitioned the design into small, doable building blocks. We discuss each subsystem in turn.

---

[7] This project originally appeared in "Arduino I: Getting Started," Morgan & Claypool Publishers, 2020. An expanded version is included here with permission as an integral part of the Greenhouse Project.

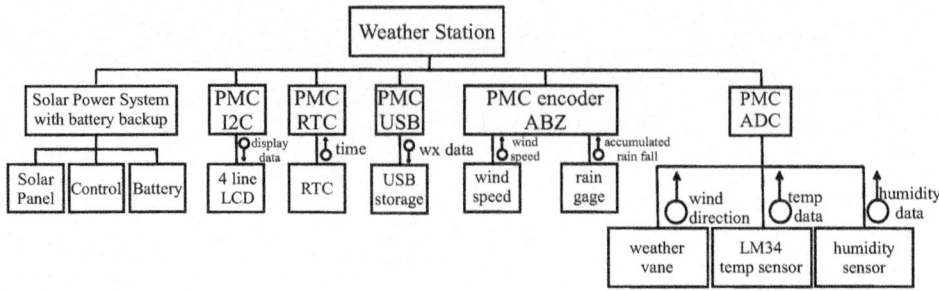

Fig. 4.48 Weather station structure chart

Liquid Crystal Display. Recall from Chap. 3, the LCD2004 (20 characters × 4 lines) provides an I2C backpack for a serial link between the Arduino PMC and the LCD and hence conserve precious digital I/O pins. Sample code was provided to display sensor data.
Temperature Sensor. The weather station is equipped with an LM34 Precision Fahrenheit Temperature Sensors. The LM34 provides 10 mV output per degree Fahrenheit. The output from the sensor is provided to Arduino PMC analog input pin AI0 for the external sensor (SNIS161D).
Humidity sensor. A Honeywell HIH-4030 sensor is used to measure humidity. The sensor provides an output voltage that may be mapped to a corresponding relative humidity (RH) value. The RH value provides a measurement of the amount of water vapor in the air. The RH is expressed as a value from 0 to 100% RH. The interface circuit for the RH sensor is shown in Fig. 4.49. The sensor provides an output voltage to indicate RH. The voltage is processed and corrected for temperature using the following equations provided by the manufacturer (Honeywell).

$$V_{out} = (V_{supply}) * (0.0062 * sensor RH) + 0.16$$

The sensor RH value is corrected for temperature:

$$TrueRH = (sensor RH/(1.0546 - 0.00216T)$$

with T expressed in degrees Centigrade.

The following sketch measures temperature and humidity. It provides a humidity reading corrected for ambient temperature.

```
//***********************************************************
//File: PMC_humidity
//
//Adapted from:
//Portenta Machine Control - Analog in 0-10 V
//
//Notes:
//Notes:
```

4.19 Application III: PMC Weather Station

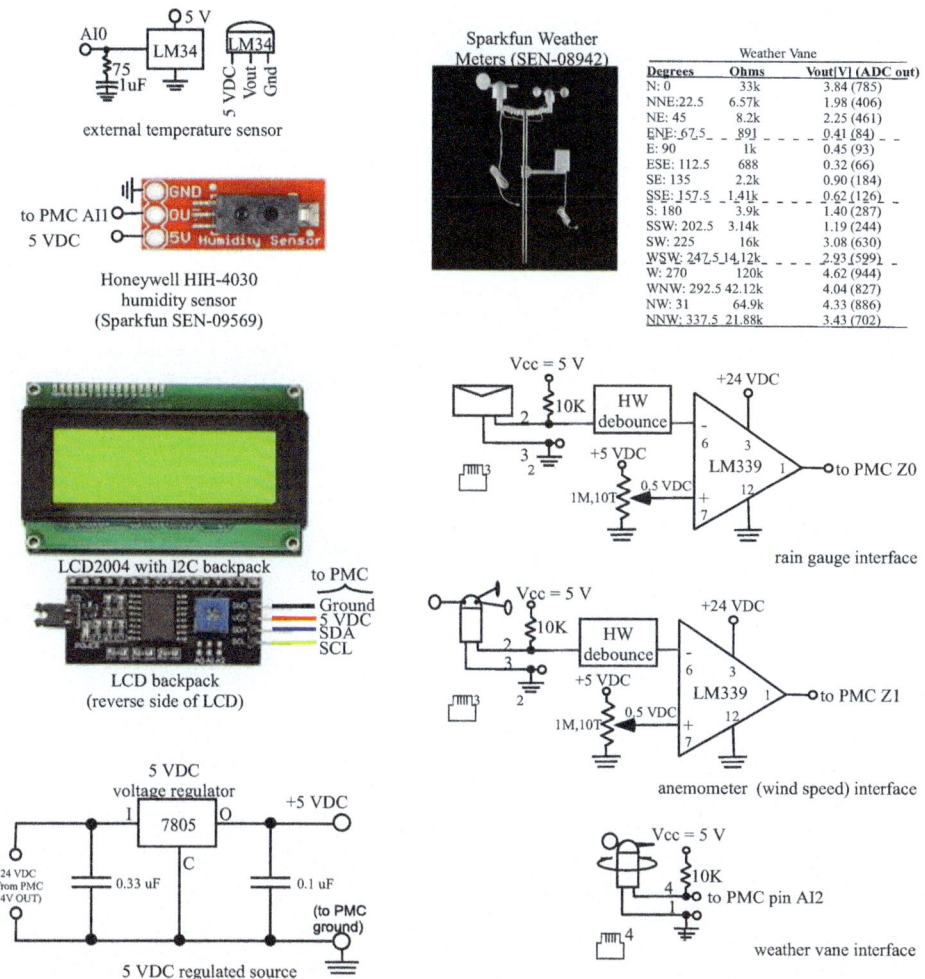

Fig. 4.49 Circuit diagram for weather station. Illustrations used with permission of Sparkfun Electronics (https://www.sparkfun.com) and Adafruit (https://www.adafruit.com)

```
//- For each channel of the ANALOG IN connector, there is a resistor
//  divider made by a 100k and 39k, the input voltage is divided by a
//  ratio of 0.28.
//- Maximum input voltage is 10V.
//- To use the 0V-10V functionality, a 24V supply on the PWR SUPPLY
//  connector is necessary.
//- Reads analog voltage at AI0 connected to LM34 IC temperature sensor.
//- The LM34 provides 10 mV of output per degree Fahrenheit.
//- Honeywell HIH-4030 humidity sensor at PMC AI1
//
//This example code is in the public domain.
//Copyright (c) 2024 Arduino, SPDX-License-Identifier: MPL-2.0
```

```
//*************************************************************

#include <Arduino_PortentaMachineControl.h>

const float RES_DIVIDER = 0.28057;
const float REFERENCE   = 3.0;
unsigned int analog_reading;

void setup()
{
Serial.begin(9600);
while(!Serial)
  {
  ; //wait for serial port connection
  }
                                        //set for 0-10 VDC in
MachineControl_AnalogIn.begin(SensorType::V_0_10);
}

void loop()
{
//read temp from LM34 at AI0
float raw_voltage_ch0 = MachineControl_AnalogIn.read(0);
float voltage_ch0 = (raw_voltage_ch0 * REFERENCE) / 65535 / RES_DIVIDER;

//ADC AI0 output
Serial.print("ADC CH0: ");
Serial.println(raw_voltage_ch0, 3);

//ADC voltage
Serial.print("LM34 voltage: ");
Serial.print(voltage_ch0, 3);
Serial.println("V");

//ADC temp
Serial.print("LM34 temperature [F]: ");
float LM_34_temp = voltage_ch0/.010;     //10 mv/degree F
Serial.println(LM_34_temp, 3);

//Read humidity at AI1
float raw_humidity_ch1 = MachineControl_AnalogIn.read(1);
Serial.print("Raw humidity CH1: ");
Serial.print(raw_humidity_ch1, 3);

float humidity_voltage_ch1 = (raw_humidity_ch1 * REFERENCE) / 65535 / RES_DIVIDER;
Serial.print("   Humidity voltage CH1: ");
Serial.print(humidity_voltage_ch1, 3);
Serial.println("V");
                                       //convert to RH per data sheet
float sensor_RH_flt  = (humidity_voltage_ch1/(0.0062 * 5.0)) - 0.16;
                                       //convert temp reading to C
float ext_temp_C = (float)((LM_34_temp - 32.0) * (5.0/9.0));
                                       //compensate for temp per data sheet
Serial.print("Sensor RH [%]: ");
Serial.println(sensor_RH_flt, 3);
Serial.print("LM34 temperature [C]: ");
Serial.println(ext_temp_C, 3);

float true_RH = sensor_RH_flt/(1.0546 - (0.00216 * ext_temp_C));
```

4.19 Application III: PMC Weather Station

```
//print result
Serial.print("Humidity [%]: ");
Serial.println(true_RH, 3);
Serial.println(" ");

delay(500);
}

//*****************************************************************
```

Weather Vane. The Sparkfun Weather Meters kit (# SEN-08942) is equipped with a weather vane, an anemometer (wind speed), and a rain gauge. The weathervane provides a voltage output from 0 to 5 VDC for different wind directions as shown in Fig. 4.49. The weathervane must be oriented to a known direction with the output voltage at this direction noted. The output voltage is provided to an RJ11 connector. Pins 1 and 4 of the RJ11 connector provide access to the vane's resistance. A 10 KΩ resistor is placed in series with the vane's resistance to provide a voltage divider circuit (https://www.sparkfun.com). The weathervane output is connected to PMC analog input AI2.

In the following sketch the analog output voltage from the weathervane is converted to wind direction and displayed on the serial monitor.

```
//*****************************************************************
//PMC_wind_direction
//
//Adapted from:
//Portenta Machine Control - Analog in 0-10 V
//
//Notes:
//- Each ANALOG IN channel is equipped with a 100k and 39k
//  resistor divider.  The input voltage is scaled by a 0.28 ratio
//- Maximum input voltage is 10V
//- To use the 0V-10V functionality, a 24V supply on the PWR SUPPLY
//  connector is necessary.
//- Reads analog voltage at AI2 connected to weather vane
//
//This example code is in the public domain.
//Copyright (c) 2024 Arduino, SPDX-License-Identifier: MPL-2.0
//*****************************************************************

#include <Arduino_PortentaMachineControl.h>

const float RES_DIVIDER = 0.28057;
const float REFERENCE   = 3.0;
unsigned int analog_reading;
float wind_from_dir;

void setup()
{
Serial.begin(9600);
while(!Serial)
  {
  ; //wait for serial port connection
  }
```

```
                                        //set for 0-10 VDC in
MachineControl_AnalogIn.begin(SensorType::V_0_10);
}

void loop()
{
float raw_voltage_ch2 = MachineControl_AnalogIn.read(2);
Serial.print("Raw Voltage CH2: ");
Serial.print(raw_voltage_ch2, 3);

float analog_dir = (raw_voltage_ch2 * REFERENCE) / 65535 / RES_DIVIDER;
Serial.print("Analog dir voltage: ");
Serial.print(analog_dir, 3);
Serial.println("V");

//Determine wind direction - vane 0 degree aligned with North
//North - ADC output: 1.16V  Low: 1.06V   High: 1.26V
if((analog_dir >= 1.06) && (analog_dir <= 1.26))
   {
   char dirstring[4] = "N  ";
   Serial.print("Direction: ");
   Serial.print(dirstring);
   wind_from_dir = 0.0;
   Serial.println(wind_from_dir);
   }

//NNE - ADC output: 3.02V  Low: 2.92V   High: 3.12V
else if((analog_dir >= 2.92) && (analog_dir <= 3.12))
   {
   char dirstring[4] = "NNE";
   Serial.print("Direction: ");
   Serial.print(dirstring);
   wind_from_dir = 22.5;
   Serial.println(wind_from_dir);
   }

//NE - ADC output: 2.75V  Low: 2.65V   High: 2.85V
else if((analog_dir >= 2.65) && (analog_dir <= 2.85))
   {
   char dirstring[4] = "NE ";
   Serial.print("Direction: ");
   Serial.print(dirstring);
   wind_from_dir = 45.0;
   Serial.println(wind_from_dir);
   }

//ENE - ADC output: 4.59V  Low: 4.49V   High: 4.69V
else if((analog_dir >= 4.59) && (analog_dir <= 4.69))
   {
   char dirstring[4] = "ENE";
   Serial.print("Direction: ");
   Serial.print(dirstring);
   wind_from_dir = 67.5;
   Serial.println(wind_from_dir);
   }

//E - ADC output: 4.55V  Low: 4.45V   High: 4.65V
else if((analog_dir >= 89) && (analog_dir <= 109))
   {
```

4.19 Application III: PMC Weather Station

```
  char dirstring[4] = "E  ";
  Serial.print("Direction: ");
  Serial.print(dirstring);
  wind_from_dir = 90.0;
  Serial.println (wind_from_dir);
  }

//ESE - ADC output: 4.68V  Low: 4.58V   High: 4.78V
else if((analog_dir >= 4.58) && (analog_dir <= 4.78))
  {
  char dirstring[4] = "ESE";
  Serial.print("Direction: ");
  Serial.print(dirstring);
  wind_from_dir = 112.5;
  Serial.println (wind_from_dir);
  }

//SE - ADC output: 4.10V  Low: 4.00V   High: 4.20V
else if((analog_dir >= 4.10) && (analog_dir <= 4.20))
  {
  char dirstring[4] = "SE ";
  Serial.print("Direction: ");
  Serial.print(dirstring);
  wind_from_dir = 135.0;
  Serial.println (wind_from_dir);
  }

//SSE - ADC output: 4.38V  Low: 4.28V   High: 4.48V
else if((analog_dir >= 4.28) && (analog_dir <= 4.48))
  {
  char dirstring[4] = "SSE";
  Serial.print("Direction: ");
  Serial.print(dirstring);
  wind_from_dir = 157.5;
  Serial.println (wind_from_dir);
  }

//S - ADC output: 3.60V  Low: 3.50V   High: 3.70V
else if((analog_dir >= 3.50) && (analog_dir <= 3.70))
  {
  char dirstring[4] = "S  ";
  Serial.print("Direction: ");
  Serial.print(dirstring);
  wind_from_dir = 180.0;
  Serial.println (wind_from_dir);
  }

//SSW - ADC output: 3.81V  Low: 3.71V   High: 3.91V
else if((analog_dir >= 3.71) && (analog_dir <= 3.91))
  {
  char dirstring[4] = "SSW";
  Serial.print("Direction: ");
  Serial.print(dirstring);
  wind_from_dir = 202.5;
  Serial.println (wind_from_dir);
  }

//SW - ADC output: 1.92V  Low: 1.82V   High: 2.02V
else if((analog_dir >= 1.82) && (analog_dir <= 2.02))
```

```
    {
    char dirstring[4] = "SW ";
    Serial.print("Direction: ");
    Serial.print(dirstring);
    wind_from_dir = 225.0;
    Serial.println (wind_from_dir);
    }

//WSW - ADC output: 2.07V  Low: 1.97V  High: 2.17V
else if((analog_dir >= 1.97) && (analog_dir <= 2.17))
    {
    char dirstring[4] = "WSW";
    Serial.print("Direction: ");
    Serial.print(dirstring);
    wind_from_dir = 247.5;
    Serial.println (wind_from_dir);
    }

//W - ADC output: 0.38V  Low: 0.28V  High: 0.48V
else if((analog_dir >= 0.28) && (analog_dir <= 0.48))
    {
    char dirstring[4] = "W  ";
    Serial.print("Direction: ");
    Serial.print(dirstring);
    wind_from_dir = 270.0;
    Serial.println (wind_from_dir);
    }

//WNW - ADC output: 0.96V  Low: 0.86V  High: 1.06V
else if((analog_dir >= 0.86) && (analog_dir <= 1.06))
    {
    char dirstring[4] = "WNW";
    Serial.print("Direction: ");
    Serial.print(dirstring);
    wind_from_dir = 292.5;
    Serial.println (wind_from_dir);
    }

//NW - ADC output: 0.67V  Low: 0.57V  High: 0.77V
else if((analog_dir >= 0.57) && (analog_dir <= 0.77))
    {
    char dirstring[4] = "NW ";
    Serial.print("Direction: ");
    Serial.print(dirstring);
    wind_from_dir = 315.0;
    Serial.println (wind_from_dir);
    }

//NNW - ADC output: 1.57V  Low: 1.47V  High: 1.67V
else if((analog_dir >= 1.47) && (analog_dir <= 1.67))
    {
    char dirstring[4] = "NNW";
    Serial.print("Direction: ");
    Serial.print(dirstring);
    wind_from_dir = 337.5;
    Serial.println (wind_from_dir);
    }

    else
```

4.19 Application III: PMC Weather Station

```
   {
   char dirstring[4] = "<->";
   Serial.print("Direction: ");
   Serial.print(dirstring);
   }

delay(500);

}
//***********************************************************************
```

Rain Gauge. The rain gauge contains an enclosed rain tipping bucket. When the bucket is full, it tips, tripping a magnetic read switch. The switch closure is routed via an RJ11 connector to PMC Encoder input Z0. An open collector LM339 op amp is used to pull the output up to 24 VDC as shown in Fig. 4.49. Each tip represents 0.011 in. (0.2794 mm) of rainfall. Note the use of a hardware debounce circuit discussed in Chap. 2 at the input to the op amp circuit.

The following sketch provides for rainfall measurement.

```
//***********************************************************************
//PMC_rain_gauge_wind_speed - measures accumulated rainfall and wind speed
//                            via Sparkfun Weather Meters (SEN-08942)
//
//Notes:
//  - PMC connected to 24 VDC power supply
//  - Digital OUTPUTS 24V IN connected to 24 VDC power supply
//  - Rain gauge: Each switch close corresponds to 0.011 inches of rain
//  - Wind speed: Switch closure of once per second = 1.492 MPH
//
//This example code is in the public domain.
//***********************************************************************

#include <Arduino_PortentaMachineControl.h>

unsigned long measurement_interval = 1000;     //1 second capture window
unsigned long time_hack;
float wind_speed;                              //in MPH

void setup()
{
Serial.begin(9600);
while (!Serial)
  {
  ;                                            //wait for serial port to connect
  }
}

void loop()
{
//Rain gauge measurements
Serial.print("Rain gauge switch: ");           //Incremented by rain guage signal
Serial.println(MachineControl_Encoders.getRevolutions(0));
Serial.print("Rain fall [in]: ");
Serial.println((float)(MachineControl_Encoders.getRevolutions(0)) * 0.011, 3);
Serial.println();
```

```
//Wind speed measurements
time_hack = millis();
while((millis()- time_hack) < measurement_interval)
  {
   wind_speed = (float)(MachineControl_Encoders.getRevolutions(1)) * 1.492;
  }

Serial.print("Wind speed switch: ");          //Incremented by wind speed indicator
Serial.println(MachineControl_Encoders.getRevolutions(1));
Serial.print("Wind speed [MPH]: ");           //Incremented by wind speed indicator
Serial.println(wind_speed);
Serial.println(" ");

delay(25);
}

//*****************************************************************************
```

Anemometer-wind speed. The anemometer is equipped with three cups to catch the wind. With each rotation of the anemometer, a reed switch is tripped. Switch closures one second apart equates to 1.492 mi per hour (2.4 Km/h) of wind speed. The switch closure is routed via an RJ11 connector to PMC Encoder input Z1. An open collector LM339 op amp is used to pull the output up to 24 VDC as shown in Fig. 4.49. Note the use of a hardware debounce circuit discussed in Chap. 2 at the input to the op amp circuit. The sketch above provides for wind speed measurement.

USB data logger with Real Time Clock We use the PMC onboard USB for data logging. The PMC also features an onboard Real Time Clock (RTC) . RTC features provide the controller the ability to track calendar time based on seconds, minutes, hours, etc.

4.19.3 UML Activity Diagram

The UML activity diagram for the main program is provided in Fig. 4.50. After initializing the subsystems, the program enters a continuous loop where temperature, wind direction, and humidity is sensed and displayed on the LCD. Encoder channels are used to capture data on wind speed and rainfall. The sensed values are then transmitted via the USB to a USB drive. The system then enters a delay to set how often the temperature and wind direction parameters are updated. We have you construct the individual UML activity diagrams for each function as an end of chapter exercise.

4.19.4 Bottom-Up Implementation

A sound implementation approach is to build up a complex system a subsystem at a time. We use this approach to assemble the weather station. Throughout the early parts of the

4.19 Application III: PMC Weather Station

Fig. 4.50 Weather station UML activity diagram

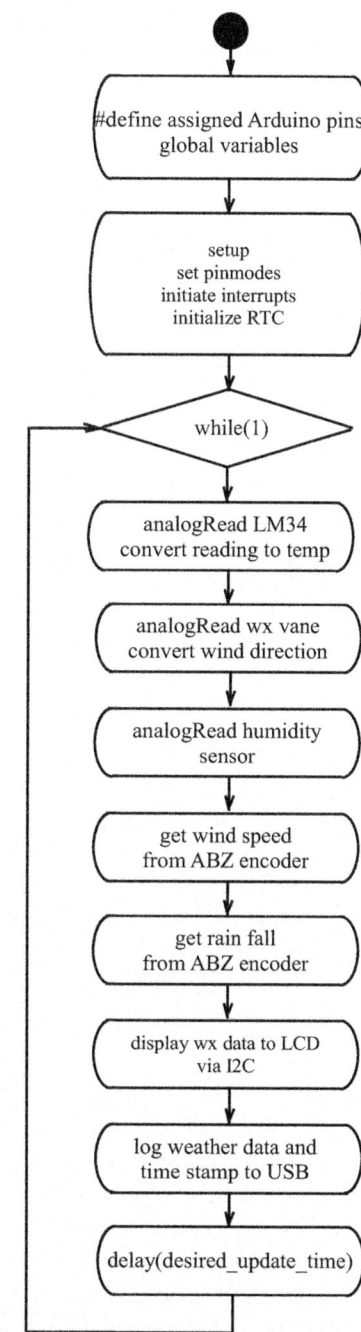

book we have provided detailed examples of many of the weather station subsystems. Use this approach to complete the weather station design.

4.20 Summary

In this chapter we explored how an Arduino PMC may be used in industrial, Internet of Things (IoT), and hybrid applications. We explored how to connect input sensors and output actuators to the PMC. We began with a review of the PMC input and output characteristics. We then explored a wide variety of digital and analog input sensors and output actuators. We also employed an operational amplifier-based transducer interface design process to interface input sensors to the PMC.

4.21 Chapter Problems

1. What will happen if a controller is used outside of its prescribed operating envelope?
2. What is switch bounce? Describe techniques to minimize switch bounce.
3. What is the difference between an incremental encoder and an absolute encoder? Describe applications for each type.
4. Describe an application for a flex sensor. Provide a supporting circuit and software with all components specified.
5. Describe an application for a fluid level sensor. Provide a supporting circuit with all components specified.
6. Why is a transimpedance amplifier typically used with a photodiode circuit?
7. Construct a table of different switch types. Provide an application for each type.
8. What is the purpose of a two channel incremental quadrature encoder?
9. In the chapter an inexpensive laser light show was introduced using two servos. Write a sketch to control the two servos from the PMC.
10. What is ultrasound? How might an ultrasound sensor be used in an industrial environment?
11. What is the advantage of the TMP36 temperature sensor over other sensors that measure negative temperatures?
12. Describe an industrial application for a tilt sensor.
13. Describe an application for an optical isolator circuit. Provide a supporting circuit with all components specified.
14. Describe an application for an environmental sensing circuit. Provide a supporting circuit with all components specified.
15. What are the ideal operational amplifier characteristics? What prevents an op amp from performing in an ideal manner?
16. In your own words describe each of the op amp nonideal parameters.

17. Derive each of the output equations for the classic operational amplifier configurations.
18. Write and test a ladder logic program for the linear actuator circuit discussed in the chapter.

References

1. Advanced Photonix Incorporated (API), *CdS Photoconductive Photocells PDV–P8001*, March 2006.
2. S. BharadwajRedd, *PLC Programming for Blinking Indicator Lights*, learn.automationcommunity.com
3. D. Egironi, *Presenting MQ sensors: low–cost gas and pollution detectors*, Open Electronics Source Electronic Projects, February 2018.
4. L.M. Faulkenberry, *An Introduction to Operational Amplifiers*, Wiley, 1977.
5. Industrial Fiber Optics, i-fiberoptics.com.
6. W. Jung, *Op Amp History*, Analog Devices www.analog.com.
7. *LM34 Precision Fahrenheit Temperature Sensors*, Texas Instruments SNIS161D, March 2000– Revised January 2016, www.ti.com.
8. *LM35 Precision Centigrade Temperature Sensors*, Texas Instruments SNIS159H, August 1999– Revised January 2017, www.ti.com.
9. *Low Voltage Temperature Sensors, TMP35/36/37*, Analog Devices, Rev. H, 2015, www.analog.com.
10. Franklin Miller, Jr. *College Physics,* fourth edition, Harcourt Brace Jovanovich, Inc. 1977.
11. Milone Technologies, *0–5 VDC Linear Resistance to Voltage Module PN–05V00199 Rev 2*, www.milonetech.com.
12. Milone Technologies, *eTape Continuous Fluid Level Sensor PN–12110215TC–X*, www.milonetech.com.
13. *MQ–2 Semiconductor Sensor for Combustible Gas*, mysensors.org
14. *MQ-3 Gas Sensor*, Hanwei Electronics Co., ltd., www.hwsensor.com.
15. *MQ-4 Gas Sensor*, Hanwei Electronics Co., ltd., www.hwsensor.com.
16. *MQ-6 Flammable Gas Sensor*, Zhengzhou Winsen Electronics Technology Co., Ltd, www.winsensor.com
17. *MQ-7 Gas Sensor*, Hanwei Electronics Co., ltd., www.hwsensor.com.
18. *MQ-8 Flammable Gas Sensor*, Zhengzhou Winsen Electronics Technology Co., Ltd, www.winsensor.com.
19. *MQ-9 Semiconductor Sensor for CO/Combustible Gas*, Hanwei Electronics Co., ltd., www.hwsensor.com.
20. *MQ-131 Gas Sensor*, mysensors.org.
21. *MQ-135 Gas Sensor*, mysensors.org.
22. *MQ-183 Gas Sensor*, Hanwei Electronics Co., ltd., www.hwsensor.com.
23. *MQ-214 Gas Sensor*, Hanwei Electronics Co., ltd., www.hwsensor.com.
24. Omega Engineering, *FAQ about Thermistors*, www.omega.com.
25. Omega Engineering, *Thermocouple FAQs*, www.omega.com.
26. *uA7800 series positive-voltage regulators*, SLVS056J, Texas Instruments, 2003.
27. J.R. Ragazzini, R.H. Randall, and F.A. Russell, *Analysis of Problems in Dynamics by Electronic Circuits*, Proceedings of the I.R.E., 444–452, May 1947.
28. A.S. Sedra and K.C. Smith, *Microelectronics*, Oxford University Press, 2004.

29. Sick/Stegmann Incorporated, Dayton, OH, (www.stegmann.com).
30. SparkFun Electronics, 6175 Longbow Drive, Suite 200, Boulder, CO 80301 (www.sparkfun.com)
31. D.F. Stout and M. Kaufman, *Handbook of Operational Amplifier Circuit Design,* McGraw–Hill Book Company, 1976.
32. Taizhou Allison Instruments Co. Ltd., *ALS–MPM–2F Water level transmitter datasheet.*

Arduino PLC IDE and Ladder Logic   5

Objectives: After reading this chapter, the reader should be able to do the following:

- Describe and configure Arduino PLC programming software and related tools;
- List and describe five different methods of programming a PLC within the IEC IEC61131--3 standard;
- Summarize the fundamental concepts of Ladder Logic (LD) PLC programming;
- Describe library instructions within the Arduino PLC IDE; and
- Employ the Arduino PLC IDE to write, compile, and execute ladder logic programs for the Portenta Machine Control.

5.1   Overview

In this chapter we explore the Arduino PLC IDE. We begin with a review of the five PLC programming languages specified within the IEC IEC61131-3 standard and available within the Arduino PLC IDE. We then narrow our focus and concentrate on ladder logic programming. We explore, download, and configure the Arduino PLC IDE and related tools. We then explore the fundamentals of ladder logic PLC programming. The fundamentals are used to complete a series of examples.

5.2 Arduino PMC Programming Tools

In the early chapters of the book we used the Arduino IDE to program various features and functions of the Portenta Machine Control (PMC). In this section we explore the Arduino PLC IDE and related software tools to program the PMC. The Arduino PLC IDE allows for programming the PMC using the five different languages within the IEC IEC61131–3 standard. These include:

- Ladder Diagram (LD)
- Sequential (SFC)
- Function Block Diagram (FBD)
- Structured Text (ST)
- Instruction List (IL).

All of these languages are powerful and important; however, we concentrate on LD programming. In the next several sections we provide a brief introduction to this powerful, graphical programming technique.

5.3 Getting Started–Arduino PLC IDE

Complete the following steps to download and configure software tools, load configuration software to the PMC, establish communication between the host laptop/PC, and activate the PLC IDE license[1]:

- **Software download and installation.** The Arduino PLC IDE and accompanying tools are available for download from `arduino.cc`. There are two different software packages required:

    - Arduino_PLC_IDE_Tools
    - Setup_Arduino_PLC_IDE

    Once downloaded, install the Arduino_PLC_IDE_Tools **first** followed by the Setup_Arduino_PLC_IDE.
- **Arduino PLC IDE.** With setup complete, open the Arduino PLC IDE by clicking on its icon. You should be greeted by the screen shown in Fig. 5.1a.
- Insert a project name, verify the target device is "Portenta Machine Control 1.0" and click "OK." A project window shown in Fig. 5.1b will appear.

---

[1] This section is adapted from "Arduino PLC IDE Setup & Device License Activation," (`arduino.cc`).

5.3 Getting Started–Arduino PLC IDE

- **Windows Device Manager.** With the PMC connected to the support laptop/PC via the USB C cable, go to the Windows Device Manager to determine the two USB COM channels used by the PMC. Two USB Serial Devices should be shown in use by the PMC. Typically the higher COM number is used by the MODBUS link and the lower number is the standard link. Note these two COM numbers for upcoming use.
- **Establish communication.** To connect the support laptop/PC to the PMC select On–line → Set up communication. This opens up a new panel called "Device Link Manager Config." Choose "MODBUS" and "Properties." Verify MODBUS is using the higher COM number noted earlier. Close the panel with "OK." Go to On–line → Connect to establish a communication link between the lost laptop/PC and the Arduino PMC.
- **Download Runtime software.** To download the PLC Runtime configuration software to the PMC, return to the Arduino PLC IDE. Go to the "PMC Configuration" panel and scroll to the bottom left corner. In the "Other" subpanel select the lower COM number used by the PMC (determined earlier using the Device Manager) and click "Download." The Runtime sketch will compile and load to the PMC. This may take a little time. Progress may be monitored via the "Output" panel.
- **License Activation.** Each PMC requires a license activation product key. This may be purchased from the Arduino website. When purchased the key is sent via e–mail. Insert the key into the license section on the main panel. Press the "Activate" button. Next, reboot the PMC using the small reset button next to the USB C connector.
- **Communication link troubleshooting.** If there are issues with the communication link between the PMC and the host PC/laptop, Arduino recommends the following troubleshooting steps[2]:

  - Download and install the latest version of the Arduino IDE.
  - Install the latest version of the Arduino Mbed OS Portenta Boards file. This file is accessible from the Arduino IDE via Tools – → Board – → Boards Manager (search on Portenta).
  - Connect PMC to host computer.
  - Run the sketch "WiFiFirmware Updater Manual" available from Examples – → Examples for Arduino Portenta H7 – → STM32H747_System – → WiFiiFirmware-Updater.
  - Partition the PMC memory using "Memory Partitioning for use with the Arduino IDE (for Opta and PMC)" (available from www.arduino.cc).
  - Re–accomplish the steps above to configure the PLC IDE and License Activation.

---

[2] Troubleshooting steps provided on the Arduino Forum.

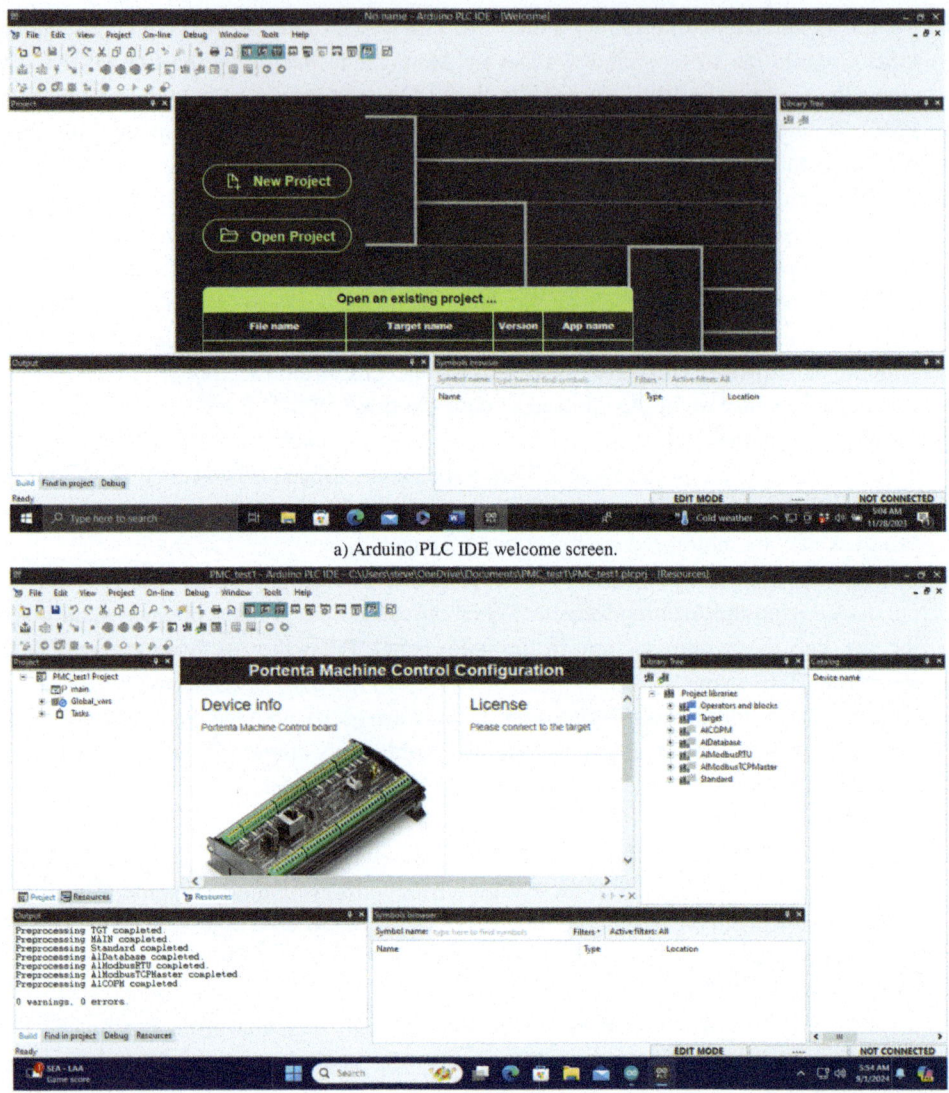

a) Arduino PLC IDE welcome screen.

b) New project screen.

Fig. 5.1 PLC IDE welcome screen. (Figure adapted and used with permission of Arduino Team (CC BY–NC–SA) (www.arduino.cc))

5.4 Structure of Arduino PLC IDE Program

Figure 5.2a demonstrates the flow of an Arduino PLC IDE program. The program executes the "Initialize" portion of the program once and then executes programs within each of the task categories (Fast, Slow, Background) at the time intervals shown.

Figure 5.2b shows program hierarchy within the Arduino PLC IDE. A project contains sections for Main, Global variables, and Tasks. The Task area is subdivided into categories previously described (Initialize, Fast, Slow, Background). Within each Task area are individual programs. A program consists of a sequence of networks (ladder rungs).

The network/ladder rung consists of a horizontal power link connecting the left power rail to the right power rail. Between the two power rails are the activities processed by the network. These include contacts (inputs) on the left and coils (outputs) on the right. The power link may also include additional activities from the Arduino IDE PLC Project Library as shown in Fig. 5.3.

A program is built up as a series of network/rungs to accomplish program steps. When the program is compiled and executed, the network/rungs are sequentially executed. Each network may be viewed as an IF–THEN statement. IF the contacts and the logic on the power link are TRUE–THEN the coil output is TRUE. Let's take a closer look at ladder program components.

5.4.1 Contacts, Coils, Branches, and Blocks

Provided in Fig. 5.4 is a partial illustration of ladder logic components. The contacts are typically a normally open (NO) or normally closed (NC) momentary contact switch. As we see in upcoming examples, the contacts may also be control signals from prior ladder rungs.

Coils serve as the output for ladder networks (rungs). As seen in Fig. 5.4, coils may be control signals, motors, lights, horns, etc.

Inputs may be configured in series/parallel combinations. A series connection of input contacts implies a logical AND statement. That is both contacts must be at logical TRUE to assert the output coil. A parallel connection of input contacts implies a logical OR statement. In the parallel configuration either contact may be logical TRUE to assert the coil. The series and parallel contact configurations may be mixed to implement complex logic configurations.

The coil outputs may also be configured in a parallel configuration. For a parallel output coil configuration both coils are asserted simultaneously (e.g. a motor and indicator lamp).

Shown in Fig. 5.4 are the extensive library functions defined within the Arduino PLC IDE. Note the extensive operations, blocks, and standard functions. We use these in upcoming examples. Let's take a closer look at counters and timers.

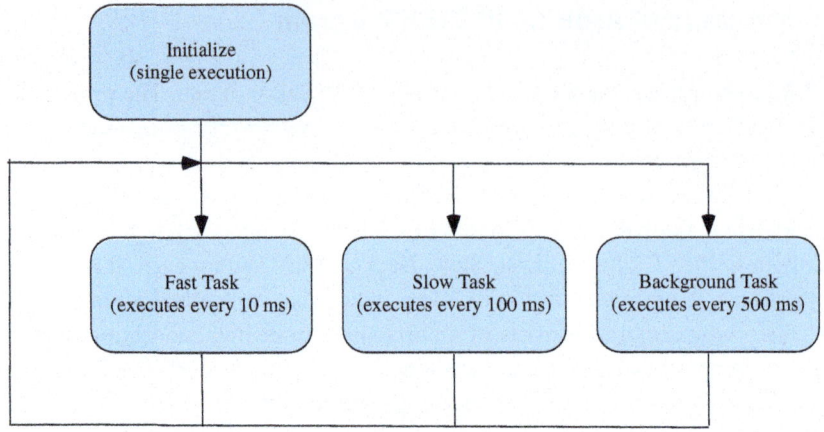

a) PLC program flow.

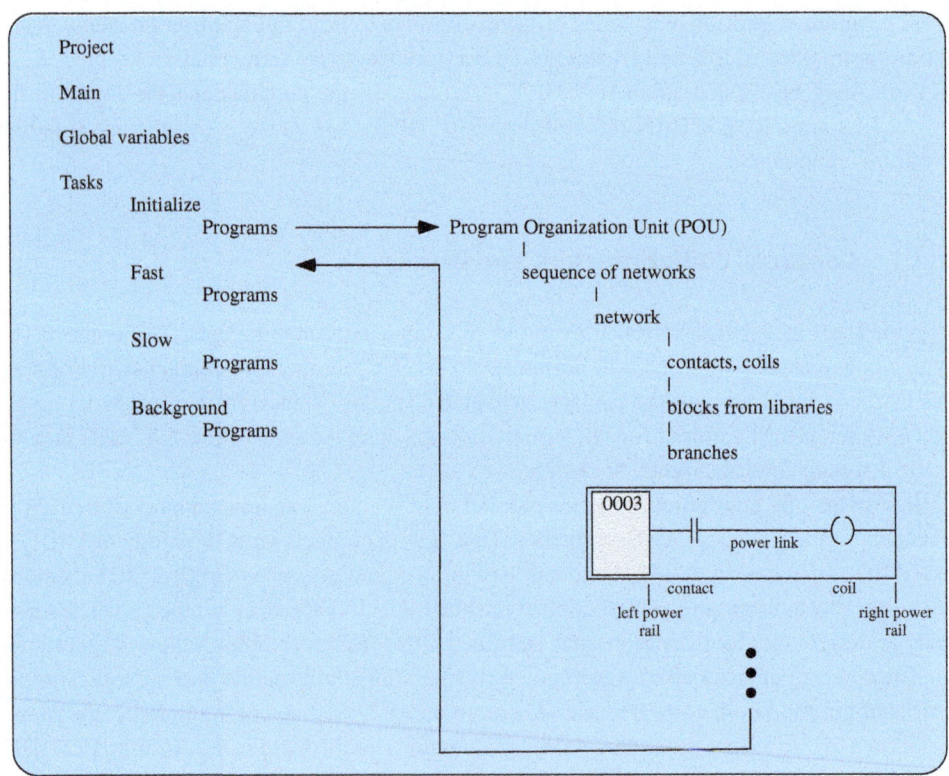

b) PLC program structure.

Fig. 5.2 Arduino PLC IDE program flow

5.4 Structure of Arduino PLC IDE Program

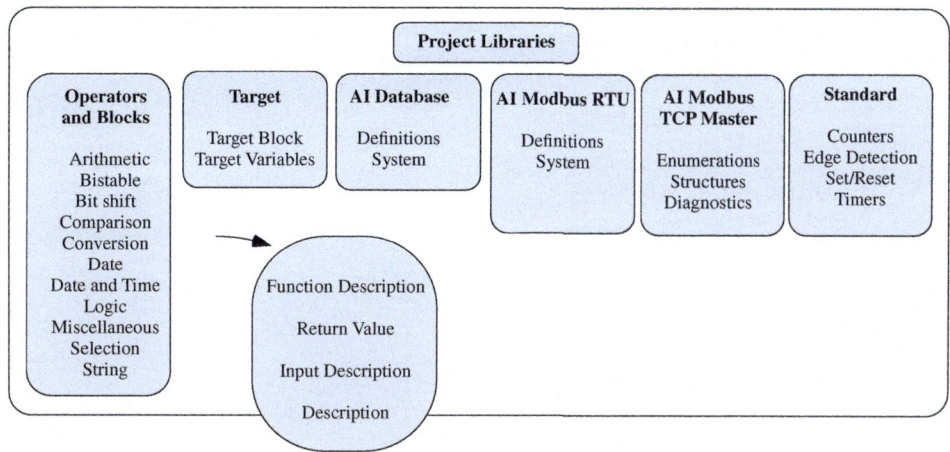

Fig. 5.3 Arduino PLC IDE Project Library

5.4.1.1 Counters

The Arduino PLC IDE features nine different predefined counters as shown in Fig. 5.5. There are counters that count down (CTD), count up (CTU), and up–down counters (CTUD). Each counter may either use signed 16–bit counters (INT), signed 32 bit counters (DINT), or unsigned 32–bit counters (DUINT). We explore the count–down timer (CTD) in some detail. Lessons learned may be applied to other counter types.[3]

The CTD timer is a signed, 16–bit counter. The CTD counts down by one (decrements) for each positive transition on the count–down (CD) input. The CTD has three inputs and two outputs:

- CD (Bool), input, count down on positive edge.
- LD (Bool), input, preset counter input, rising edge on LD sets Counter Value (CV) to Preset Value (PV).
- PV (Int), input, Preset Value, desired initial count of timer for count down.
- Q (Bool), output, count–down timer output. Set to logic one when counter value (CV) equals zero. Remains at zero until LD positive edge.
- CV (Int), output, current value of counter.

Reference Fig. 5.5 for a timing diagram illustrating counter operation.

---

[3] Detailed information for each counter type is accessible from within the Arduino PLC IDE Library.

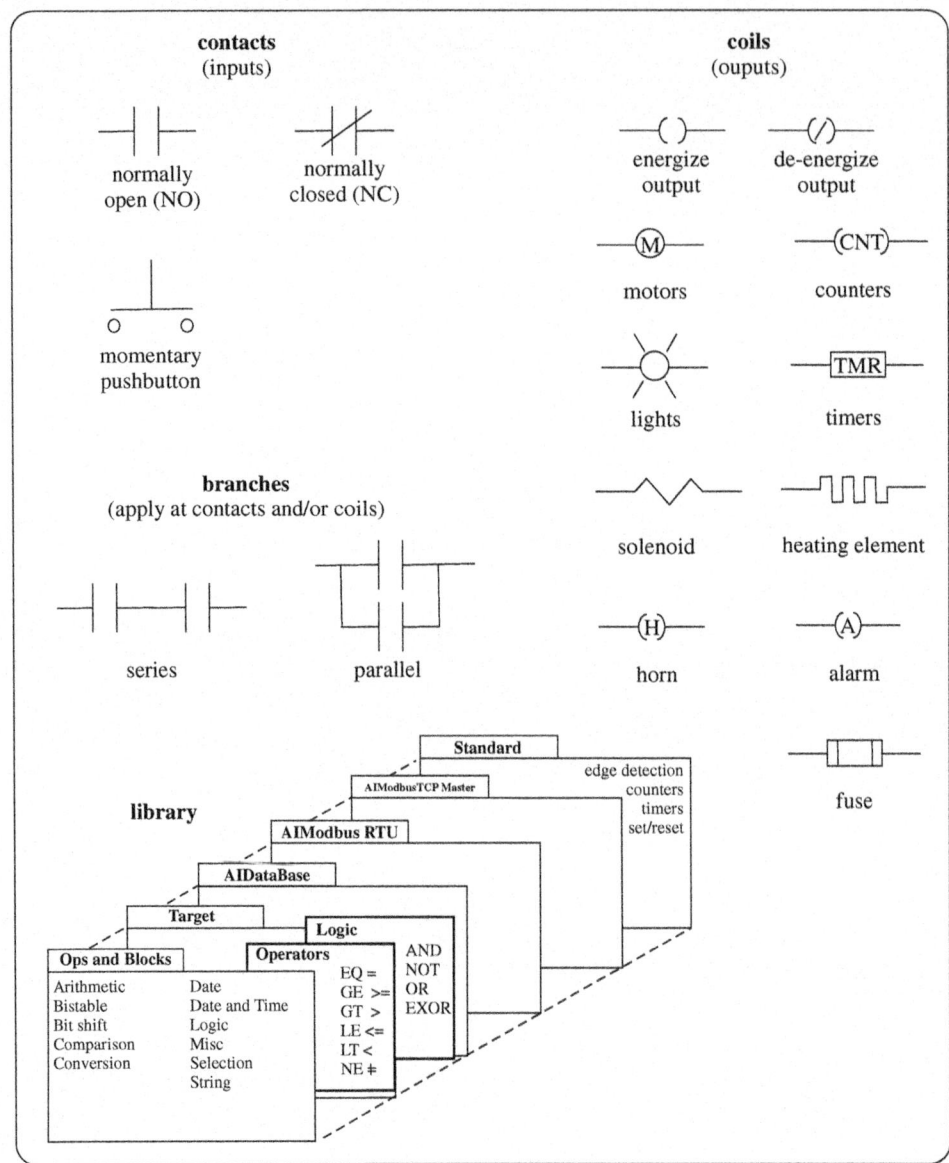

Fig. 5.4 Ladder logic network components

5.4 Structure of Arduino PLC IDE Program

Counter Type	Counter Current Value Type		
	INT	DINT	UDINT
down	CTD	CTD_DINT	CTD_UDINT
up	CTU	CTU_DINT	CTU_UDINT
up-down	CTUD	CTUD_DINT	CTUD_UDINT

Notes:
INT: signed, 16 bits
DINT : signed, 32 bits
DUINT: unsigned 32 bits

CTD - signed , 16 bit down counter
Counts down by one for each positive transition on the count down (CD) input.

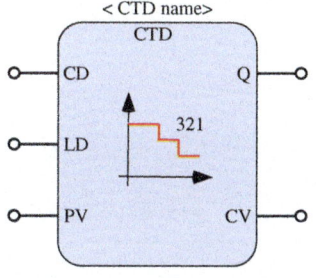

Q (BOOL): count-down output
PV (INT): preset value
CV (INT): counter current
LD (BOOL): preset counter input
CD (BOOL): count-down on positive edge

Notes:
1. Rising edge LD sets CV to value of PV.
2. Rising edge CD decrements CV.
3. When CV = 0, output Q set.
4. Q remains 0 until LD positive edge.

Fig. 5.5 Arduino PLC IDE counters

5.4.1.2 Timers

The Arduino PLC IDE features two different predefined timers: Off–delay Timer (TOF) and On–delay Timer (TON) as shown in Fig. 5.6. The TOF timer delays the deactivation of output Q by the preset time (PT) value in milliseconds. The TON timer provides a delay of PT milliseconds after IN becomes True before Q becomes True.[4]

5.4.2 LD Editor

Provided in Fig. 5.7 is a summary of launching a new ladder logic (LD) program within the Arduino PLC IDE and how to add and edit program rungs.

---

[4] Detailed information for each timer type is accessible from within the Arduino PLC IDE.

TOF - Off-delay Timer: provides delayed output (Q) deactivation (milliseconds) with respect to input IN.

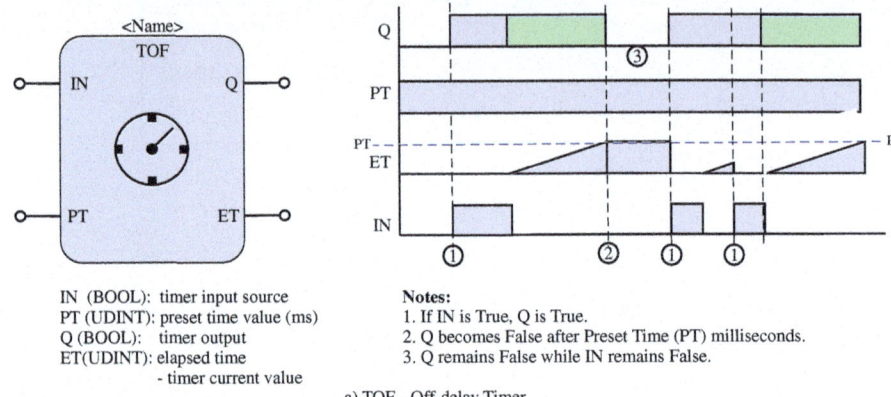

IN (BOOL): timer input source
PT (UDINT): preset time value (ms)
Q (BOOL): timer output
ET(UDINT): elapsed time
 - timer current value

Notes:
1. If IN is True, Q is True.
2. Q becomes False after Preset Time (PT) milliseconds.
3. Q remains False while IN remains False.

a) TOF - Off-delay Timer

TON - On-delay Timer: provides delayed output (Q) (milliseconds) of the input IN.

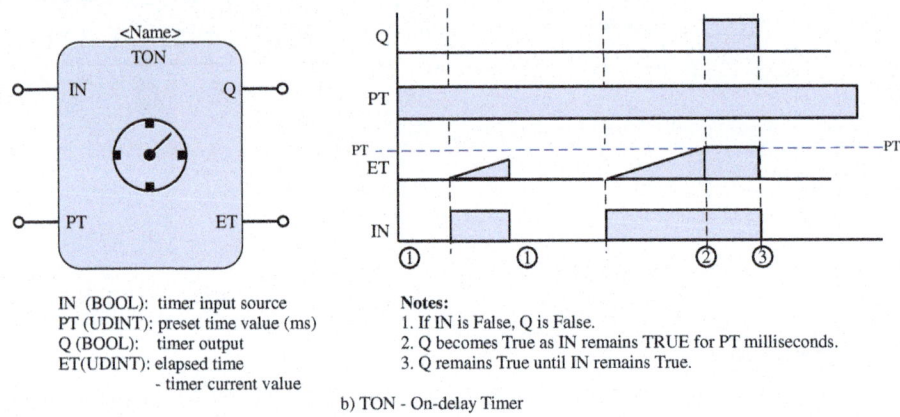

IN (BOOL): timer input source
PT (UDINT): preset time value (ms)
Q (BOOL): timer output
ET(UDINT): elapsed time
 - timer current value

Notes:
1. If IN is False, Q is False.
2. Q becomes True as IN remains TRUE for PT milliseconds.
3. Q remains True until IN remains True.

b) TON - On-delay Timer

Fig. 5.6 Arduino PLC IDE timers

5.5 LD Program Examples

In this section we provide several examples to illustrate the use of ladder logic programming to provide basic control. Figure 5.8 provides the schematic used in the upcoming examples. In the Application section of the chapter we assemble the circuit into a DIN rail mounted Test Fixture.

5.5 LD Program Examples

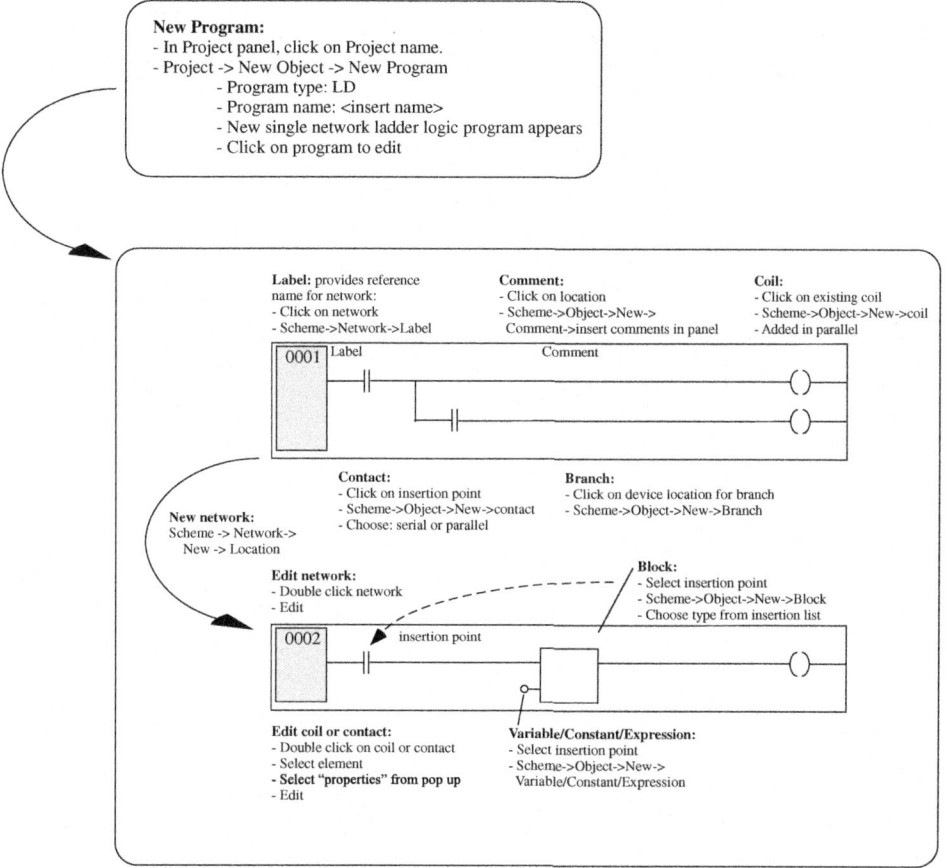

Fig. 5.7 Ladder logic network editing

Figure 5.9 provides the steps to launch an LD program. Follow the UML diagram to build the first ladder logic program. In the program, when the red pushbutton is pressed the red LED will illuminate as shown in Fig. 5.11a. The ladder logic is provided in Fig. 5.10.

Extend this basic example to implement and test the ladder logic circuits provided in Fig. 5.11b, c.

Figure 5.12 provides a basic counter configuration. In the example a count–down counter (CTD) is configured to start at the count of seven. The count is initially loaded by pressing PB1 connected to the preset counter input (LD). The CTD decrements for every push button press of PB2. When the counter reaches zero, output Q goes to logic high and illuminates the LED connected to Output_1.

Figure 5.13a provides a basic Off–delay Timer (TOF) configuration. In the example the TOF is configured with a two second (2000 ms) delay. When pushbutton PB3 is pressed the Q output goes logic high illuminating the LED at Output_2. When PB3 is released the

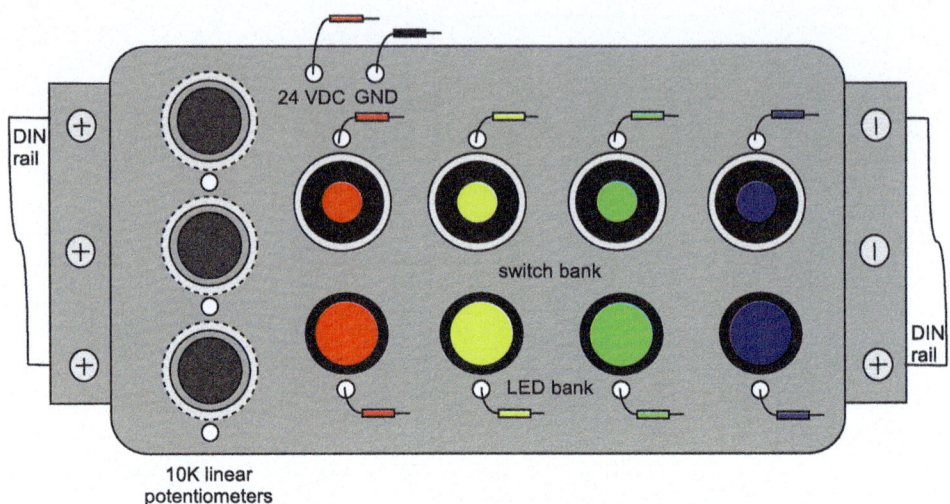

a) PMC test fixture.

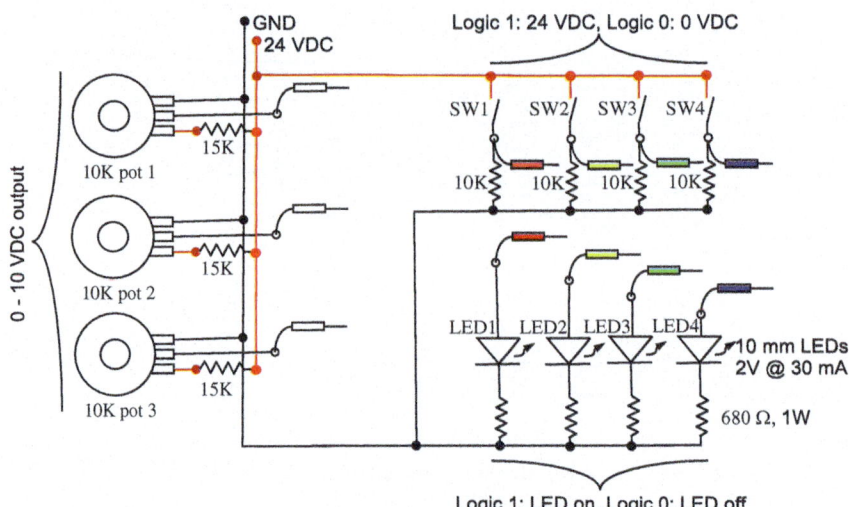

b) PMC test fixture schematic.

Fig. 5.8 Ladder logic test circuit

5.5 LD Program Examples

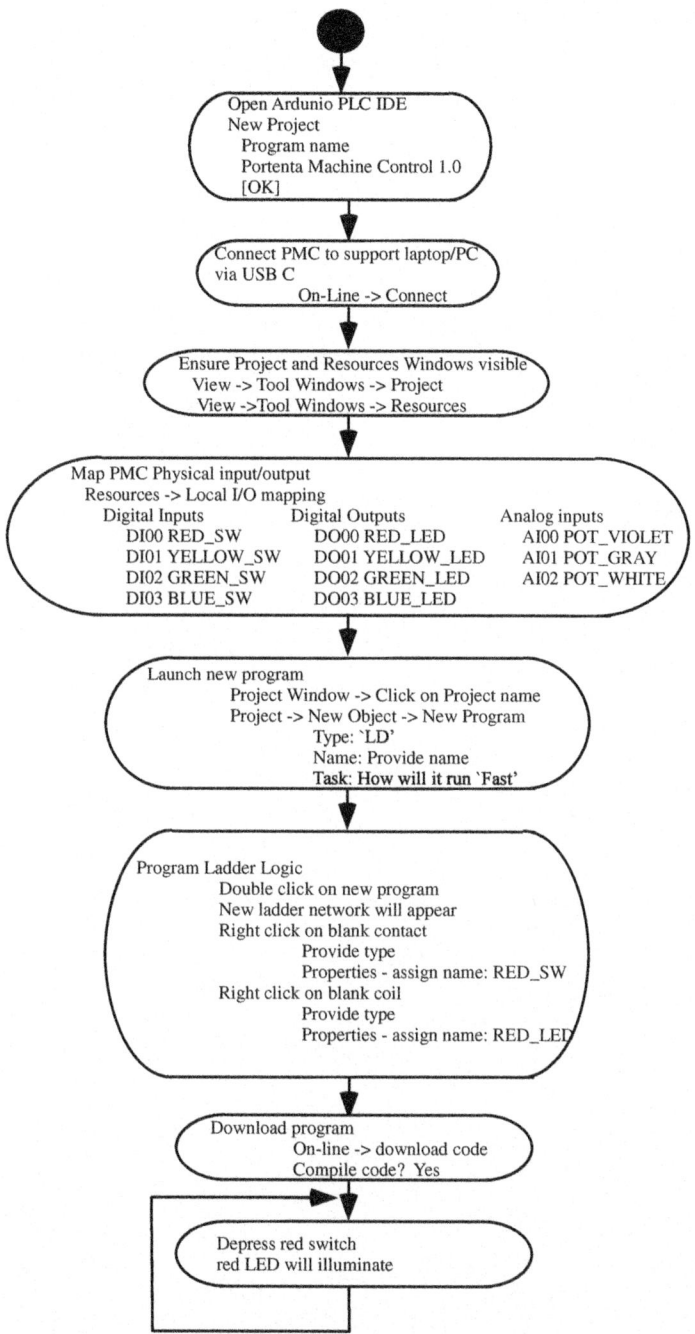

Fig. 5.9 Ladder logic program

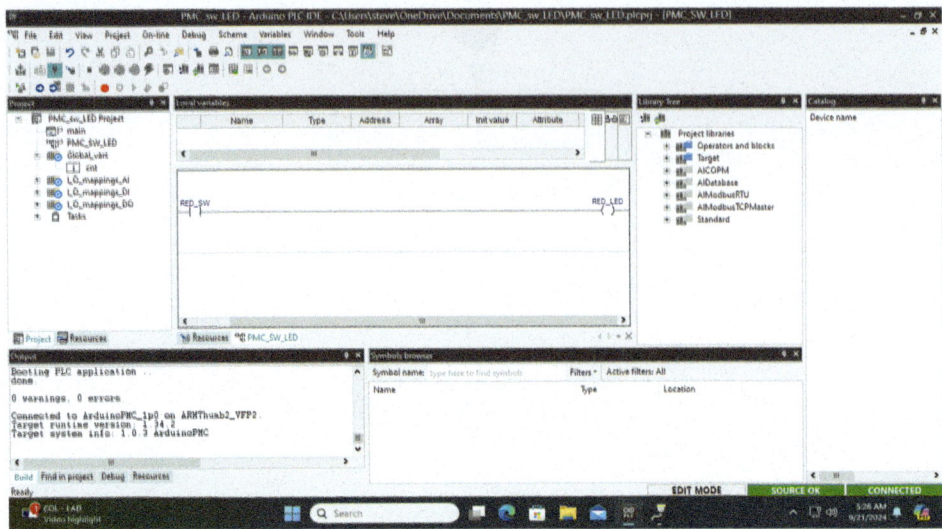

Fig. 5.10 Ladder logic program

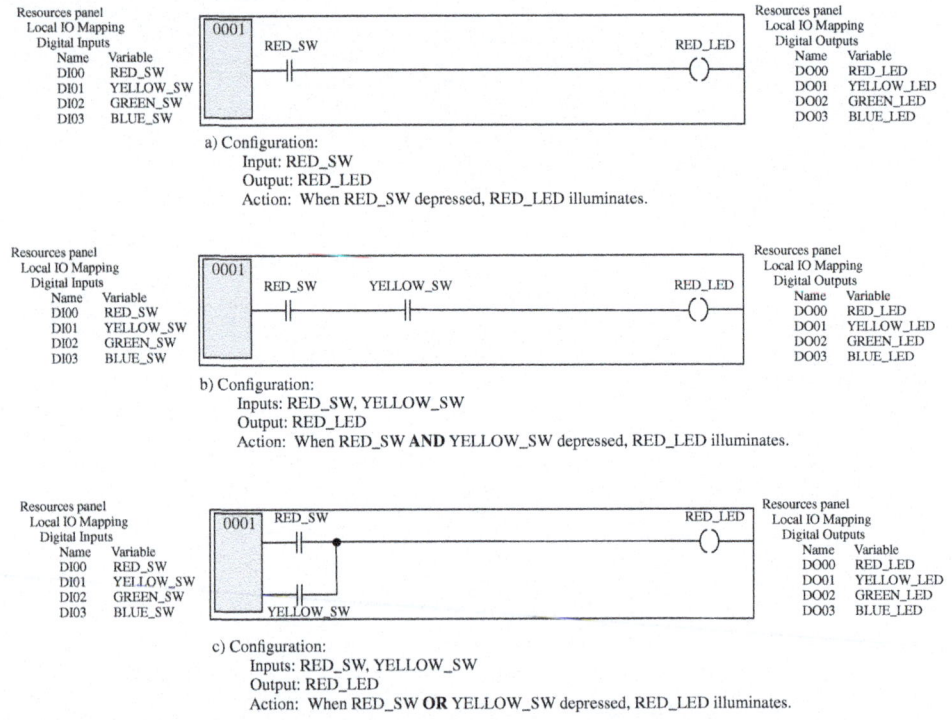

Fig. 5.11 Basic ladder logic examples

5.6 Ladder Logic Examples

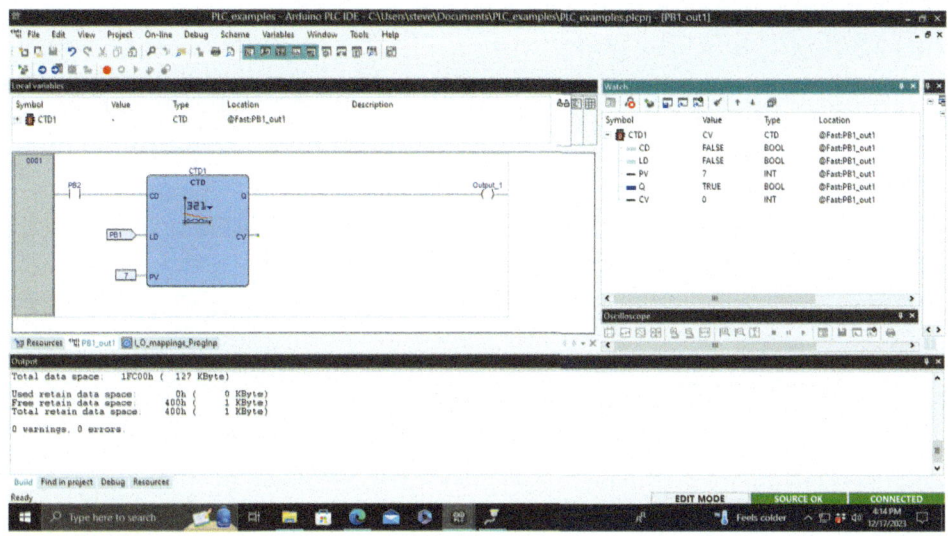

Fig. 5.12 Basic ladder logic counter example

LED stays illuminated for another two seconds. In Fig. 5.13b note how the output Q from the count–down timer CTD1 serves as the enable signal for the TOF timer in place of PB3.

Figure 5.14 provides analog–to–digital (ADC) conversion example. A greater than (>) comparison operator is used to compare the analog voltage provided by a potentiometer connected to Input 5 (Pot 2) to a threshold value of 30,000. Input 5 is configured for 16–bit analog conversion so values range from 0 to 65,535 ($0 \; to \; 2^{(b-1)}$) depending on the position of the potentiometer knob. When the Input 5 value exceeds 30,000 to LED connected to Output_2 illuminates. The Pot2 value may be observed using the Watch Window. The potentiometer represents any number of analog sensors (e.g. light, temperature, etc.). We discuss a wide variety of analog sensors in the next chapter.

5.6 Ladder Logic Examples

5.6.1 LM34 Temperature Sensor

In this example an LM34 is used to measure ambient temperature. When the temperature exceeds 90 °F, a fan turns on. The fan remains on until the ambient temperature drops below 70 °F. The difference between the fan on setting and off setting is called hysteresis. The two values are different to ensure the fan is not constantly turning on and off near the set point. The circuit diagram is provided in Fig. 5.15 and the accompanying ladder logic program in Fig. 5.16.

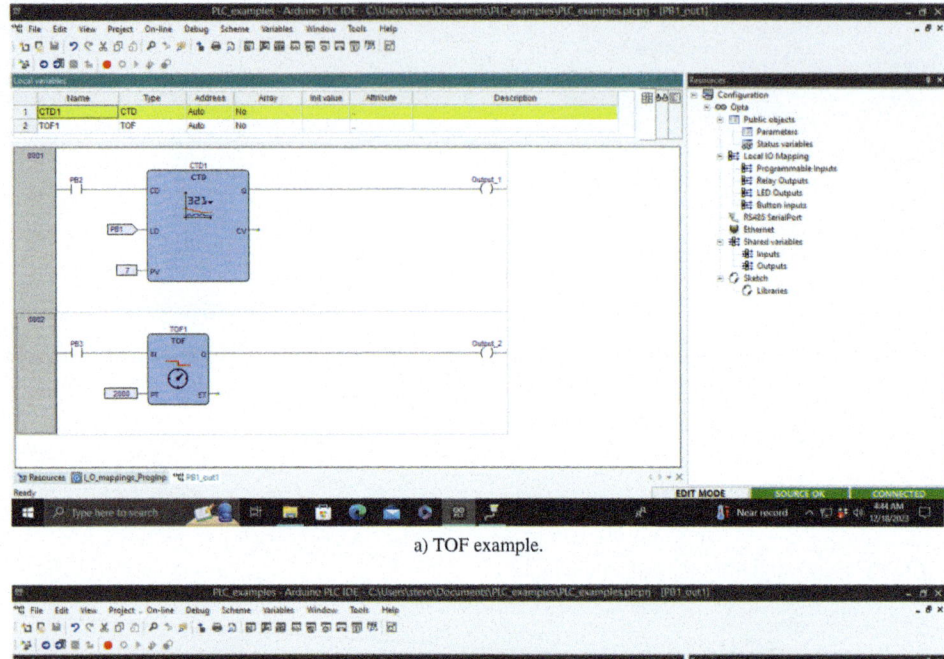

a) TOF example.

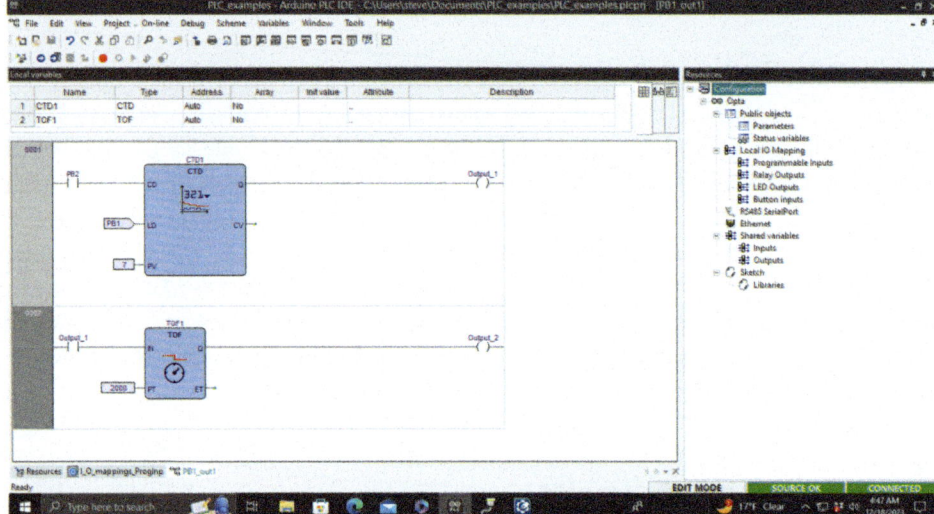

b) Count down counter output as input for TOF.

Fig. 5.13 Basic ladder logic timer examples

5.6 Ladder Logic Examples

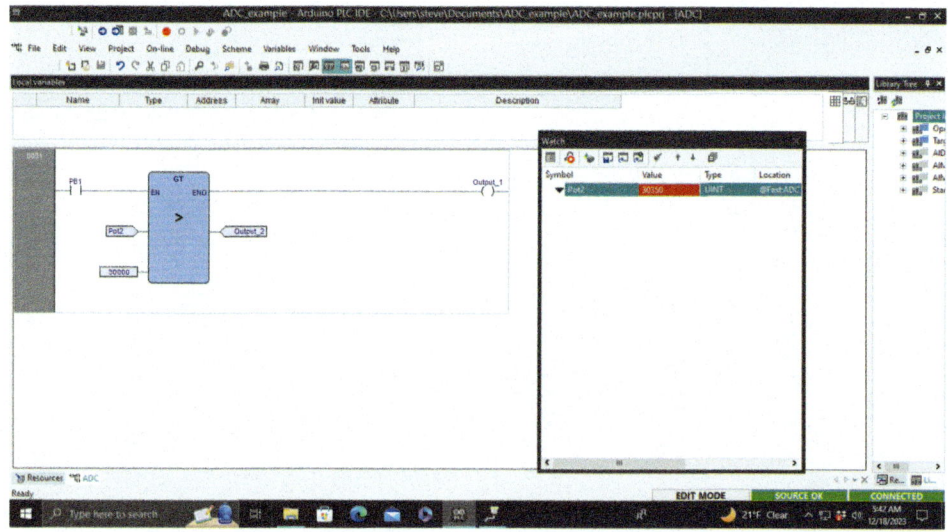

Fig. 5.14 Basic ladder logic ADC example

The LM34 temperature sensor output is fed directly to PMC input AI0. The fan and LEDs (Y, G, B) are connected to PMC outputs DO 00 to 03. The thresholds to turn the fan on and off are determined by:

- Equating the temperature reading to LM34 output (e.g. 90°C provides 900 mV).
- Determining the output from a 12–bit ADC for this voltage.
- The ADC output at the specified voltage is the threshold.

$$threshold = (0.900/10.0) \times 4096) = 370$$

5.6.2 Smoke Detector

The ladder logic sketch provided in 5.17 senses Vout and activates a Sonalert when smoke is detected at a set sensitivity level. The threshold value for activation is set for Vout at 1.0 VDC.

The MQ series of sensors may be calibrated to provide a reading in gas parts per million, or "PPM." The interested reader is referred to procedures in the appropriate data sheet (Egironi 2018).

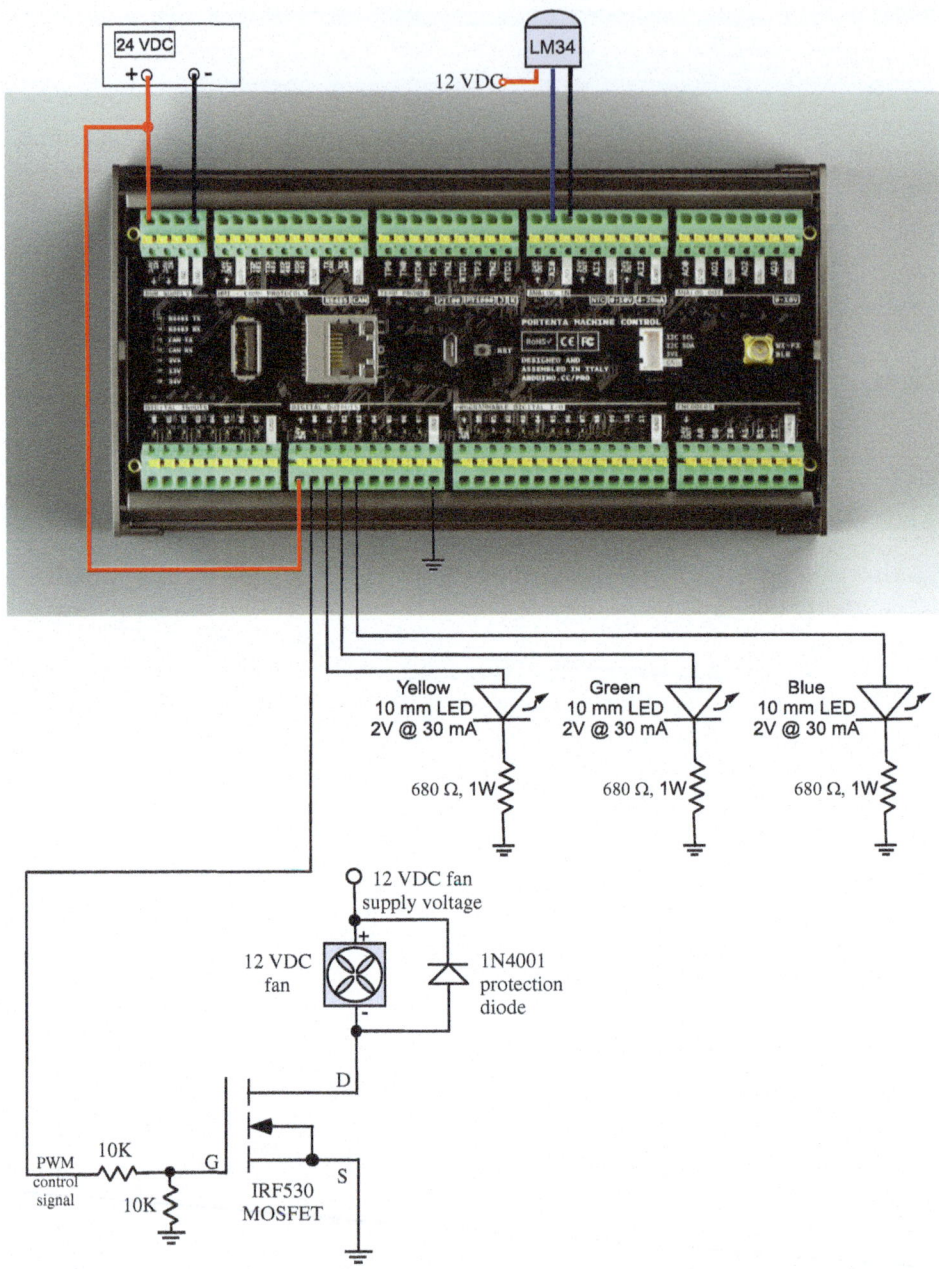

Fig. 5.15 LM34 fan controller. Images used courtesy of the Arduino Team (CC BY–NC–SA) (www.arduino.cc)

5.7 Application I: Sequencer Control Logic

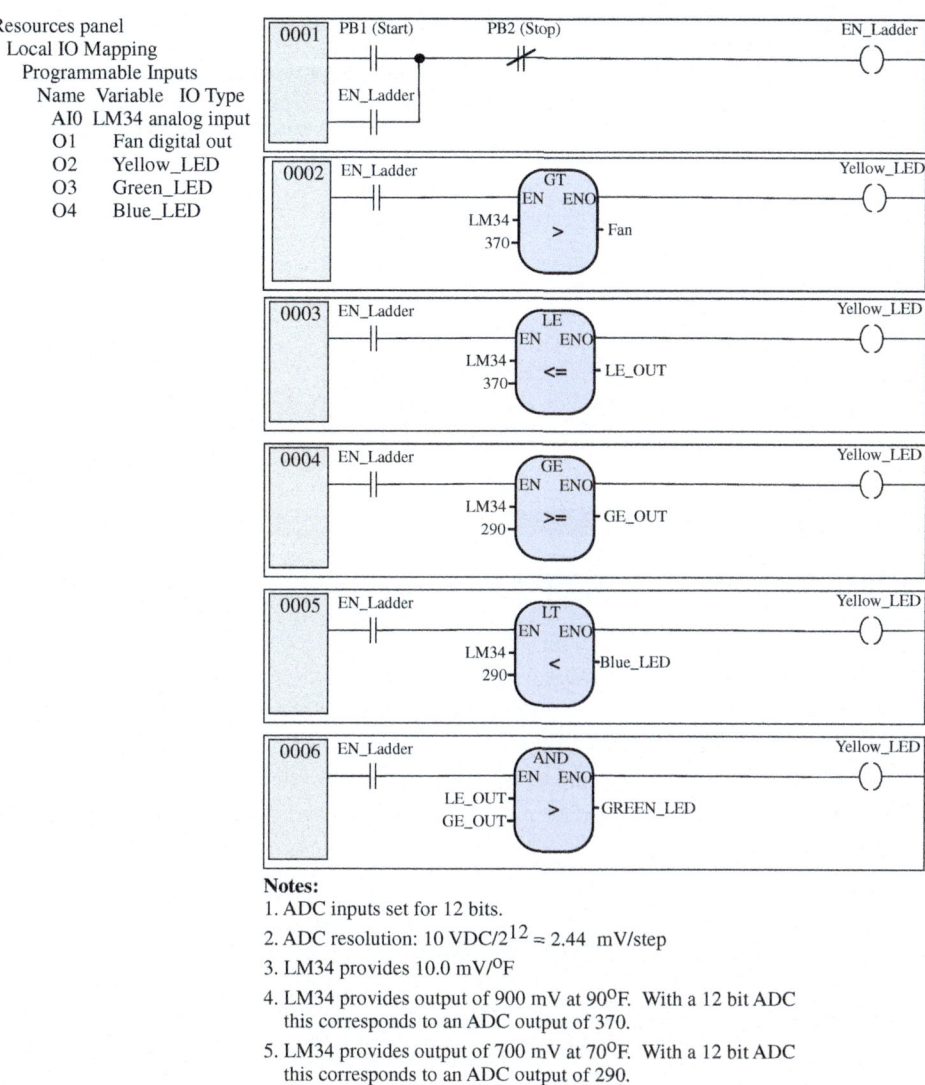

Notes:
1. ADC inputs set for 12 bits.
2. ADC resolution: 10 VDC/$2^{12} \approx 2.44$ mV/step
3. LM34 provides 10.0 mV/°F
4. LM34 provides output of 900 mV at 90°F. With a 12 bit ADC this corresponds to an ADC output of 370.
5. LM34 provides output of 700 mV at 70°F. With a 12 bit ADC this corresponds to an ADC output of 290.

Fig. 5.16 LM34 fan controller ladder logic

5.7 Application I: Sequencer Control Logic

In many control applications, a sequencer is used to implement a repeatable fixed sequence of control signals as shown in Fig. 5.18a.

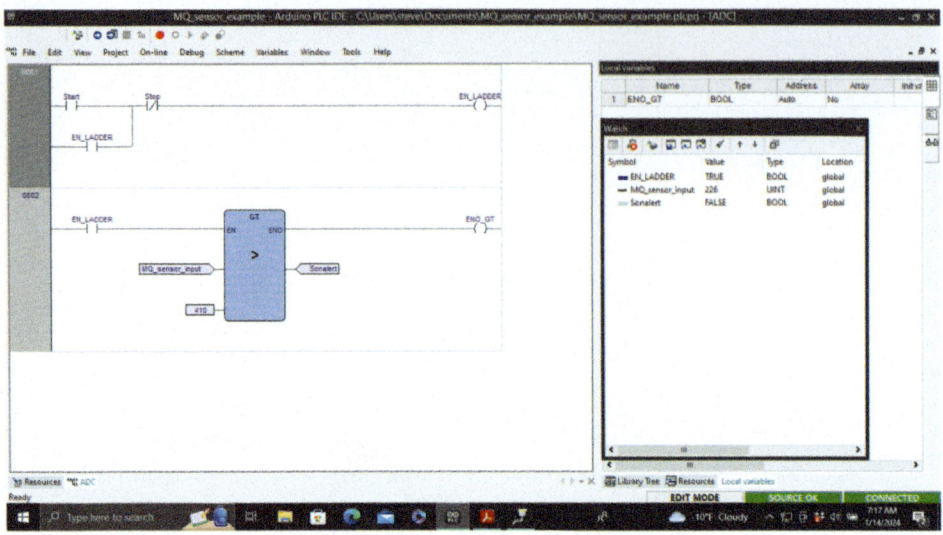

Fig. 5.17 MQ sensor ladder logic sketch

A sequencer is implemented using a series of binary counters operating at different clock frequencies. The output from the counters are provided to combinational logic circuitry to generate the desired control signal sequence as shown in Fig. 5.18b.

To implement a binary counter in ladder logic, cross coupled timers (e.g. TON) are used. The output of the configuration provides a square wave output as shown in Fig. 5.18c (BharadwajReddy).

5.7.1 Stepper Motor Control–Ladder Logic Sequencer

To implement a stepper motor controller in ladder logic, two counters (A_Light and B_Light) are used. The first counter operates at twice the frequency of the second counter. The two counters with their accompanying inverted outputs (A_Light' and B_Light') provide the output sequence shown in Fig. 5.19a.

To achieve the desired stepper motor control signal as shown in Fig. 5.19c, the outputs from the counters are combined using combinational logic as shown in Fig. 5.19b. The example implemented with the Arduino PLC IDE is shown in Figs. 5.20 and 5.21.

5.7 Application I: Sequencer Control Logic

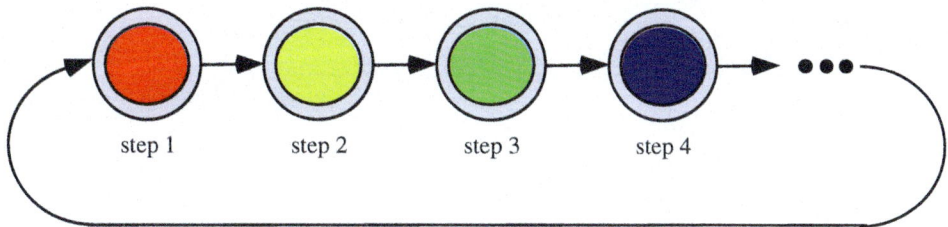

a) Sequence.

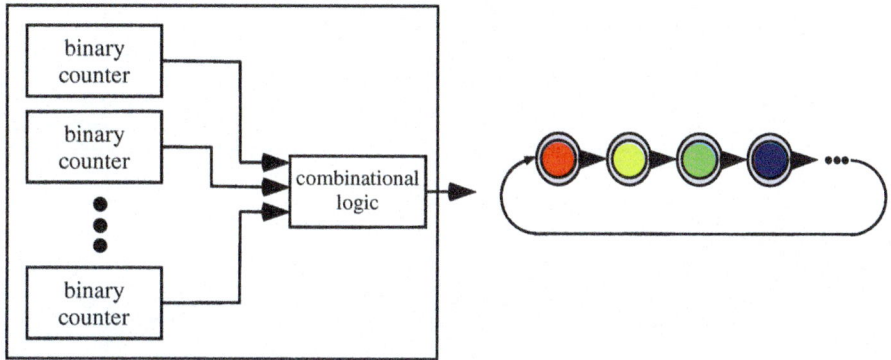

b) Sequence generator.

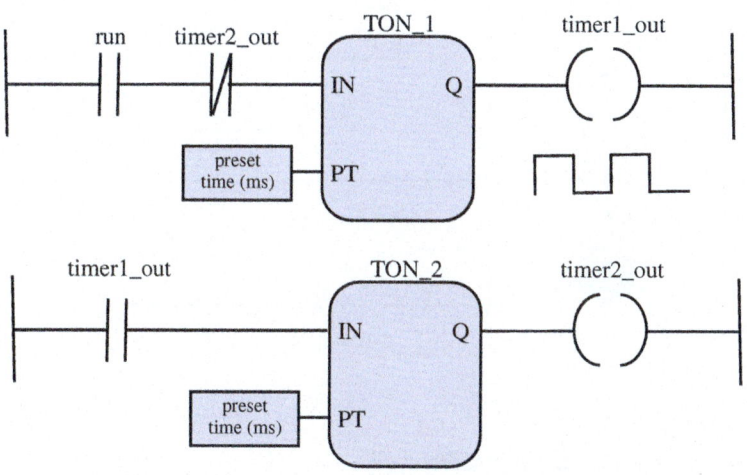

c) Cross coupled TON timers.

Fig. 5.18 Sequencer control logic

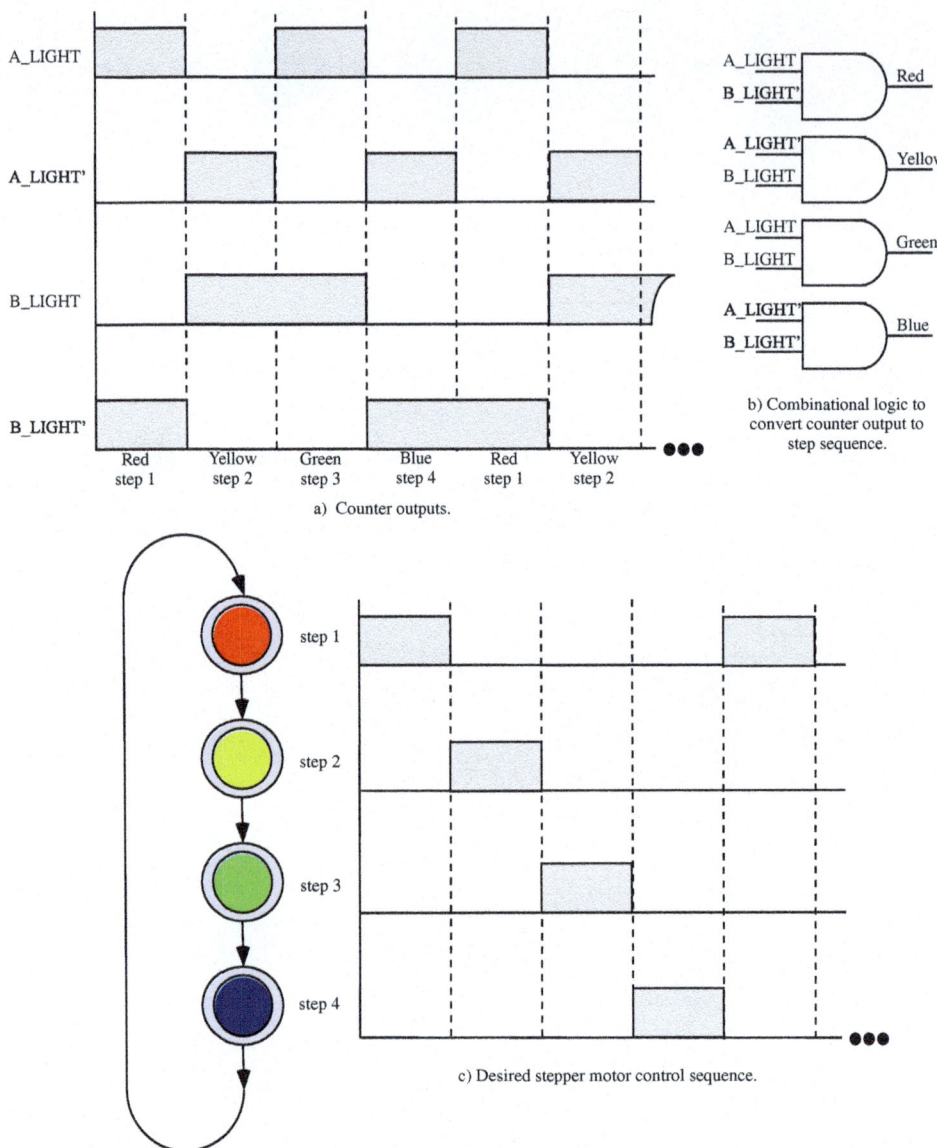

Fig. 5.19 Stepper motor controller in ladder logic

5.8 Application II: Test Fixture

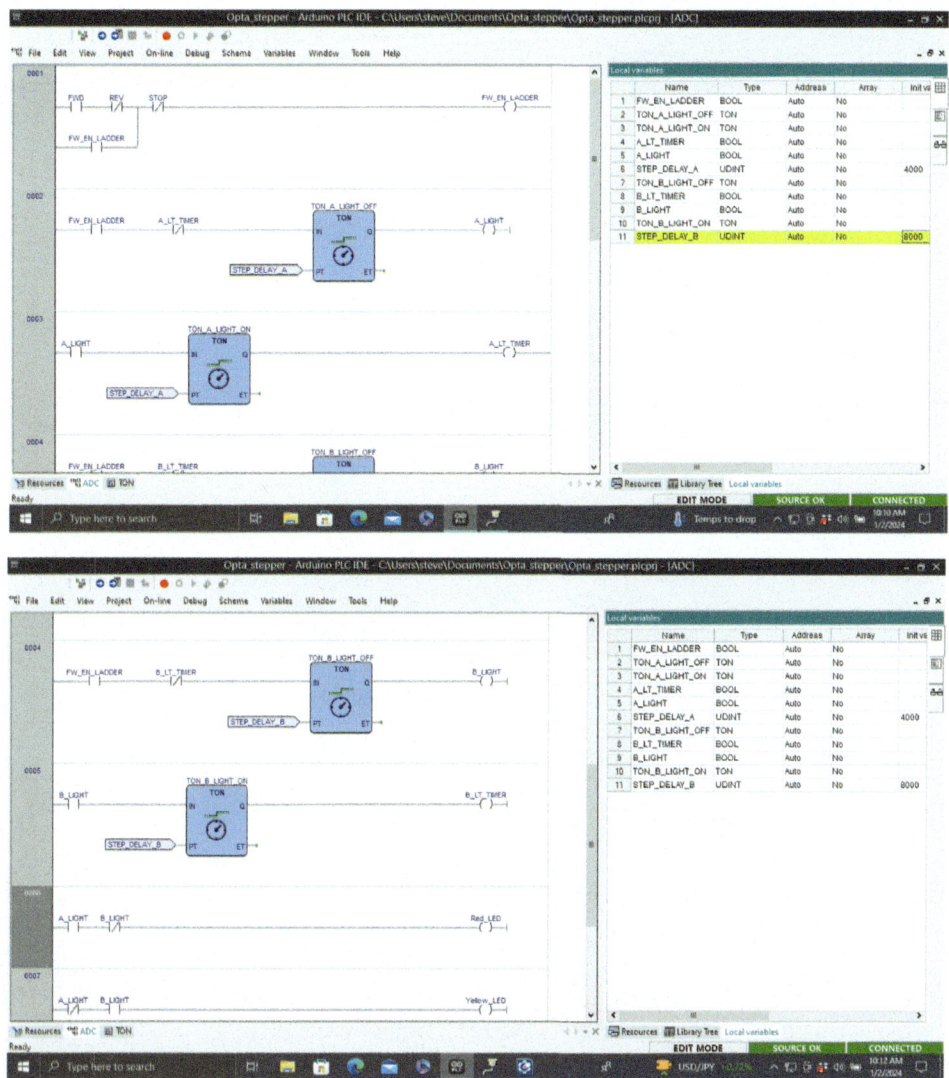

Fig. 5.20 Stepper motor controller in ladder logic. Using PLC IDE

5.8 Application II: Test Fixture

Provided in Fig. 5.8a is the test fixture to easily inject input signals and track outputs from a PMC executing a ladder logic program. The test fixture consists of a series of three 10K Ohm potentiometers, four light emitting diodes (LEDs), and four pushbuttons. The circuit is housed within a plastic chassis box mounted to a DIN rail via DIN rail mounting adaptors. The completed fixture is shown in 5.22.

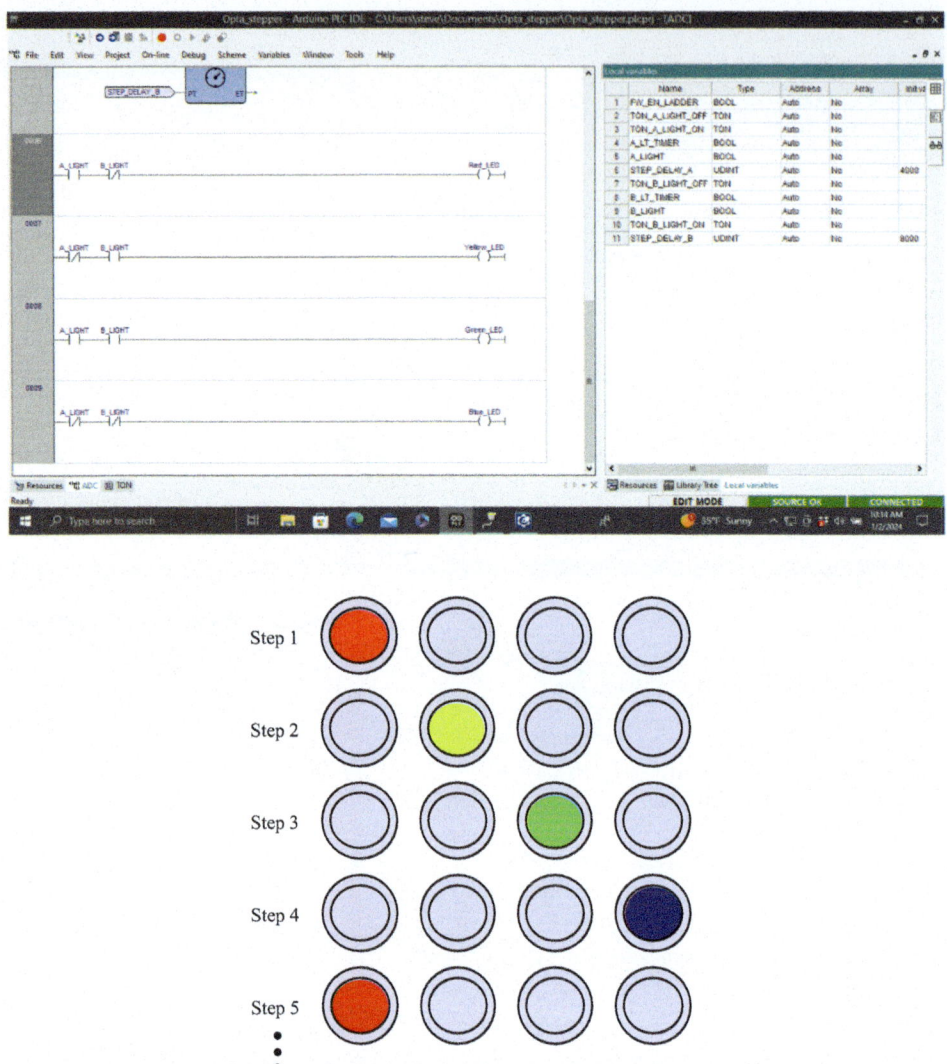

Fig. 5.21 Stepper motor controller in ladder logic. Using PLC IDE

5.9   Application III: Greenhouse Temperature Sensing System

Figure 5.23 provides a ladder logic implementation of a system to detect when a given physical variable (e.g. temperature, water level, etc.) is too high, too low, or within a desired range. We use the Test Fixture to test the ladder logic program.

5.9 Application III: Greenhouse Temperature Sensing System

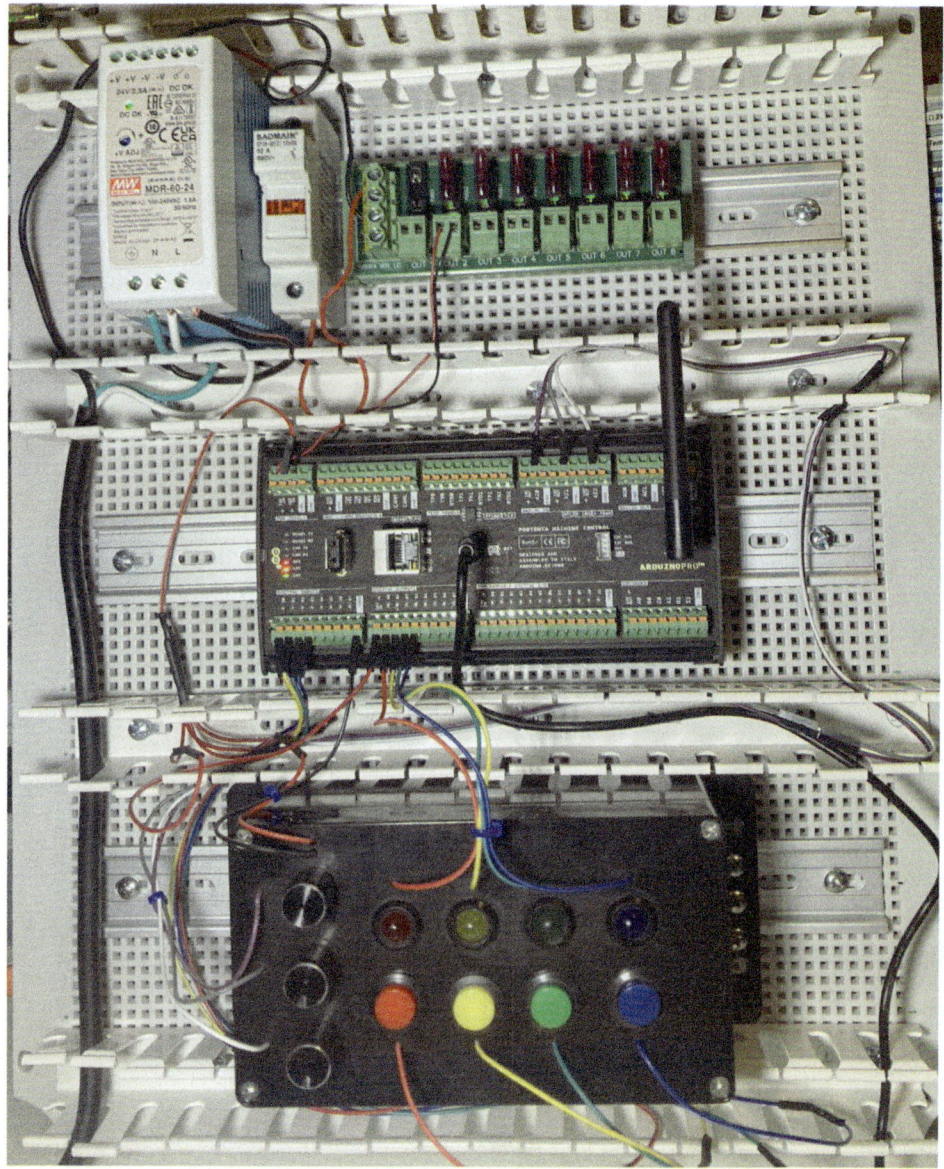

Fig. 5.22 Ladder logic test fixture on DIN rail

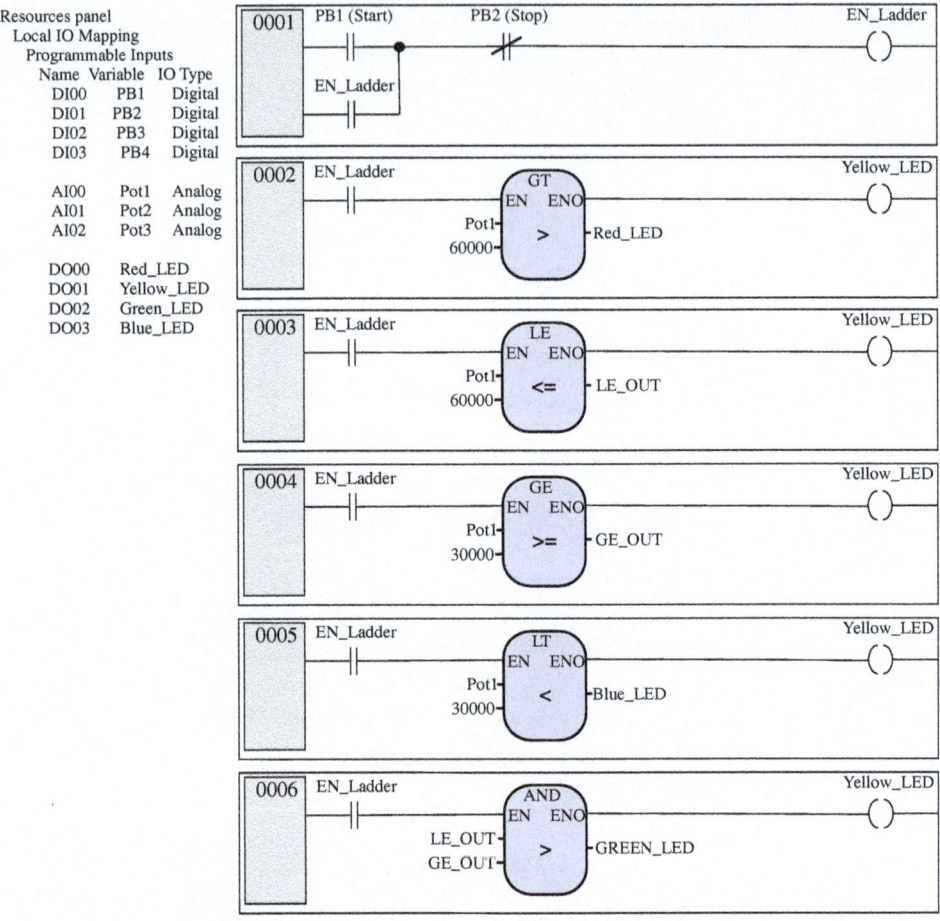

Fig. 5.23 Ladder logic sensing system

Here is a summary of ladder logic operation:

- Resource definitions for inputs and outputs are shown at the top left of the figure. Also, local variables of type Bool are declared for EN_Ladder, LE_OUT, and GE_OUT.
- Rung/network 0001 contains two contacts PB1 (NO) and PB2 (NC). The PB1 input serves as the Start PB for the process while PB2 serves as the STOP PB. When PB1 is pressed, the coil EN_Ladder is asserted and seals the PB1 switch ON via the parallel branch containing EN_Ladder. The process remains active until PB2 STOP is pushed.
- Rung/Network 0002 contains the greater than (>) comparison operator. It is used to compare the analog voltage provided by a potentiometer connected to AI00 (Pot 1) to a threshold value of 60,000. AI00 is configured for 16–bit analog conversion so values

range from 0 to 65,535 ($0 \ to \ 2^{(b-1)}$). When the AI00 value exceeds 60,000 the Red LED illuminates. The Pot 1 value may be observed using the Watch Window. It is important to note the potentiometer represents any number of analog sensors.
- Rung/Network 0003, 0004, and 0006 work together to illuminate a Green LED if the potentiometer connected to AI00 (Pot 1) provides an output between 60,000 and 30,000. Note how the LE_OUT and the GE_OUT signals are ANDed together to form the Green_LED output.
- Rung/Network 0005 illuminate a Blue LED if the potentiometer connected to AI00 (Pot 1) provides an output less than 30,000.

5.10 Summary

In this chapter we explored the Arduino PLC IDE. We began with a review of the five PLC programming languages specified within the IEC IEC61131–3 standard and available within the Arduino PLC IDE. We then narrowed our focus and concentrated on ladder logic programming. We explored, downloaded, and configured the Arduino PLC IDE and related tools. We then explored the fundamentals of ladder logic PLC programming. The fundamentals were used to complete a series of examples.

5.11 Chapter Problems

1. The Arduino PLC IDE allows for programming of PLCs using the five different languages within the IEC IEC61131–3 standard. Prepare a table listing each type of programming language along with its pros and cons.
2. Describe the difference between ladder logic contacts, coils, branches, and blocks. Provide an example of each.
3. Modify the ladder logic program above so the LEDs sequentially cycle left to right and then back right to left.
4. There are many predefined operations, blocks, and standard functions defined with the PLC IDE Library. Provide a summary table of these library elements.
5. Describe the difference between the TOF and the TON timers. Write a ladder logic program to demonstrate the difference between the two timers.
6. Design a ladder logic program providing a signal when 12 cans on a conveyor has passed a sensor. The signal is used to start a new box. The boxes are packed 144 per palette. Provide a signal when 144 boxes have been filled with cans.
7. Develop a ladder logic circuit to indicate the fluid level in a vat. Provide output signals when the vat is nearing empty or nearing full. An analog sensor is used to indicate vat fluid level.

8. A small corporation is owned by four individuals. The distribution of the corporation shares is as follows: Mr. Quine owns four shares, Mrs. Karnaugh owns three shares, Mr. Boole owns two shares, and Ms. McCluskey owns one share. You are to build a voting machine that will be used when the owners vote on corporate issues. Each of the owners has a switch that is closed to record a 'yes' vote and opened to record a 'no' vote. When one of the owners casts a 'yes' vote, that vote is weighted by the number of the owner's shares. For example, when Mrs. Karnaugh votes 'yes', her vote counts three times that of Ms. McCluskey. The output of the voting machine will be a visual indication–maybe a LED that is lit? When the result of any vote is a 'yes' by the majority of the shareholders. If the issue does not get a majority vote, the indicator LED will not light. If the issue results in a tie vote, a separate LED will be lit to indicate this tie result. Develop a ladder logic program to implement the voting machine.[5]

References

1. *Arduino Opta Collective Datasheet,* Product Reference Manual, SKU: AFX00001– AFX00002– AFX00003, arduino.cc, June 2023.
2. *Arduino PLC User's Manual,* arduino.cc (Accessible via the Arduino PLC IDE help tab.).
3. P. Marquinez, *Arduino PLC IDE Setup & Device License Activation,* arduino.cc.
4. *Programming Introduction with Arduino PLC IDE–Create Programs with all the IEC–61131–3 languages on the Arduino PLC IDE,* Finder S.p.A., opta.findernet.com.
5. J. Bagur and J.C. Linares, *Opta User Manual,* arduino.cc.

---

[5] Hint: Employ Boolean logic simplification techniques (e.g. Karnaugh maps) to determine equations for a successful vote and tie. Implement the equations with PLC IDE logical operators. Scenario courtesy of Alex Dwellis, Colonel, USAF retired.

Application: IoT Greenhouse   6

Objectives: After reading this chapter, the reader should be able to do the following:

- Apply instrumentation, IoT, and PLC concepts to design a control system;
- Describe tools used to systematically design hardware and software tools for a control project; and
- Implement the control system with both the Arduino IDE and PLC IDE.

6.1 Objective

The objective of this chapter is to demonstrate the concepts discussed in this book. Simply put, our goal is to provide the theory, design, and construction of a passively heated greenhouse. We equip the greenhouse with instrumentation to monitor and control key parameters using the Arduino Portenta Machine Control. The concepts may be adapted for many other non–greenhouse systems.[1]

We begin the chapter with the theory of greenhouse design. Prior to this project, the author had no background in greenhouse design and construction. However, it was a project considered for some time. The reader is also assumed to have no background in this area. This section was compiled from a number of excellent sources listed at the end of the chapter. We then present the design and construction of a passively heated greenhouse. In our example an existing eight–by–eight–foot garden shed is converted into a greenhouse. This is a do it yourself (DIY) project.

---

[1] This chapter was adapted for the Arduino PMC with permission from "Arduino III: Internet of Things," S. Barrett, Springer, 2021.

© The Author(s), under exclusive license to Springer Nature Switzerland AG 2026
S. F. Barrett, *Arduino VIII*, Synthesis Lectures on Digital Circuits & Systems,
https://doi.org/10.1007/978-3-031-85944-1_6

The chapter then provides details on an Arduino PMC based greenhouse control system. Finally, key greenhouse parameters are made available for monitoring and control using IoT concepts. Prior to continuing, the reader is encouraged to review microcontroller–based system design concepts provided in Appendix B.

6.2 Greenhouse Theory

There is considerable information available to guide the design of a passive greenhouse. A passive greenhouse uses solar energy to either extend the growing season for plants or to grow plants year round. Equally important is to provide for greenhouse cooling and ventilation during hot summer months. The information provided here is a compilation of the excellent sources listed at the end of this chapter. A thorough review of these sources is recommended.

Passive heating uses the energy of the sun to heat the interior of the greenhouse during the day. The heat energy is stored using a large thermal mass such as barrels filled with water as shown in Fig. 6.1. When the temperature drops at night, the energy stored in the water is released into the greenhouse to mitigate internal temperature fluctuations.

The British Thermal Unit (BTU) is the energy required to raise one pound of water a degree Fahrenheit. During the day, the water barrels absorb energy. At night a drop of one degree per pound of water would release one BTU of heat energy.

To efficiently capture the solar energy, greenhouse windows should face south. Windows may be included on the east and west sides as well. Typically, the north facing wall is not equipped with windows but is thoroughly insulated.

The south facing windows ideally should be inclined at an angle. A general rule of thumb is to take your location's latitude and add ten degrees. I live in Laramie, Wyoming (41.3114° N, 105.5911° W); therefore, ideally the south facing windows should be inclined at approximately 51°.

To determine the amount of water required for passive energy storage several rules of thumb are used:

- To extend the growing season, 2.5 gallons of water needs to be stored for every square foot of glazing material (windows).
- To grow plants year round, five gallons of water needs to be stored for every square foot of glazing material.
- It is worth noting that a gallon of water weighs 8.3 pounds.

To maximize the collection of solar energy, it was decided to place windows on the south, east, and west walls of the greenhouse. In this specific example, an eight–by–eight–foot existing garden shed was converted to a greenhouse.

6.2 Greenhouse Theory

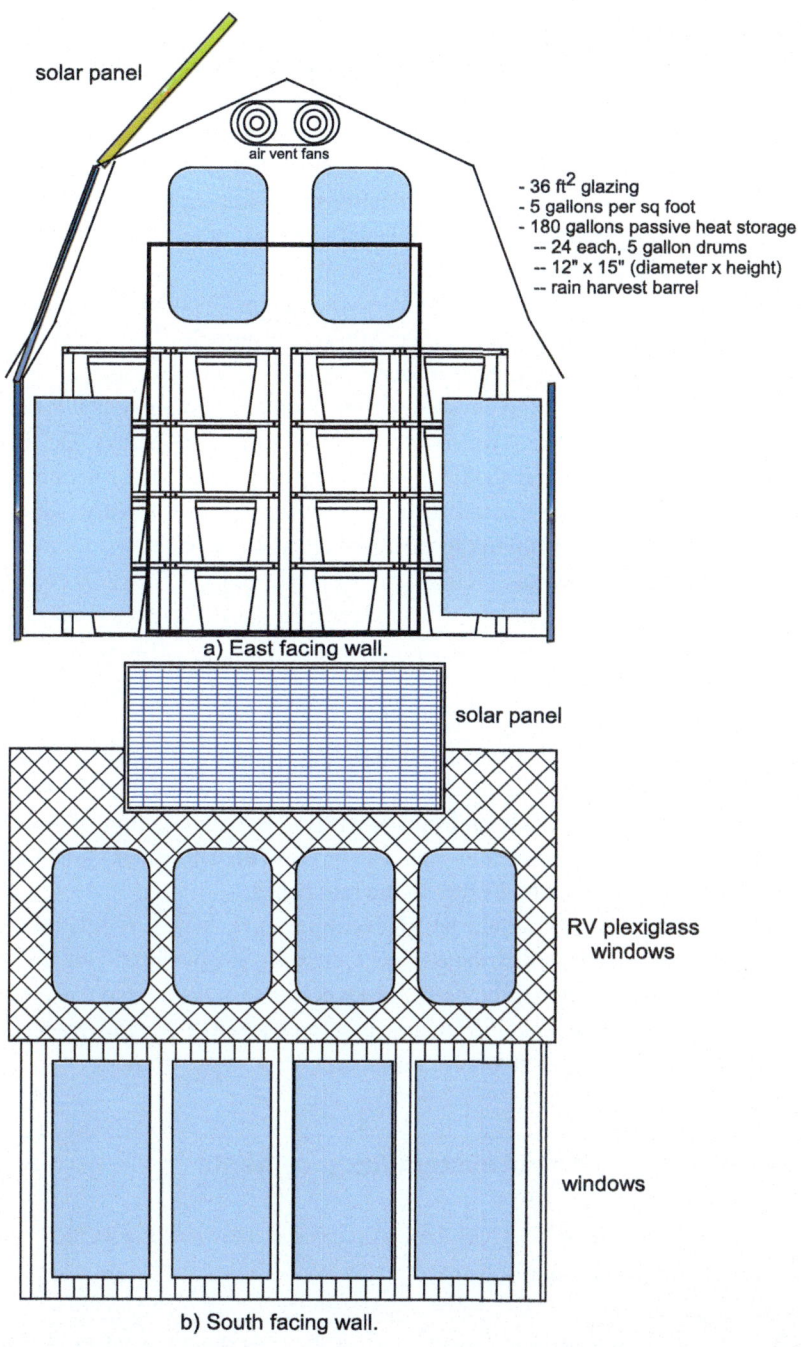

Fig. 6.1 Greenhouse concepts

A combination of mobile home glass windows and plexiglass bubble (RV skylight) windows were installed. The total glazing area was 36.2 square feet. To provide for year–round growing, 180 gallons of water is required for passive storage. A total of 24 five–gallon, black plastic buckets were used. The balance of water storage was accomplished using a 55–gallon rain barrel. The buckets were placed on shelves within the greenhouse. Realize the greenhouse floor will be supporting 180 gallons (1500 pounds) of water. Ensure the floor is adequately supported to do this. I included some additional $2'' \times 4''$ (inch) support framing under the floor for this purpose.

The shed has traditional $2'' \times 4''$ (inch) framing covered by half inch thick exterior wallboard. The roof of the shed was peaked and covered with roofing shingles as shown in Fig. 6.2 top left. The north facing wall of the shed is insulated with R–13 insulation.

The shed's interior is insulated and covered with wallboard in non–window locations. The interior roof was also insulated and covered with $1'' \times 4''$ (inch) pine planks. The interior was painted white for reflectivity. The shed floor was covered with a quarter inch black rubber mat. A 12 VDC exhaust fan was installed in the east facing eave. Finally, windows were covered on the shed interior with eighth inch plexiglass to provide an additional thermal barrier. The greenhouse exterior and interior are shown in Fig. 6.2.[2]

6.3 Water Harvesting

A rain barrel is used to capture water for greenhouse use. Typically, the roof of the greenhouse or a nearby structure is used to capture rain and direct it to a containment barrel. A rule of thumb for determining how much roof area is needed is a half–gallon of water may be collected per square foot of roof area for one inch of rain fall. The gutter system on the roof is helpful for directing the captured water to the rain barrel.

Water entering the rain barrel should be filtered to remove leaves, twigs, roof material, etc. Once collected the water is distributed for greenhouse use by employing a small water pump. A gravity feed system may be used; however, the rain barrel may need to be set at a prohibitive height to obtain suitable water pressure (Watson).[3] For this project we use a commercially available 50–gallon, plastic rain barrel for water catchment.

6.4 Greenhouse Control System Requirements

The Greenhouse Control System (GCS) block diagram is shown in Fig. 6.3.
The GCS has the following requirements:

---

[2] Springer is a global publisher. Due to the wide range of applicable building codes and standards and permitting requirements, please check with local requirements concerning permitting and installation requirements.

[3] Check local rules and codes to determine if water harvesting is permitted in your location.

Fig. 6.2 Greenhouse project

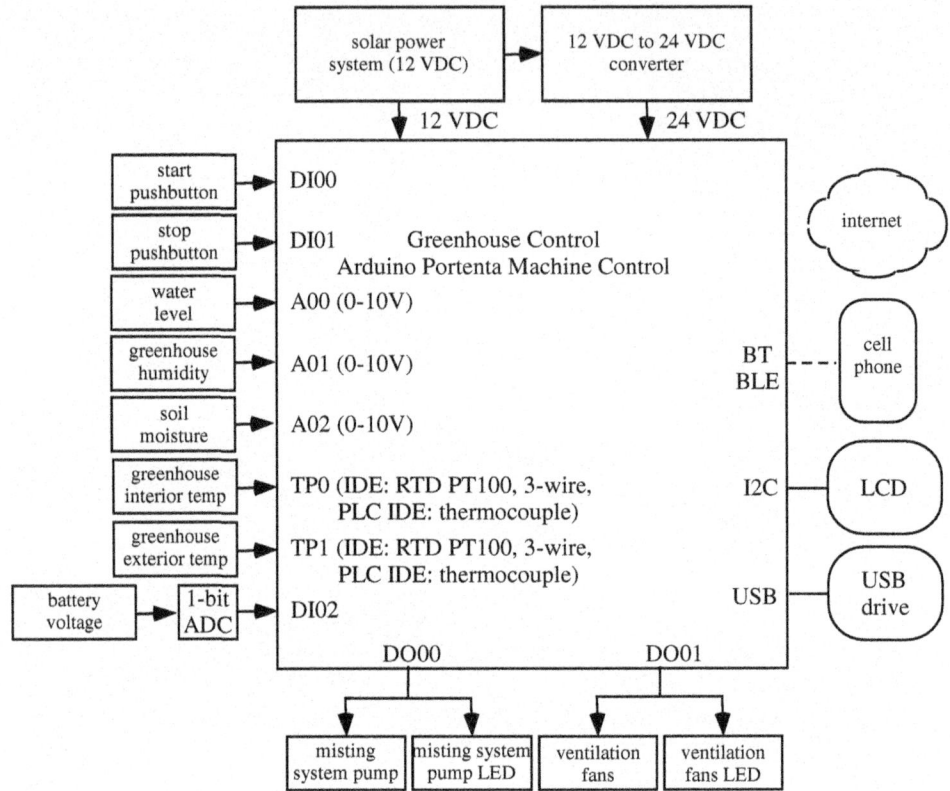

Fig. 6.3 Greenhouse control system

- Self–contained solar power with a 12 VDC battery backup;
- A 12–24 VDC converter Meanwell DDR–120A–24) to power the Portenta Machine Control;
- Compartmentalized layout for systematic design and expansion;
- Input sensors to measure rain barrel water level, humidity, soil moisture, temperatures, and battery voltage level;
- Arduino PMC based technology for greenhouse instrumentation and control;
- Output control for a misting system and ventilation fan;
- Cell phone and/or internet connectivity for greenhouse monitoring; and
- Control software written both with the Arduino IDE and the PLC IDE.

6.6 Greenhouse Control Subsystem

In keeping with a compartmentalized design, the GCS has three subsystems:

- Solar power subsystem;
- Greenhouse Control Subsystem; and
- Arduino PMC PLC based interface subsystem.

Each subsystem is discussed in turn.

6.5 Solar Power Subsystem

For the greenhouse project several power supply options were considered as shown in Fig. 6.4. In Fig. 6.4a a DIN rail mounted, 24 VDC power supply operating, from 115 VAC provides system power. Additional voltage regulators are used to provide 12 VDC and 5 VDC. This is the power system used during project development.

In Fig. 6.4b, c, solar power is used as the primary source. In Fig. 6.4b a 24 VDC solar power system is used to provide 24 VDC, 12 VDC and 5 VDC for the project. In Fig. 6.4c a 12 VDC solar power system is used. A 12 VDC—to–24 VDC DIN rail mounted converter is used to provide 24 VDC for the Portenta Machine Control. This is the option implemented for project deployment. This system has been operating in the greenhouse for over three years.

The solar power system consists of a solar panel, a solar power manager, a rechargeable battery, and fuses for circuit protection. In this project we use the DFRobot DFR0580 Solar Power Manager for a 12 VDC lead–acid battery. With an 18 VDC, 100 W solar panel and a 12 VDC lead–acid battery; the DFR0580 can provide regulated output voltages of 5 VDC at 5 A and 12 VDC at 8 A. This is suitable for the GCS project (www.DFRobot.com). A diagram of the solar power system is shown in Fig. 6.5. The 12–to–24 VDC converter is part of the Arduino PMC PLC based interface subsystem.

A distribution panel was designed for the power system and is shown in Fig. 6.6. The various components are connected together as shown in the figure with an automotive style fuse block and automotive style blade fuses. The panel is housed in a QILIPSU plastic, hinged $16.1'' \times 12.2'' \times 7.1''$ enclosure.

6.6 Greenhouse Control Subsystem

As shown in Fig. 6.3, the following requirements have been set for the Greenhouse Control portion of the GCS system:

- Panel mount start and stop pushbuttons (PMC inputs DI00 and DI 01);
- Monitor the water level in the rain barrel (analog sensor PMC ADC A00);

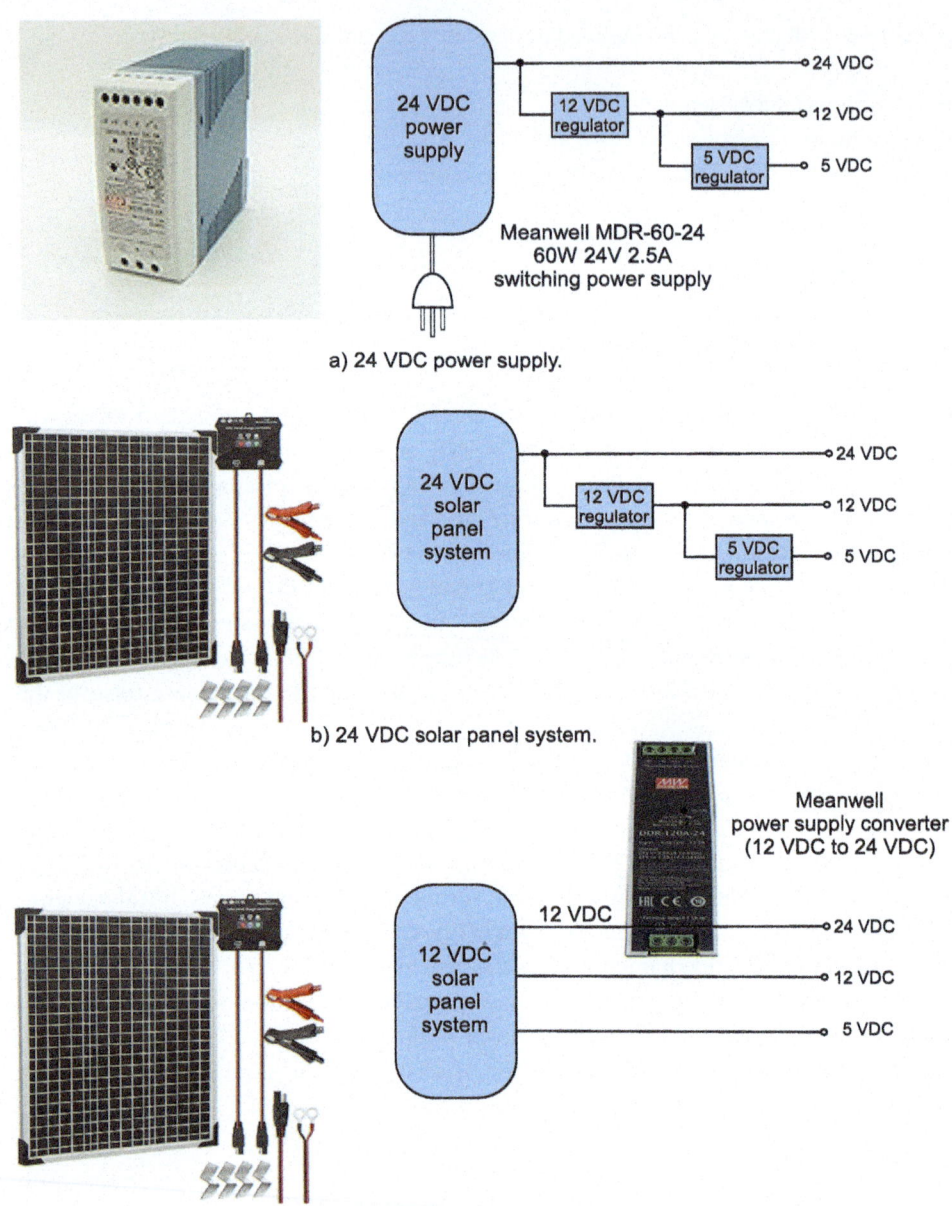

Fig. 6.4 Greenhouse power supply options

6.6 Greenhouse Control Subsystem

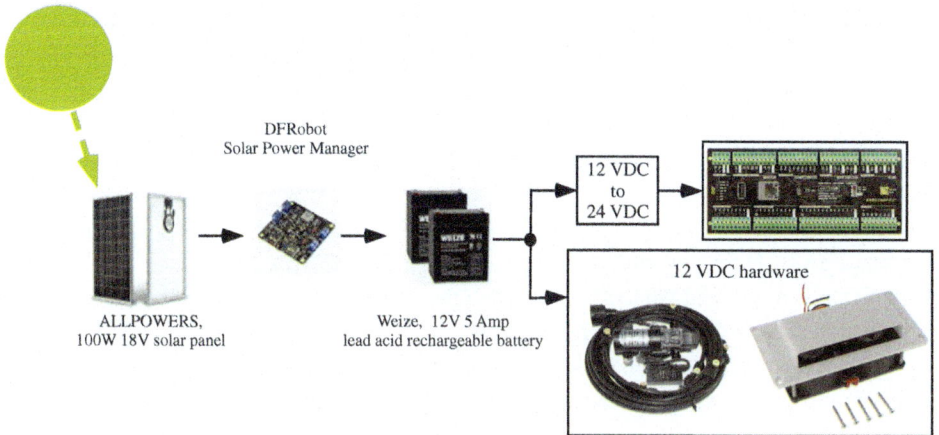

Fig. 6.5 Solar power system. Images courtesy of AllPowers, DFRobot, Weize, and Arduino

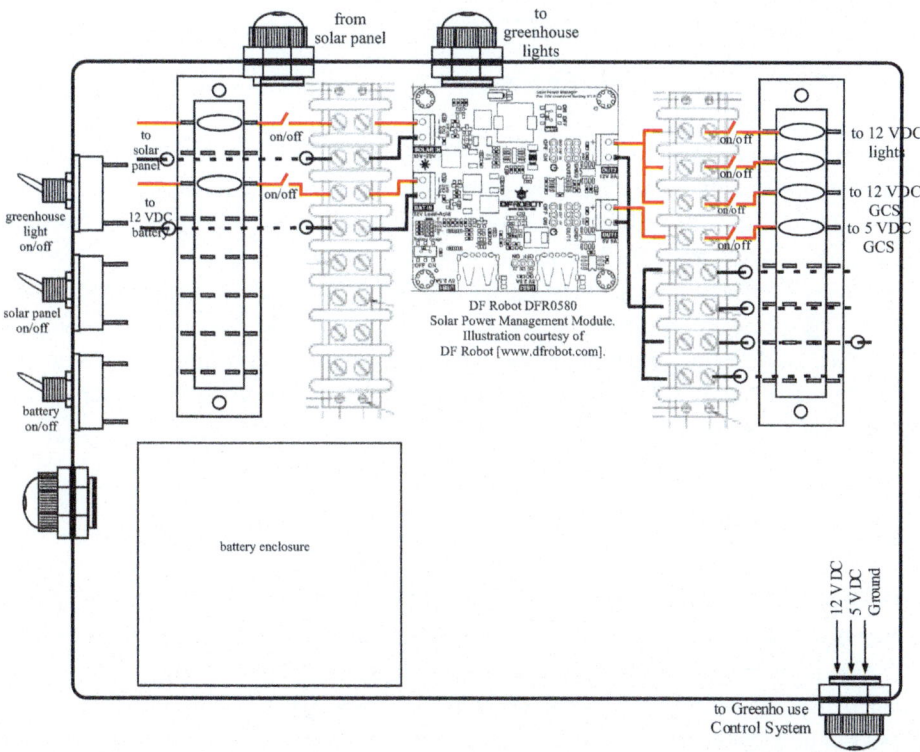

Fig. 6.6 Solar power distribution panel

- Monitor humidity level within the greenhouse (analog sensor PMC ADC A01);
- Monitor plant soil moisture (analog sensor to PMC ADC A02);
- Monitor interior greenhouse temperature (IDE: RTD 3–wire; PLC IDE: thermocouple to PMC TP0 and TN0);
- Monitor exterior greenhouse temperature (IDE: RTD 3–wire; PLC IDE: thermocouple to PMC TP1 and TN1);
- Monitor battery voltage level (battery voltage to one–bit ADC to PMC DI02) and illuminate an LED for low battery voltage;
- Activate vent fan and vent van LED when the internal greenhouse temperature is above a desired value (digital output DI00); and
- Activate misting system and misting system LED when the internal humidity level falls below the desired value and/or plant soil moisture is low (DI01).

To meet these requirements the instrumentation system shown in Fig. 6.7 is used. Each component is discussed in turn.

6.6.1 Milone E–Tape Fluid Sensor

Milone Technologies manufacture a line of continuous fluid level sensors. The sensor resembles a ruler and provides a near linear response. The sensor reports a change in resistance to indicate the distance from sensor tube bottom to the fluid surface. To convert the resistance change to a voltage change, the Milone 0–5 VDC Resistance to Voltage Module is used. The module shown in Fig. 6.8 (left) is powered from 12 VDC. The output from the module ranges up to 5 VDC (www.milonetech.com) to indicate fluid level. The sensor's analog output is fed to PMC ADC input A00. The Milone sensor is used to monitor rain barrel water level. A sample sketch and PLC ladder logic program to collect data from an analog sensor were provided in Chaps. 4 and 5 respectively.

6.6.2 Humidity Sensor

A Honeywell HIH–4030 sensor is used to measure greenhouse humidity. The sensor provides an output voltage that may be mapped to a corresponding relative humidity (RH) value. The RH value provides a measurement of the amount of water vapor in the air (see Fig. 6.8 (center)). The RH is expressed as a value from 0 to 100% RH.

The sensor provides an output voltage to indicate RH. The voltage is processed and corrected for temperature using the following equations provided by the manufacturer (Honeywell).

$$V_{out} = (V_{supply}) * (0.0062 * sensor\ RH) + 0.16$$

The sensor RH value is corrected for temperature:

6.6 Greenhouse Control Subsystem

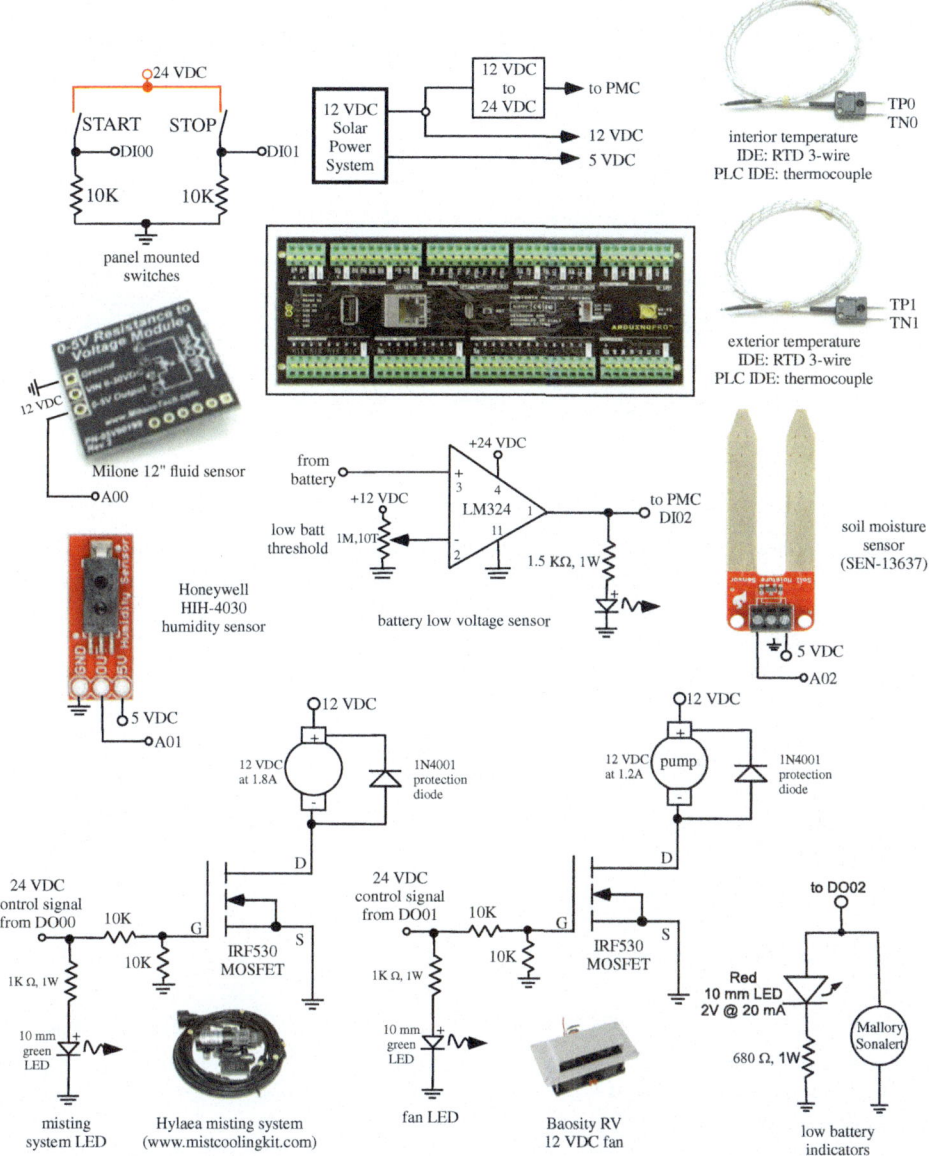

Fig. 6.7 Greenhouse control system. Arduino illustrations used with permission of the Arduino Team (CC BY–NC–SA) (www.arduino.cc)

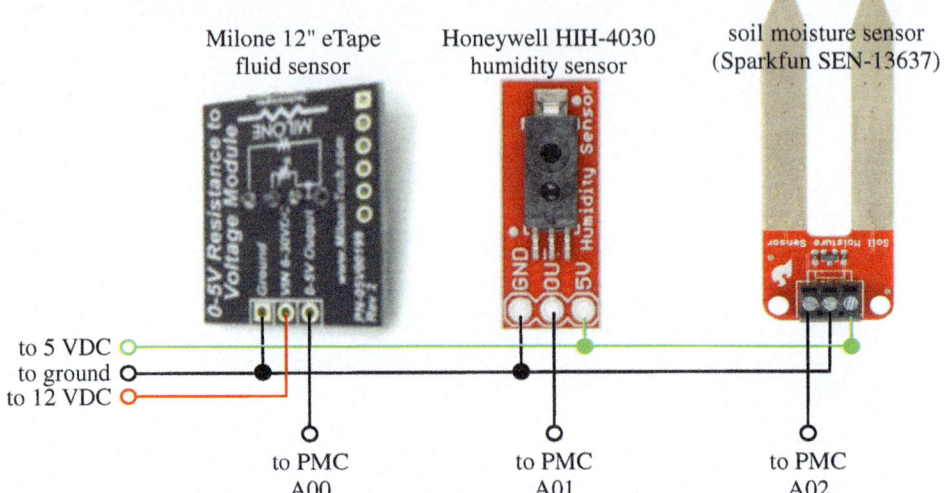

Fig. 6.8 (left) Milone Technologies eTape liquid level sensor. Image courtesy of Milone Technology (www.milonetech.com). (center) Honeywell HIH–4030 sensor. Image courtesy of Sparkfun (CY BY 2.0) (www.sparkfun.com). (right) Sparkfun soil moisture sensor. Image courtesy of Sparkfun (CY BY 2.0) (www.sparkfun.com)

$$TrueRH = (sensor\ RH/(1.0546 - 0.00216T)$$

with T expressed in degrees Centigrade. A sample sketch to collect data from this sensor is provided in Chap. 4.

6.6.3 Soil Moisture Sensor

A Sparkfun soil moisture sensor (SEN–13637) is used to monitor plant soil moisture content. The sensor is powered by a 5 VDC source. The interface circuit is shown in Fig. 6.8 (right). A sample sketch and PLC ladder logic program to collect data from an analog sensor were provided in Chaps. 4 and 5 respectively.

6.6.4 Thermocouple/RTD Interior/exterior Greenhouse Temperature Sensors

To monitor the interior and exterior greenhouse temperature a pair of K–type thermocouples are used with the PLC IDE. A sample ladder logic network is provided in Fig. 6.9.

6.6 Greenhouse Control Subsystem

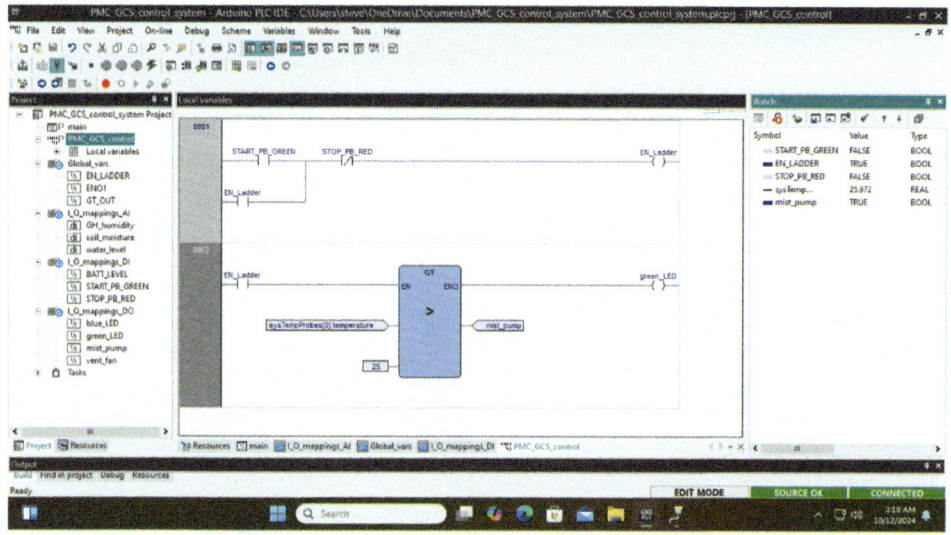

Fig. 6.9 PLC IDE ladder logic for thermocouple temperature measurement

In the example the thermocouple output (sysTempProbes[0].temperature) is compared to a threshold of 25°C. If the thermocouple output exceeds this value, the mist pump is asserted. For the IDE, a three–wire RTD (PT100, Adafruit #3290) is used to measure temperature. An example was provided in Chap. 4 to measure RTD temperature using the Arduino IDE.

6.6.5 Misting System and LED

A Hylaea misting system consisting of a 12 VDC fluid pump and misting delivery hardware is provided in a kit (www.mistcoolingkit.com). A MOSFET– based interface circuit is provided in Fig. 6.7. A 1N4001 diode serves as protection for the inductive load. The 12 VDC source is supplied by the DF Robot DFR0580 Solar Power Management Module OUT 2 rated at 12 V, 8 A. An LED indicator circuit is also provided to indicate when the pump is running.

6.6.6 Vent Fan and LED

A Baosity recreational vehicle 12 VDC vent fan is mounted in the east eave of the greenhouse roof. It provides for safe venting of the greenhouse interior when it becomes too hot. A MOSFET– based interface circuit is provided in Fig. 6.7. A 1N4001 diode serves as protection for the inductive load. The 12 VDC source is supplied by the DF Robot DFR0580 Solar Power Management Module OUT 2 rated at 12 V, 8 A. An LED indicator circuit is also provided to indicate when the vent is running.

6.6.7 GCS System Code

The UML activity diagram for the GCS system code is provided in Fig. 6.10. A first draft of an Arduino PLC IDE is provided in Application I. The Arduino IDE–based sketch follows. The activity diagram provides a template for basic control system operation. It may be customized for your local climate conditions.

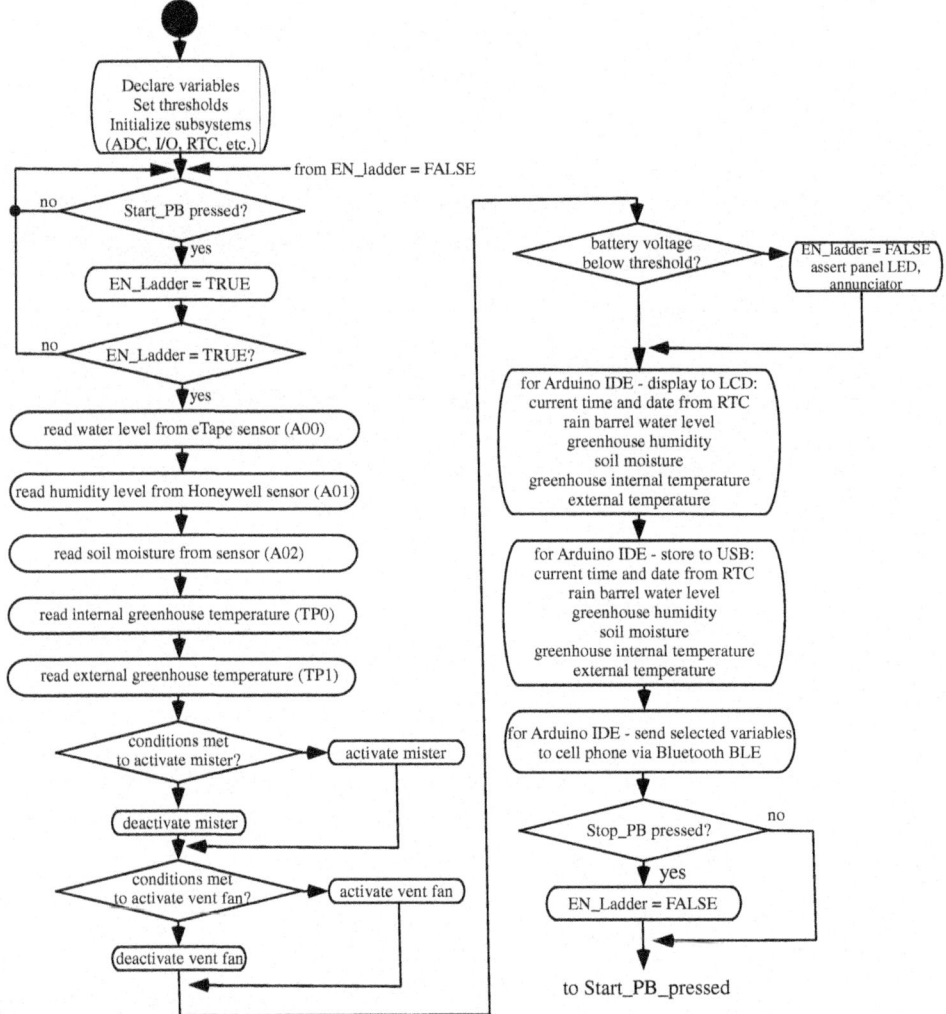

Fig. 6.10 Greenhouse UML activity diagram

6.6 Greenhouse Control Subsystem

It may seem a bit overwhelming to write the entire control system. We use a systematic step–by–step approach to guide our efforts:

- Construct a detailed UML activity diagram as a map to guide code development,
- Develop and insert comments into the code to establish the general program flow,
- Use prior examples throughout the book as tools to code each individual process on the UML activity diagram, and
- Bring each individual process individually online and test for accuracy before moving on to other processes.

```
//****************************************************************
//PMC_greenhouse_control
//
//This sketch demonstrates how to monitor and control
//environmental conditions within a greenhouse using
//the Arduino Portenta Machine Control.
//
//Author:   S. Barrett, Nov - Dec, 2024
//
//This sketch is in the public domain.
//****************************************************************

//include
#include <Arduino_PortentaMachineControl.h>

//logical values
#define  YES   1
#define  NO    0
#define  TRUE  1
#define  FALSE 0

//RTD values
//Rref resistor - PT100: 430.0 ; PT1000: 4300.0
#define RREF         400.0
//Nominal 0-degrees-C sensor resistance
//PT100: 100.0; PT1000: 1000.0
#define RNOMINAL    100.0

//declare global variables for ADC processing
const float RES_DIVIDER = 0.28057;
const float REFERENCE   = 3.0;
unsigned int analog_reading;

//water level threshold [inches] - determined experimentally
float water_level;           //level read from rain harvest barrel
float water_level_threshold = 4.0; //set level for 4"

//temperature sensor variables
float temp_ch0_C, temp_ch1_C;
float temp_ch0_F, temp_ch1_F;

//soil sensor threshold - determined experimentally
float soil_dry_threshold = 100.0;
```

```cpp
//internal temperature thresholds - determined experimentally
float temp_ch0_F_mister_threshold = 40.0;
float temp_ch0_F_fan_threshold = 80.0;

//humidity threshold - determined experimentally
unsigned int sensor_RH_threshold = 40;

//battery level threshold
unsigned int readings;
unsigned int batt_threshold = 58982;   //9 VDC
bool battery_OK = FALSE;

//control signals
uint16_t start_pb = NO;        //start pb connected to DI00
uint16_t  stop_pb = NO;        //stop pb  connected to DI01
bool EN_ladder = FALSE;        //seals start switch
uint16_t soil_dry = NO;        //soil sensor control signals
uint16_t water_level_good = NO; //water level in harvest barrel

void setup()
{
Serial.begin(9600);
while(!Serial)
  {
  ; //wait for serial port connection
  }

Wire.begin();

//ADC - set for 0-10 VDC in
MachineControl_AnalogIn.begin(SensorType::V_0_10);

//Digital In - configure
if(!MachineControl_DigitalInputs.begin())
  {
  Serial.println("Digital input GPIO expander initialization fail!!");
  }

//RTD
MachineControl_RTDTempProbe.begin(THREE_WIRE);

//RTC - set Real Time Clock (RTC)
//left as homework assignment

} //end setup

//***************************************************************************

void loop()
{
                            //start_PB connected to DI00
                            //if start_pb pressed, EN_ladder = TRUE
start_pb = MachineControl_DigitalInputs.read(DIN_READ_CH_PIN_00);
if(start_pb == TRUE)
  {
  EN_ladder = TRUE;
  }
Serial.print("EN_ladder: ");
Serial.println(EN_ladder);
```

6.6 Greenhouse Control Subsystem

```
Serial.println(" ");

                                      //process/control greenhouse environment
if(EN_ladder == TRUE)
  {

  //***read internal greenhouse temperature (TP0) - PT100 RTD***
  //Read temperatures first - required for humidity calculations
  MachineControl_RTDTempProbe.selectChannel(0);
  Serial.println("CHANNEL 0 SELECTED");
  uint16_t rtd = MachineControl_RTDTempProbe.readRTD();
  float ratio = rtd;
  ratio /= 32768;

  //Check and report any faults via Serial Monitor
  uint8_t fault = MachineControl_RTDTempProbe.readFault();
  if(fault)
     {
     Serial.print("Fault 0x"); Serial.println(fault, HEX);
     if(MachineControl_RTDTempProbe.getHighThresholdFault(fault))
        {
        Serial.println("RTD High Threshold");
        }
     if(MachineControl_RTDTempProbe.getLowThresholdFault(fault))
        {
        Serial.println("RTD Low Threshold");
        }
     if(MachineControl_RTDTempProbe.getLowREFINFault(fault))
        {
        Serial.println("REFIN- > 0.85 x Bias");
        }
     if(MachineControl_RTDTempProbe.getHighREFINFault(fault))
        {
        Serial.println("REFIN- < 0.85 x Bias - FORCE- open");
        }
     if(MachineControl_RTDTempProbe.getLowRTDINFault(fault))
        {
        Serial.println("RTDIN- < 0.85 x Bias - FORCE- open");
        }
     if(MachineControl_RTDTempProbe.getVoltageFault(fault))
        {
        Serial.println("Under/Over voltage");
        }
     MachineControl_RTDTempProbe.clearFault();
     } //end if(faulT)
  else
     {
     Serial.print("RTD int temp value: "); Serial.println(rtd);
     Serial.print("Ratio = "); Serial.println(ratio, 8);
     Serial.print("Resistance = "); Serial.println(RREF * ratio, 8);
     Serial.print("Temperature [C] = ");
     Serial.print(MachineControl_RTDTempProbe.readTemperature(RNOMINAL, RREF));

     //Centigrade
     temp_ch0_C = MachineControl_RTDTempProbe.readTemperature(RNOMINAL, RREF);

     //Centigrade to Fahrenheit conversion
     temp_ch0_F = (temp_ch0_C * 9.0/5.0) + 32;
     Serial.print("   Temperature CH0 [F]: ");
```

```
    Serial.println(temp_ch0_F);
    } //end else
Serial.println();
delay(1000);
//end RTD

//***read external greenhouse temperature (TP1)- PT100 RTD***
MachineControl_RTDTempProbe.selectChannel(1);
Serial.println("CHANNEL 1 SELECTED");
uint16_t rtd1 = MachineControl_RTDTempProbe.readRTD();
float ratio1 = rtd1;
ratio1 /= 32768;

//Check and report any faults via Serial Monitor
uint8_t fault1 = MachineControl_RTDTempProbe.readFault();
if(fault1)
    {
    Serial.print("Fault 0x"); Serial.println(fault, HEX);
    if(MachineControl_RTDTempProbe.getHighThresholdFault(fault))
        {
        Serial.println("RTD High Threshold");
        }
    if(MachineControl_RTDTempProbe.getLowThresholdFault(fault))
        {
        Serial.println("RTD Low Threshold");
        }
    if(MachineControl_RTDTempProbe.getLowREFINFault(fault))
        {
        Serial.println("REFIN- > 0.85 x Bias");
        }
    if(MachineControl_RTDTempProbe.getHighREFINFault(fault))
        {
        Serial.println("REFIN- < 0.85 x Bias - FORCE- open");
        }
    if(MachineControl_RTDTempProbe.getLowRTDINFault(fault))
        {
        Serial.println("RTDIN- < 0.85 x Bias - FORCE- open");
        }
    if(MachineControl_RTDTempProbe.getVoltageFault(fault))
        {
        Serial.println("Under/Over voltage");
        }
    MachineControl_RTDTempProbe.clearFault();
    } //end if(fault)
else
    {
    Serial.print("RTD ext temp value: "); Serial.println(rtd1);
    Serial.print("Ratio = "); Serial.println(ratio1, 8);
    Serial.print("Resistance = "); Serial.println(RREF * ratio1, 8);
    Serial.print("Temperature [C] = ");
    Serial.print(MachineControl_RTDTempProbe.readTemperature(RNOMINAL, RREF));

    //Centigrade
    temp_ch1_C = MachineControl_RTDTempProbe.readTemperature(RNOMINAL, RREF);

    //Centigrade to Fahrenheit conversion
    temp_ch1_F = (temp_ch1_C * 9.0/5.0) + 32;
    Serial.print("   Temperature CH1 [F]: ");
    Serial.println(temp_ch1_F);
```

6.6 Greenhouse Control Subsystem

```
    } //end else
Serial.println(" ");
Serial.println(" ");
delay(1000);
//end RTD

//***read water level from eTape sensor (A00)***
//The eTape sensor is placed in the water harvest barrel.
//It is used to ensure there is adequate water available
//before turning on the water mister system.
//The threshold to turn on the misters is experimentally
//determined.
//Sensor provides 5V out for 12" water

float raw_voltage_ch0 = MachineControl_AnalogIn.read(0);
float voltage_ch0 = (raw_voltage_ch0 * REFERENCE) / 65535 / RES_DIVIDER;

//ADC output
Serial.print("ADC CH0 - harvested water level: ");
Serial.print(raw_voltage_ch0, 3);
Serial.println(" ");

//ADC voltage
Serial.print("Water level voltage: ");
Serial.print(voltage_ch0, 3);
Serial.println("V");

//Convert to water level
//Conversion factor: 0 to 12" -> 0 to 5 VDC
float water_level = voltage_ch0 * (12.0/5.0);
Serial.print("Water level [inches]: ");
Serial.println(water_level, 3);

//determine water level status
if(water_level > water_level_threshold)
  {
  Serial.println("Water level sufficient for misting");
  water_level_good = YES;
  }
else
  {
  Serial.println("Water level not sufficient for misting");
  water_level_good = NO;
  }
Serial.println(" ");

//***read humidity level from Honeywell sensor (A01)***
//internal (TP0) and external (TP1) temperature previously read
//variables: temp_ch0_C, temp_ch1_C

//Read humidity at AI1
float raw_humidity_ch1 = MachineControl_AnalogIn.read(1);
Serial.print("Raw humidity CH1: ");
Serial.print(raw_humidity_ch1, 3);
Serial.println(" ");

float humidity_voltage_ch1 = (raw_humidity_ch1 * REFERENCE) / 65535 / RES_DIVIDER;
Serial.print("Humidity voltage CH1: ");
Serial.print(humidity_voltage_ch1, 3);
```

```
  Serial.println("V");
                                            //convert to RH per data sheet
  float sensor_RH_flt = (((humidity_voltage_ch1/5.0) - 0.16) * (1/0.0062));
                                            //compensate for temp per data sheet
  float true_RH = sensor_RH_flt/(1.0546 - (0.00216 * temp_ch0_C));
  unsigned int sensor_RH = (unsigned int)(true_RH);       //cast to int for display

  //print result
  Serial.print("Humidity [%]: ");
  Serial.println(sensor_RH);
  Serial.println(" ");
  Serial.println(" ");

  //***read soil moisture from sensor (A02)***
  float raw_voltage_ch2 = MachineControl_AnalogIn.read(2);
  float voltage_ch2 = (raw_voltage_ch2 * REFERENCE) / 65535 / RES_DIVIDER;

  //ADC output
  Serial.print("ADC CH2: ");
  Serial.println(raw_voltage_ch2, 3);

  //ADC voltage
  Serial.print("Soil moisture: ");
  Serial.print(voltage_ch2, 3);
  Serial.println("V");

  //determine threshold experimentally
  if(voltage_ch2 < soil_dry_threshold)
    {
    Serial.println("Soil dry");
    soil_dry = YES;
    }
  else
    {
    Serial.println("Soil moist");
    soil_dry = NO;
    }
  Serial.println(" ");

  //***battery level > threshold?***
  //Battery level below set threshold?
  //Threshold set with external hardware
  //HIGH: battery above threshold
  //LOW:  battery below threhold

  readings = MachineControl_DigitalInputs.read(DIN_READ_CH_PIN_02);
  if(readings == TRUE)
    {                              //battery okay
    Serial.println("battery okay");
    battery_OK = TRUE;
    }
  else
    {
    Serial.println("battery low");
    battery_OK = FALSE;
    EN_ladder = FALSE;
    }
  Serial.println();
```

6.6 Greenhouse Control Subsystem

```
Serial.println("Current Status:");
Serial.print("int temp [F]: ");
Serial.println(temp_ch0_F);
Serial.print("RH [%]:");
Serial.println(sensor_RH);
Serial.print("Soil dry?");
Serial.println(soil_dry);
Serial.print("water level good?");
Serial.println(water_level_good);
Serial.print("battery OK?");
Serial.println(battery_OK);

//***conditions met to activate mister?***
//Depends on local conditions
//Demo values used here:
//- Internal greenhouse temp above threshold
//- Greenhouse humidity below threshold
//- Soil moisture below threshold
//- Adequate water available
//- Battery above threshold
if((temp_ch0_F > temp_ch0_F_mister_threshold)&& //GH internal temp
   (sensor_RH < sensor_RH_threshold)&&          //GH humidity
   (soil_dry==YES)&&                            //soil moisture
   (water_level_good==YES)&&                    //water level
   (battery_OK==TRUE))                          //battery level
   {                                            //mister on
   Serial.println("Mister on");
   MachineControl_DigitalOutputs.write(0, HIGH);
   }
else                                            //mister off
   {
   Serial.println("Mister off");
   MachineControl_DigitalOutputs.write(0, LOW);
   }

//***conditions met to activate vent fan?***
//Depends on local conditions
//Demo values used here:
//- Internal greenhouse temp above threshold
//- Battery above threshold
if((temp_ch0_F > temp_ch0_F_fan_threshold)&&   //GH internal temp
   (battery_OK==TRUE))                          //battery level
   {                                            //fan on
   Serial.println("Fan on");
   MachineControl_DigitalOutputs.write(1, HIGH);
   }
else                                            //fan off
   {
   Serial.println("Fan off");
   MachineControl_DigitalOutputs.write(1, LOW);
   }

//***display measured values to LCD***
//***store to USB drive
//current time, date
//rain barrel water level
//greenhouse humidity
//soil moisture
//greenhouse internal temperature
```

```
    //greenhouse external temperature
    //left as homework assignment

    //***Bluetooth BLE: send selected variables to cell phone***
    //left as homework assignment

  } //end if (EN_ladder == True)
                                //stop_PB connected to DI01
                                //if stop_pb pressed, EN_ladder = FALSE
  stop_pb = MachineControl_DigitalInputs.read(DIN_READ_CH_PIN_01);
  if(stop_pb == TRUE)
    {
    EN_ladder = FALSE;
    }
  Serial.print("EN_ladder: ");
  Serial.println(EN_ladder);
  Serial.println("   ");
  delay(1000);
  }//end loop

//**************************************************************************
```

6.6.8 GCS Printed Circuit Board

The layout for the GCS printed circuit board (PCB) and the actual circuit board design are provided in Fig. 6.11. The PCB is mounted to a DIN rail using DIN rail mounting adaptors (Molence C45 35 mm × 15 mm bracket). The PCB provides an interface between the PMC and the peripheral components.

6.6.9 Enclosure

The completed GCS System is mounted within a QILIPSU hinged cover, stainless steel latch, junction box ($20'' \times 16.1'' \times 7.9''$) with mounting plate. The layout of the junction box is provided in Fig. 6.12.

6.7 Testing

The final project step is to thoroughly test all system features. A test plan is developed to test and document the proper operation of each system feature and the overall system. The beginning of a test plan for the Greenhouse Control System is provided in Fig. 6.13. Should a test fail, the software is corrected. The test plan is then restarted.

In developing and testing hardware and software for a system, it is not always possible or desirable to have close access to the system hardware. For example, in the development of

6.7 Testing

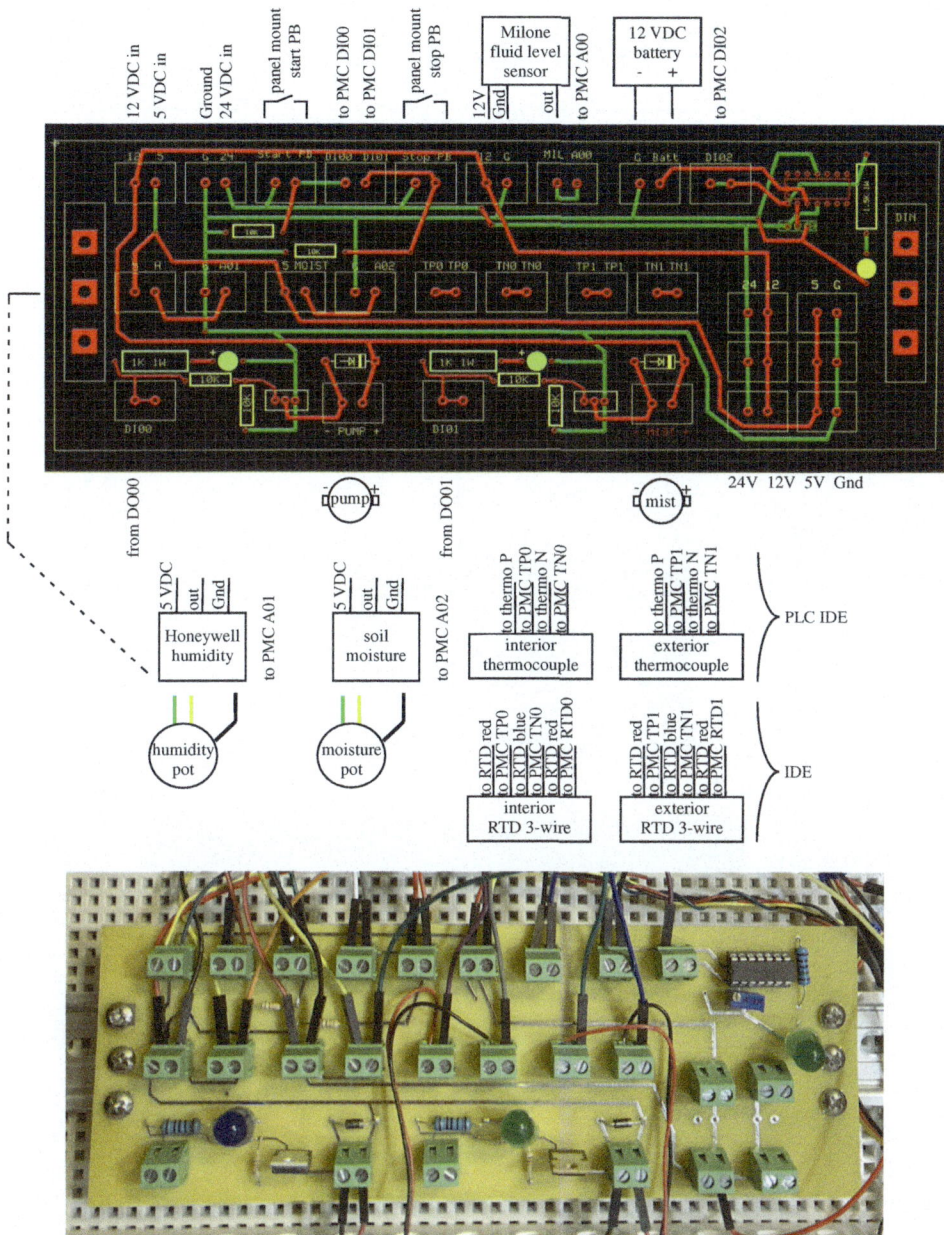

Fig. 6.11 Greenhouse printed circuit board

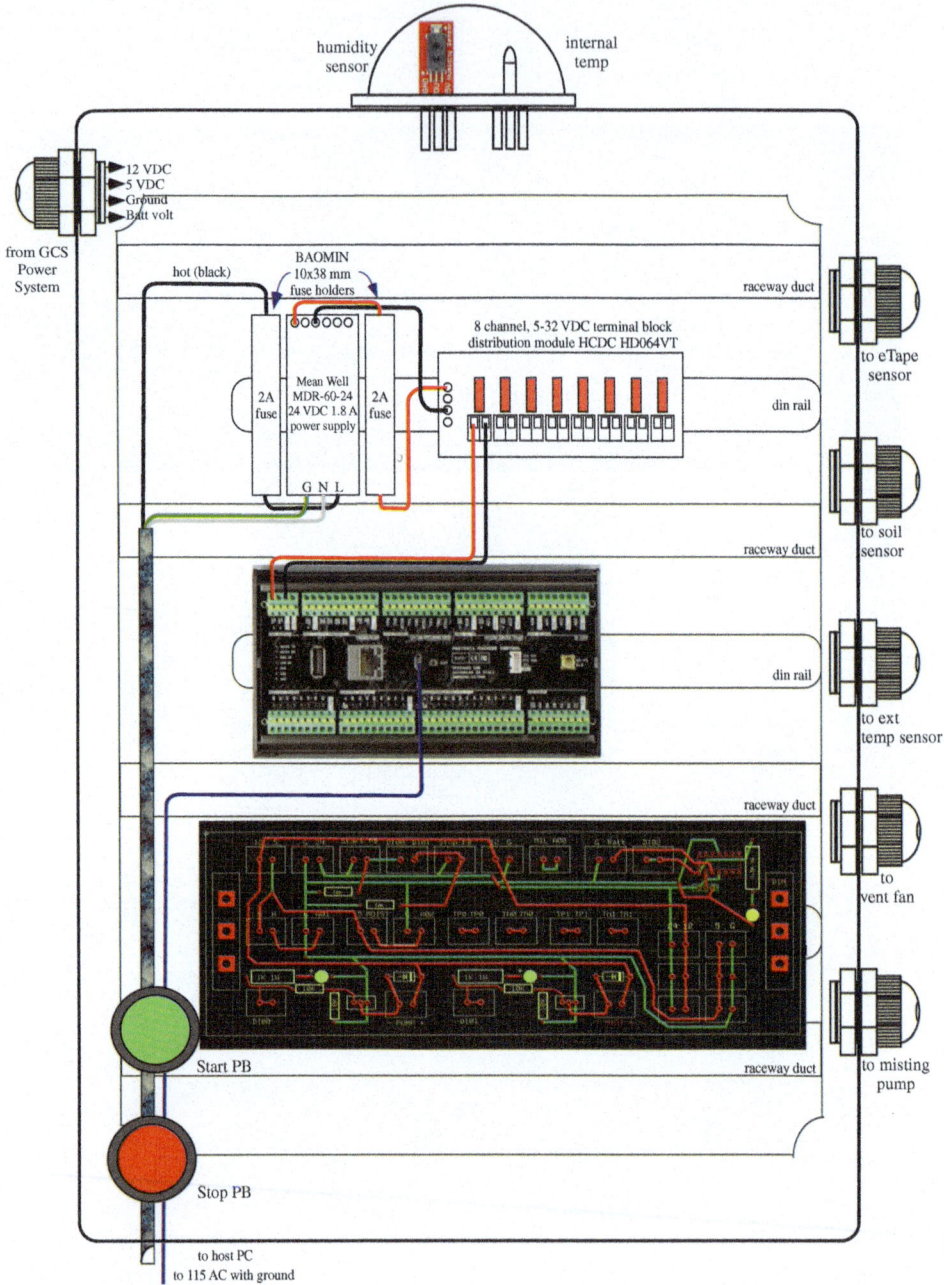

Fig. 6.12 GCS panel layout

6.7 Testing

Test #	Description	Expected Status	Results
1	Start program	Program should bypass main loop until Start PB pressed. When Start PB pressed, program should continuously process main loop.	
2	Data collection.	Within main loop, program should **collect data from "sensors"** (potentiometers). Program should respond to changes in **"sensor" values.**	
3	Vent fan	When internal temperature measurement exceeds threshold setting, vent fan and LED will come on. Fan and LED turns off when internal temperature falls below threshold setting.	
4	Mist pump	When soil moisture measurement falls below threshold setting, mist pump and LED will come on. Mist pump and LED turns off when soil moisture is above threshold setting.	
5	Battery status	When battery voltage measurement falls below threshold setting, battery LED 1 will illuminate. Battery LED 1 turns off when battery voltage is above threshold setting.	
6	Stop program	When Stop PB is pressed, program stops **data collection and responding to "sensor"** values changes.	

Fig. 6.13 GCS test plan

control software for a greenhouse, the greenhouse is not always readily available as shown in Fig. 6.14. Also, different weather conditions are not conveniently available to test the control algorithm under a variety of conditions. In these situations, a simulator may be used to substitute for the system. The simulator provides the necessary inputs and signals in place of the system so that software may be developed. I have used the technique when developing control systems for a high–end expensive audio amplifier system and also for controlling large ($12' \times 12'$) industrial door controllers.

In this vein, a greenhouse simulator was developed as shown in Fig. 6.15. For the simulator, potentiometers were used as sensor simulators and LEDs were used for the pump and fan. The completed test circuit is shown in Fig. 6.16. The simulator allows exhaustive testing of different weather conditions and control system operation.

Fig. 6.14 Greenhouse availability. Mother Nature had other plans

6.8 Application: Greenhouse Control System–Ladder Logic

In this application activity we implement a Greenhouse Control System using PLC ladder logic. We use the same hardware configuration as used earlier in the chapter except the internal and external temperature sensors are thermocouples. As mentioned earlier, it is a bit overwhelming to create, test, and implement a complex control system. Again, we use a systematic step–by–step approach to guide our efforts:

- Construct a detailed UML activity diagram as a map to guide code development,
- Use prior examples throughout the book as tools to code each individual process on the UML activity diagram with ladder logic,
- Construct an overall ladder logic diagram of the control system,
- Review the process for constructing an Arduino PLC IDE–based program provided in Fig. 5.9.
- Bring each individual process individually online and test for accuracy before moving on to other processes, and

6.8 Application: Greenhouse Control System–Ladder Logic

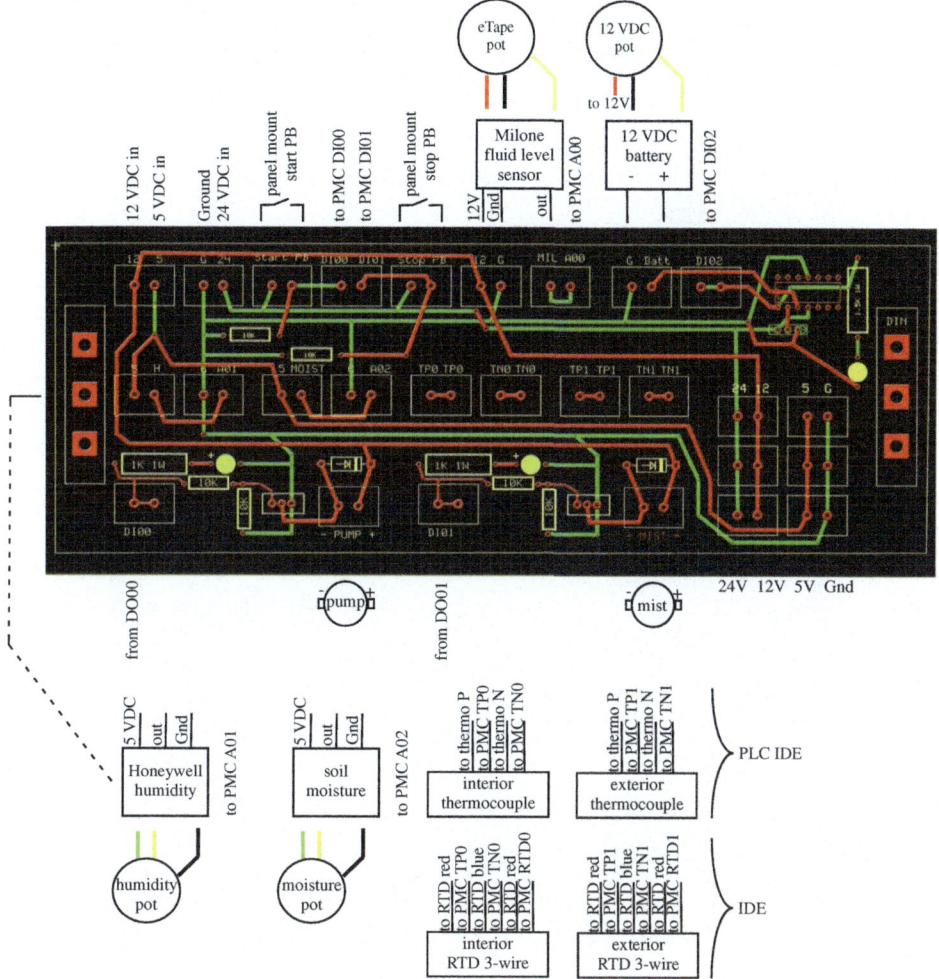

Fig. 6.15 GCS text fixture layout

- Thoroughly test the system with a detailed, documented plan.

A detailed UML activity diagram was provided in Fig. 6.10. Provided in Figs. 6.17 and 6.18 is the diagram of a Greenhouse Control System in ladder logic.
To implement the diagram follow the steps provided in Fig. 5.9. The results are provided in Figs. 6.19, 6.20, and 6.21.
The system is thoroughly tested using a test plan such as provided in Fig. 6.13 and the Arduino PLC IDE built–in testing features such as the Watch and Oscilloscope.

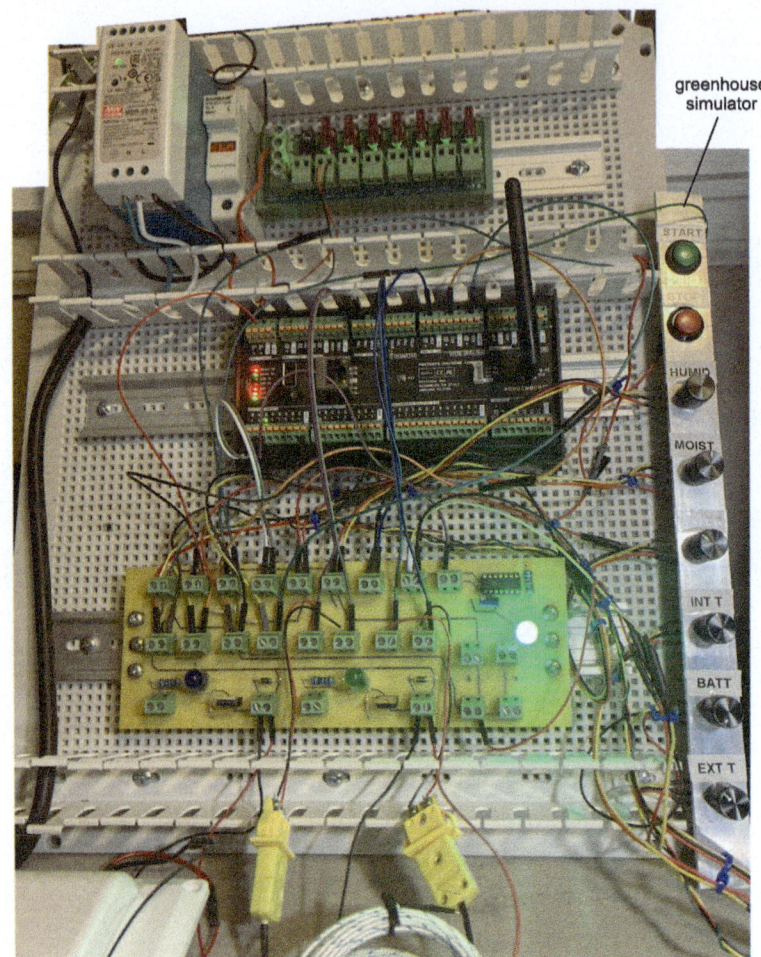

Fig. 6.16 GCS text fixture

6.9 Summary

The goal of this chapter was to demonstrate in action the concepts discussed in this book. Simply put, our goal is to provide the theory, design, and construction of a passively heated greenhouse. We equip the greenhouse with instrumentation to monitor and control key parameters.

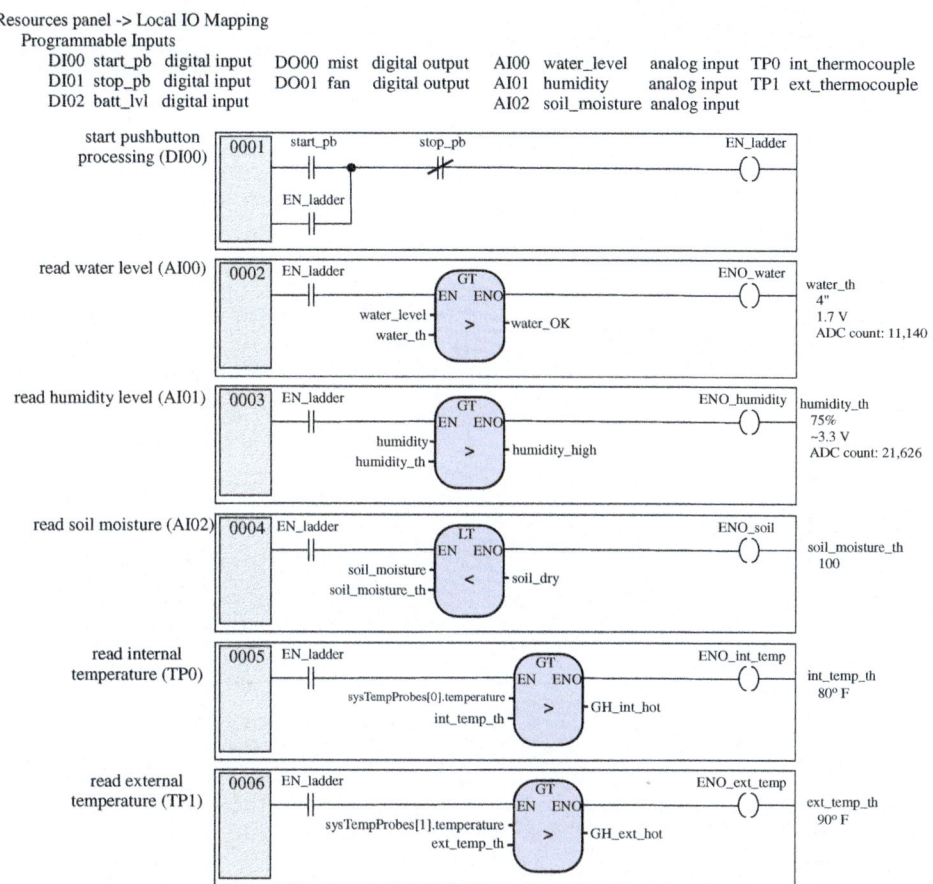

Fig. 6.17 Greenhouse control system in ladder logic

6.10 Chapter Problems

1. Provide a detailed structure chart for the GCS system code.
2. Develop a detailed test plan to insure requirements have been met for the GCS system.
3. What are the critical variables that should be regularly monitored?
4. Earlier in the chapter we provided a PMC_greenhouse_control sketch. Extend the features of the sketch to include:

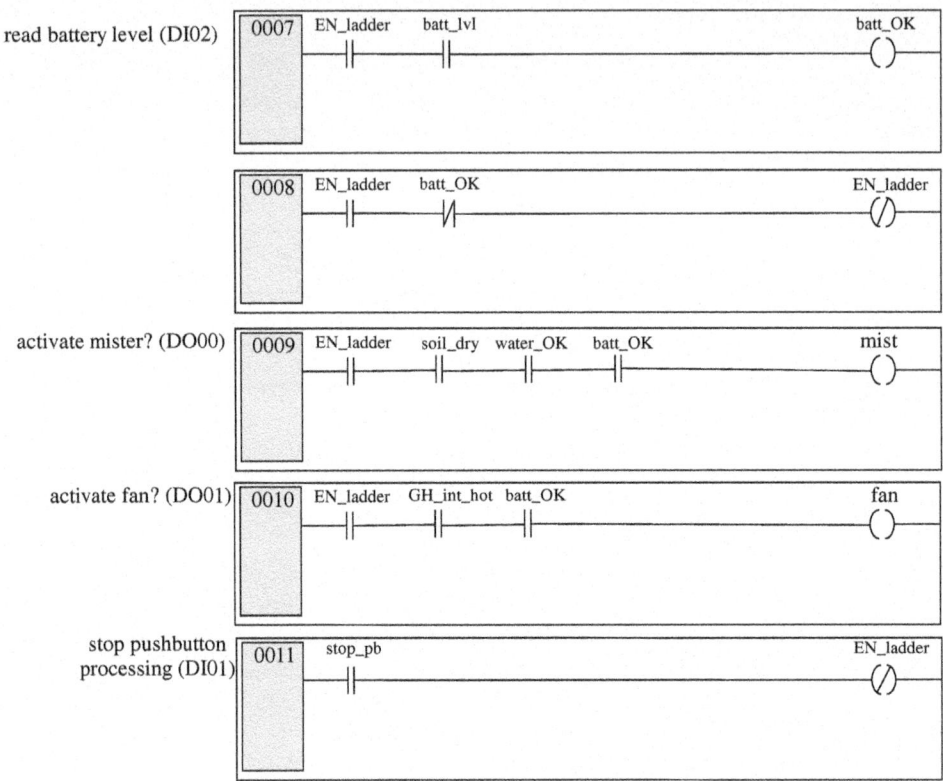

Fig. 6.18 Greenhouse control system in ladder logic

- Provide the sketch with a Real Time Clock,
- Equip the sketch with an LCD to display key greenhouse parameters,
- Provide for storage of key greenhouse parameters on a USB drive, and
- Develop a Bluetooth BLE application to report key greenhouse parameters for viewing on a client cell phone.

5. In the PLC ladder logic control system provided in the Application section, the humidity sensor reading was not compensated for temperature. Add this compensation to the ladder logic diagram and code.

6.10 Chapter Problems

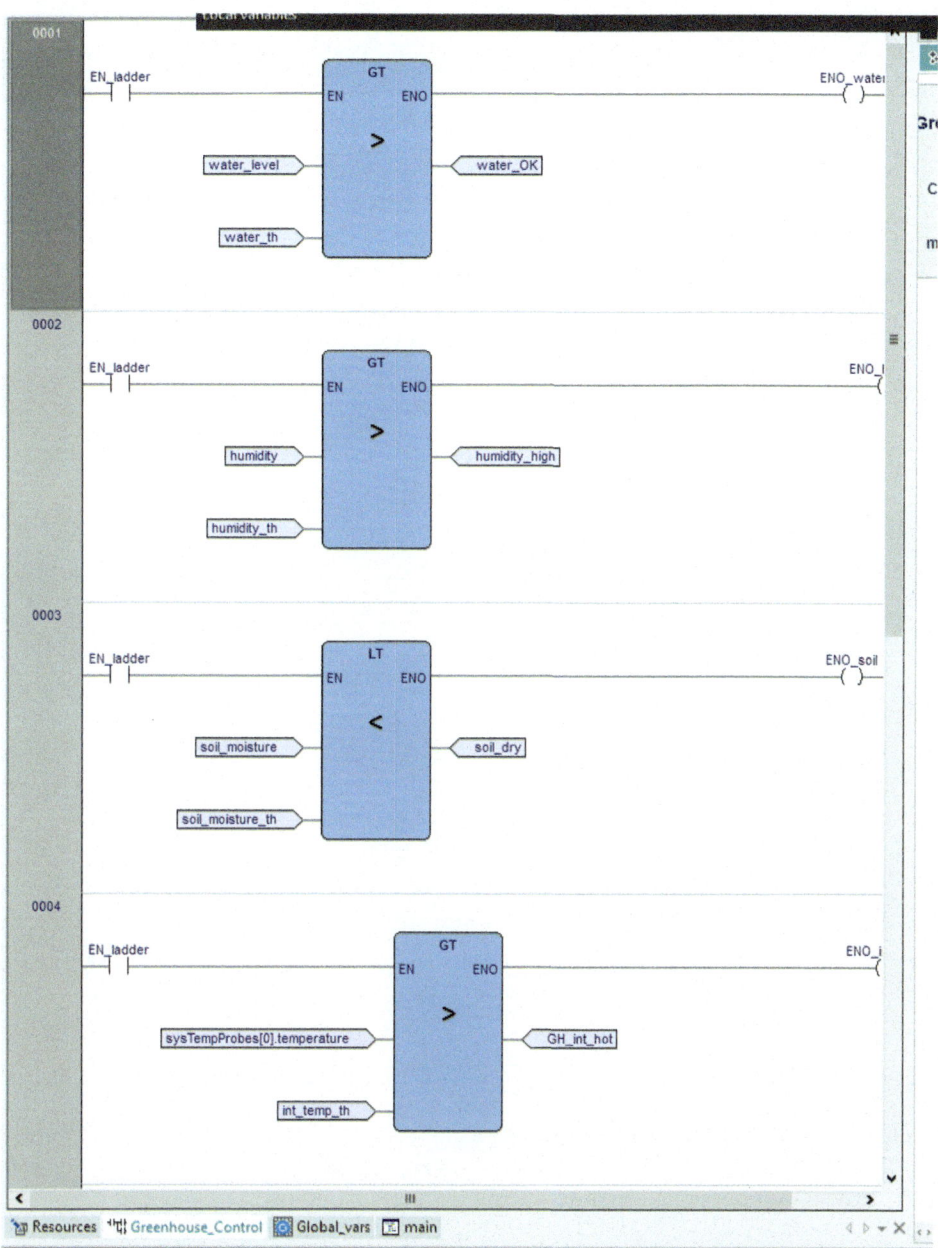

Fig. 6.19 Greenhouse control system in ladder logic

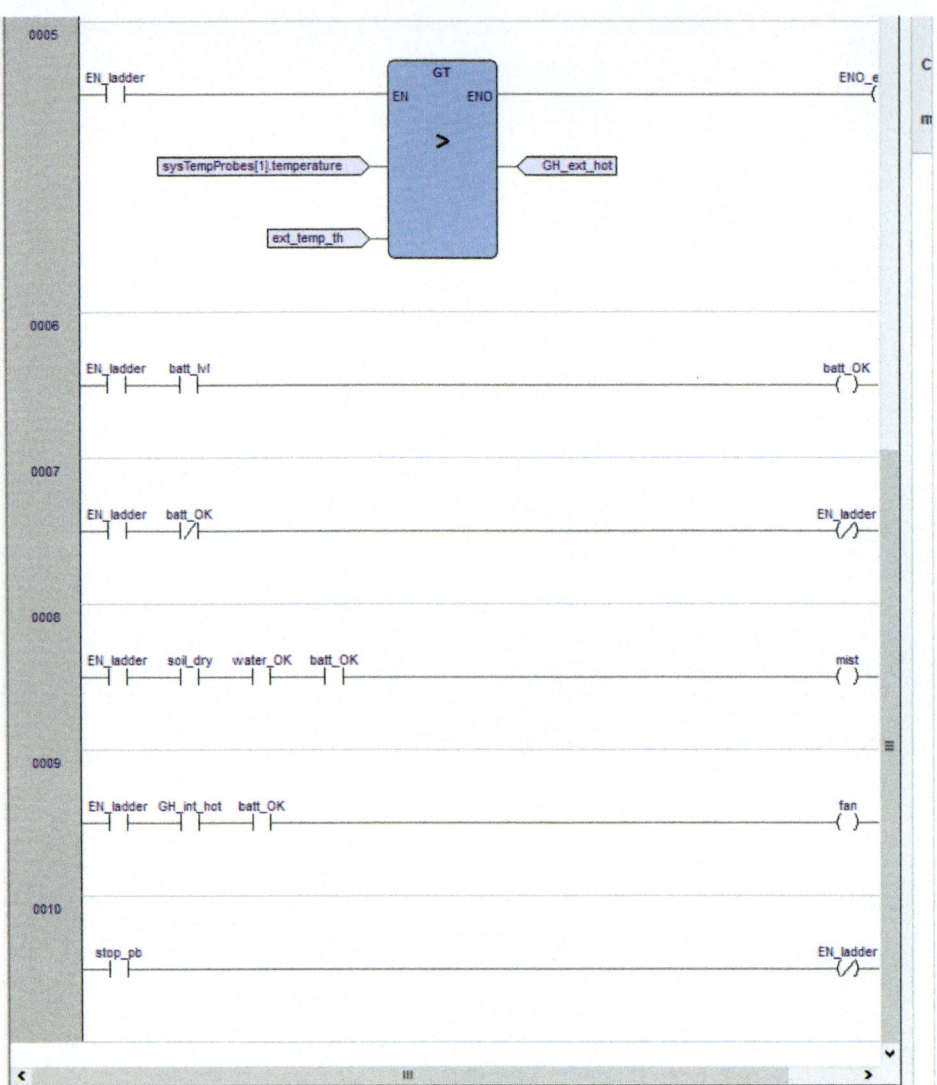

Fig. 6.20 Greenhouse control system in ladder logic

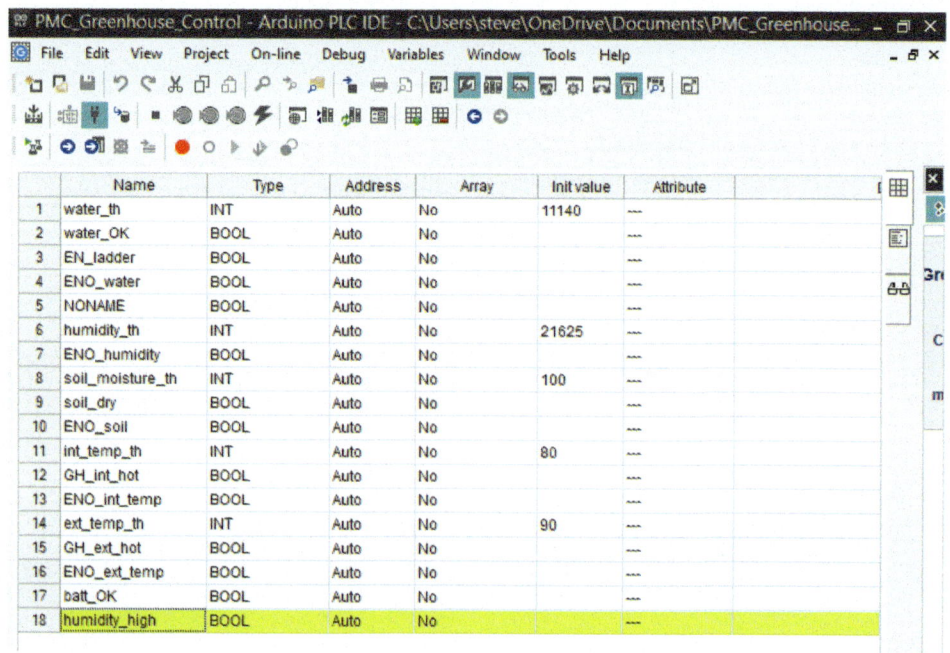

Fig. 6.21 Greenhouse control system in ladder logic

References

1. Baird C. (2011) *The Complete Guide to Building Your Own Greenhouse*, Atlantic Publishing.
2. Doxon, L. *How to Calculate Greenhouse Heating*, https://homeguides.sfgate.com.
3. Encinias, V. *Efficient Greenhouse Design*, https://gpnmag.com.
4. *Five Low–Tech Winter Greenhouse Heating Techniques*, https://www.rimolgreenhouses.com.
5. Honeywell, *HIH–4030/31 Series Humidity Sensors*, Honeywell Sensing and Control, www.honeywell.com/sensing.
6. Lindsey, C. and M. Plinke (2016) *The Year–Round Solar Greenhouse*, New Society Publishers.
7. Marshall, R. (2006) *How to Build Your Own Greenhouse*, Storey Publishing.
8. Milone Technologies, *0–5 VDC Linear Resistance to Voltage Module PN–05V00199 Rev 2*, www.milonetech.com.
9. Milone Technologies, *eTape Continuous Fluid Level Sensor PN–12110215TC–X*, www.milonetech.com.
10. National Semiconductor, *LM34 Precision Fahrenheit Temperature Sensors*, DS006685, National Semiconductor Corporation, www.national.com, 2000.
11. Oehler, M. (2007) *The Earth–Sheltered Solar Greenhouse Book*, Mole Publishing Company.
12. Schmidt, P. (2011) *The Complete Guide to Greenhouses & Garden Projects*, Creative Publishing International.
13. *Passive Solar Greenhouse*, Bradford Research Center, University of Missouri, 2017, http://bradford.cafnr.org/passive---solar---greenhouse.

14. Schiller, L. *Three Methods for Heating Greenhouses for Free*, https://www.motherearthnews.com.
15. Schiller, L. *How to Design a Year–Round Solar Greenhouse*, https://www.motherearthnews.com.
16. Thoma, M. *Seven Useful Features You Need in a Passive Solar Greenhouse*, `tranquilurbanhomestead.com`.
17. Watson, G. *Rain Barrels–A Homeowner's Guide*, SW Florida Water Management District, `WaterMatters.org`.

Appendix
Safety

In this Appendix[1] a review of safety concepts are provided for ready reference. The information provided is condensed from the Electrical Safety Foundation International (ESFI) and several other sources. "The Electrical Safety Foundation International is the premier non–profit organization dedicated exclusively to promoting electrical safety at home and in the workplace. Since 1994, ESFI has led the way in promoting electrical safety across North America. Over the years, ESFI has become highly regarded by industry, media and consumer safety partners alike by constantly reinvigorating the way electrical safety is addressed. ESFI creates unique awareness and educational resources designed to meet the diverse needs of a variety of at–risk groups. The Electrical Safety Foundation International is dedicated exclusively to promoting electrical safety at home and in the workplace through education, awareness, and advocacy(esfi.org)."

In the next several sections we review the background information on the effects of electricity on the human body and present electrical safety procedures for safe equipment operation. You must strictly adhere to these procedures when working on book associated activities. With this in mind, we provide laboratory safe operating procedures and also electric shock treatment procedures.

Safety is paramount and is everybody's business!

A.1 Physiological Effects of Electricity

For an electrically induced physiological effect to occur, your body must become part of an electrical circuit. Current must enter at some point on your body and exit via another point. Through safe laboratory procedures and proper equipment design, current is prevented from entering the body.

---

[1] This Appendix was adapted with permission from "Arduino VI: Bioinstrumentation," S. Barrett, Springer, 2024.

© The Editor(s) (if applicable) and The Author(s), under exclusive license to Springer Nature Switzerland AG 2026
S. F. Barrett, *Arduino VIII*, Synthesis Lectures on Digital Circuits & Systems, https://doi.org/10.1007/978-3-031-85944-1

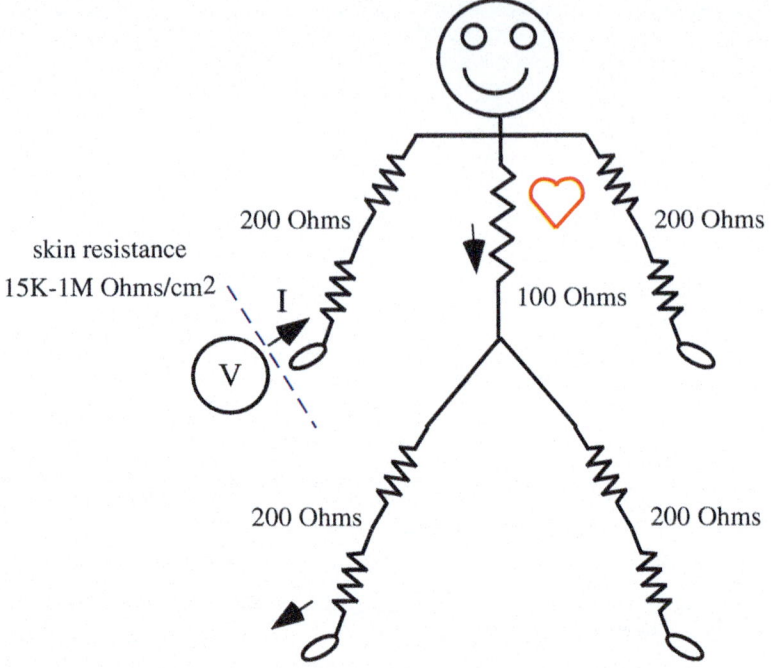

Fig. A.1 Body resistance

From an electrical point of view the body is a resistor. The resistance of skin ranges from 15K to 1M Ω per square centimeter. However, the resistance may decrease to one percent of its dry value when it is wet. Beyond the skin barrier, the internal resistance is approximately 200 Ω per limb and 100 Ω per trunk as shown in Fig. A.1. Application of Ohm's Law ($voltage = current \times resistance$) shows a voltage of 300 volts applied across the body with dry skin will result in a current of approximately 10 mA. The same current flow will result at much lower values of voltage when your skin is moist or wet (Webster 2020).

When the body comes in contact with voltage, it is current flow through the body that causes physiological damage and potentially death. The body may be exposed to harmful currents via macroshock or microshock. In macroshock the body is exposed to a high voltage level. Microshocks provide low level, yet harmful currents to the heart. Microshocks result from leakage current within an AC powered circuit. Leakage currents are small currents that flow between two conductors at different potentials. Even if the conductors are well insulated, leakage currents may occur due to capacitive and resistive effects. Leakage currents flow into the body and the heart due to catheters or internal conductive monitoring leads (Webster 2020).

Shock of either type is prevented via a combination of proper power system design and distribution, proper equipment grounding, proper equipment design, and safe operating

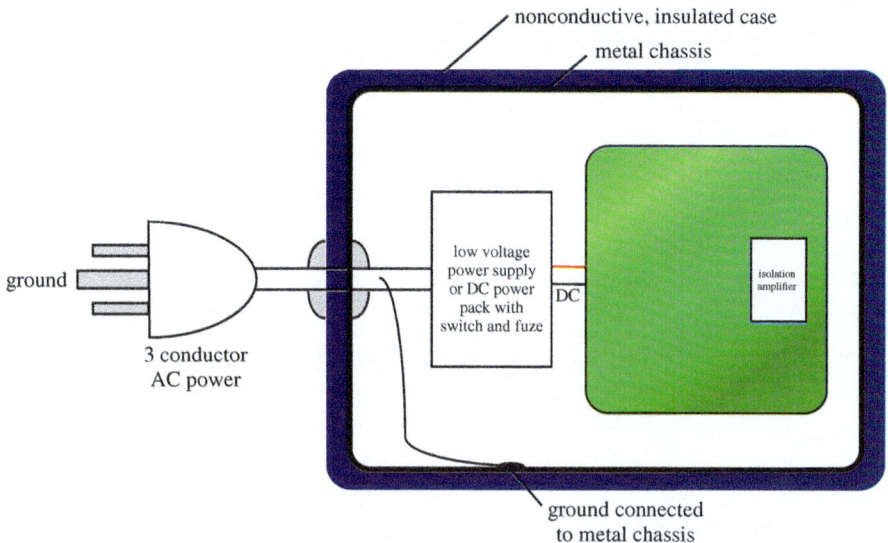

Fig. A.2 Safety design

practices and processes. As shown in Fig. A.2, a person is protected from both types of shock via these equipment design concepts (Webster 2020, Baretich 2015):

- A reliable equipment chassis ground to ground;
- Reduction of leakage current through properly insulated equipment conductors;
- Equipment double insulation with a nonconductive external covering or chassis;
- Equipment operation at the lower AC or DC voltage practical; and
- Use of isolation amplifiers when using conductive leads.

Equipment powered from an AC source must use a three conductor power cord. The ground conductor is connected to the equipment's metal chassis and also to Earth ground via a sound and reliable electrical distribution system. Also, equipment operation at the lowest possible AC or DC voltage further reduces shock hazard. The lower voltage coupled with a high, intact skin resistance reduce the potential from harmful current (Webster 2020).

Not much current flow is required to produce a physiological effect. The table below shows that currents as low as 0.5 mA illicit a perceptible physiological response. Note that as more current flows through the body the physiological effect is more severe (Webster 2020, ANSI/AAMI ESI–1993).

Low range [mA]	High range [mA]	Physiological response
0.5	3	Start to feel the energy, tingling sensation
3	10	Experience pain, muscle contraction
10	40	Grip paralysis threshold–called "let–go" current
30	75	Respiratory systems shuts down
100	200	Experience heart fibrillation
200	500	Heart clamps tight
Over 1500		Tissue and organs burn

"Let–go current" is the maximum current at which you can voluntarily withdraw from the current source. At higher current levels the muscles involuntarily contract and you can't let go of the current source. At current levels as low as 10 mA respiratory paralysis may result.

At approximately 50 mA ventricular fibrillation may result. Ventricular fibrillation is a condition where the ventricles of the heart quiver instead of pump. It is caused by an interruption of the normal electrical cardiac conduction cycle. Defibrillation action is usually required to bring the heart to a complete stop so that the normal cardiac conduction cycle may resume. At even higher levels of current sustained myocardial contraction and severe burns may occur. Currents as low as 10 mA through the body may result in death.

A.2 Electrical Safety Principles

The following is a list of electrical safety "do's" and "don'ts" compiled from several different electrical safety sources.

- Two or more individuals must be present in a laboratory at a given time to monitor one another's safety.
- Remove jewelry (rings, watches, necklaces, long earrings, etc.) when working with electrical or mechanical equipment.
- Watch out for loose wires.
- Work with one hand behind your back. This will prevent electrical current from entering one hand passing through your heart and out the other hand.
- Be aware of capacitors. The retain charge even when equipment is turned off.
- Use a three conductor electrical system with an Earth ground.
- Keep fingers out of "live" chassis.

Appendix A: Safety

- Don't install components when the power is on.
- Don't work with wet or oily hands.
- Don't cut live wires.
- Double check circuit wiring before energizing a circuit.
- Make sure that all members of your laboratory group realize when a circuit is being energized.
- Repair or replace leads with loose connectors.
- Avoid touching rotating parts or allowing wires to touch them.
- **Safety is everybody's business!**

A.3 Shock Rescue Procedures

In response to an electrical accident, follow these procedures immediately:

- Call for help.
- De–energize the circuit.
- Separate the person from the energy source.
- Make sure you and the victim are in a safe zone – not in contact with any electrical source, away from downed or broken wires.
- Never grab the person or pull the person off the current source with your hands; you might become part of the circuit and become injured as well.
- Use a dry wood broom, leather belt, plastic rope or something similar that is non–conductive such as wood or plastic cane with hook on the end to free the person from the energy source.
- Administer first aid, apply mouth–to–mouth resuscitation and/or CPR, know what to do.
- Keep the victim lying down, warm and comfortable to maintain body heat until help arrives. Do not move the person in case of injury to neck or back.
- If the victim is unconscious, put him/her on side to let fluids drain.
- Make sure the victim receives professional medical attention (person shocked could have heart failure hours later).

A.3.1 References

ANSI/AAMI ESI–1993, *Safe Current Limits for Electromechanical Apparatus*, American National Standard, 1993.

M.F. Baretich, *Electrical Safety Manual 2015–A Comprehensive Guide to Electrical Safety Standards for Healthcare Facilities,* American Safety in Medical Technology (AAMI), www.aami.org.

Electrical Safety Foundation International (ESFI), esfi.org.

J.G. Webster and A.J. Nimunkar, *Medical Instrumentation Application and Design,* fifth edition, John Wiley and Sons, Inc, 2020.

Appendix B
Embedded Systems Design

Objectives: After reading this appendix, the reader should be able to do the following:
- Define an embedded system;
- List all aspects related to the design of an embedded system;
- Provide a step–by–step approach to embedded system design;
- Discuss design tools and practices related to embedded systems design; and
- Apply embedded system design practices in the design of a microcontroller system employing several interacting subsystems.

B.1 Overview

In this appendix,[2] we begin with a definition of just what is an embedded system. We then explore the process of how to successfully (and with low stress) develop an embedded system prototype that meets established requirements.

B.2 What Is An Embedded System?

An embedded system contains a microcontroller to accomplish its job of processing system inputs and generating system outputs. The link between system inputs and outputs is provided by a coded algorithm stored within the processor's resident memory. What makes embedded

---

[2] The information on embedded system design first appeared in "Microcontroller Fundamentals for Engineers and Scientists," Morgan and Claypool Publishers, 2006. It has been adapted with permission. Although first developed for embedded systems design, concepts provided here apply to PLC system design.

systems design so interesting and challenging is the design must also take into account the proper electrical interface for the input and output devices, limited on–chip resources, human interface concepts, the operating environment of the system, cost analysis, related standards, and manufacturing aspects (Anderson). Through careful application of this material you will be able to design and prototype embedded systems.

B.3 Embedded System Design Process

In this section, we provide a step–by–step approach to develop the first prototype of an embedded system that will meet established requirements. There are many formal design processes that we could study. We concentrate on the steps that are common to most. We purposefully avoid formal terminology of a specific approach and instead concentrate on the activities that are accomplished as a system prototype is developed. The design process we describe is illustrated in Fig. B.1 using a Unified Modeling Language (UML) activity diagram. We discuss the UML activity diagrams later in the appendix.

B.3.1 Project Description

The goal of the project description step is to determine what the system is ultimately supposed to do. To achieve this step you must thoroughly investigate what the system is supposed to do. Questions to raise and answer during this step include but are not limited to the following:

- What is the system supposed to do?
- Where will it be operating and under what conditions?
- Are there any restrictions placed on the system design?

To answer these questions, the designer interacts with the client to ensure clear agreement on what is to be done. If you are completing this project for yourself, you must still carefully and thoughtfully complete this step. The establishment of clear, definable system requirements may require considerable interaction between the designer and the client. It is essential that both parties agree on system requirements before proceeding further in the design process. The final result of this step is a detailed listing of system requirements and related specifications.

B.3.2 Background Research

Once a detailed list of requirements has been established, the next step is to perform background research related to the design. In this step, the designer will ensure they understand

Fig. B.1 Embedded system design process

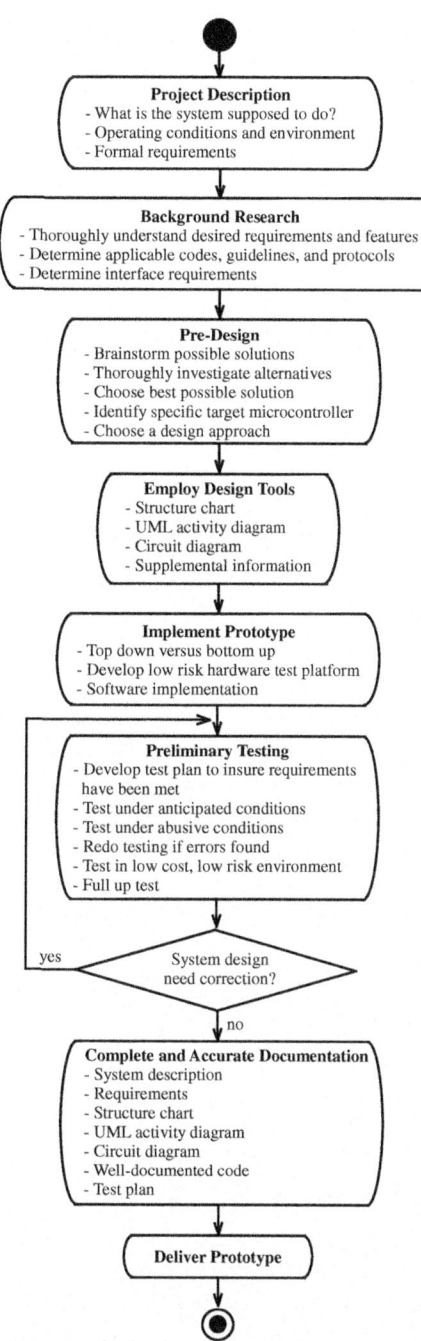

all requirements and features required by the project. This will again involve interaction between the designer and the client. The designer will also investigate applicable codes, guidelines, protocols, and standards related to the project. This is also a good time to start thinking about the interface between different portions of the project particularly the input and output devices peripherally connected to the microcontroller. The ultimate objective of this step is to have a thorough understanding of the project requirements, related project aspects, and any interface challenges within the project.

B.3.3 Pre–design

The goal of the pre–design step is to convert a thorough understanding of the project into possible design alternatives. Brainstorming is an effective tool in this step. Here, a list of alternatives is developed. Since an embedded system typically involves both hardware and/or software, the designer can investigate whether requirements could be met with a hardware only solution or some combination of hardware and software. Generally, speaking a hardware only solution executes faster; however, the design is somewhat fixed once fielded. On the other hand, a software implementation provides flexibility and a typically slower execution speed. Most embedded design solutions will use a combination of both hardware and software to capitalize on the inherent advantages of each.

Once a design alternative has been selected, the general partition between hardware and software can be determined. It is also an appropriate time to select a specific hardware device to implement the prototype design. If a microcontroller technology has been chosen, it is now time to select a specific controller. This is accomplished by answering the following questions:

- What microcontroller systems or features (i.e., ADC, PWM, timer, etc.) are required by the design?
- How many input and output pins are required by the design?
- What is the maximum anticipated operating speed of the microcontroller expected to be?
- How much memory is required by the algorithm?

B.3.4 Design

With a clear view of system requirements and features, a general partition is determined between hardware and software, and a specific microcontroller chosen, it is now time to tackle the actual design. It is important to follow a systematic and disciplined approach to design. This will allow for low stress development of a documented design solution that meets requirements. In the design step, several tools are employed to ease the design process. They include the following:

- Employing a top–down design, bottom up implementation approach,
- Using a structure chart to assist in partitioning the system,
- Using a Unified Modeling Language (UML) activity diagram to work out program flow, and
- Developing a detailed circuit diagram of the entire system.

Let's take a closer look at each of these. The information provided here is an abbreviated version of the one provided in "Microcontrollers Fundamentals for Engineers and Scientists." The interested reader is referred there for additional details and an in–depth example (Barrett and Pack).

Top down design, bottom up implementation. An effective tool to start partitioning the design is based on the techniques of top–down design, bottom–up implementation. In this approach, you start with the overall system and begin to partition it into subsystems. At this point of the design, you are not concerned with how the design will be accomplished but how the different pieces of the project will fit together. A handy tool to use at this design stage is the structure chart. The structure chart shows the hierarchy of how system hardware and software components will interact and interface with one another. You should continue partitioning system activity until each subsystem in the structure chart has a single definable function.

UML Activity Diagram. Once the system has been partitioned into pieces, the next step in the design process is to start working out the details of the operation of each subsystem we previously identified. Rather than beginning to code each subsystem as a function, we will work out the information and control flow of each subsystem using another design tool: the Unified Modeling Language (UML) activity diagram. The activity diagram is simply a UML compliant flow chart. UML is a standardized method of documenting systems. The activity diagram is one of the many tools available from UML to document system design and operation. The basic symbols used in a UML activity diagram for a microcontroller based system are provided in Fig. B.2 (Fowler).

To develop the UML activity diagram for the system, we can use a top–down, bottom–up, or a hybrid approach. In the top–down approach, we begin by modeling the overall flow of the algorithm from a high level. If we choose to use the bottom–up approach, we would begin at the bottom of the structure chart and choose a subsystem for flow modeling. The specific course of action chosen depends on project specifics. Often, a combination of both techniques, a hybrid approach, is used. You should work out all algorithm details at the UML activity diagram level prior to coding any software. If you can not explain system operation at this higher level, first, you have no business being down in the detail of developing the code. Therefore, the UML activity diagram should be of sufficient detail so you can code the algorithm directly from it (Dale).

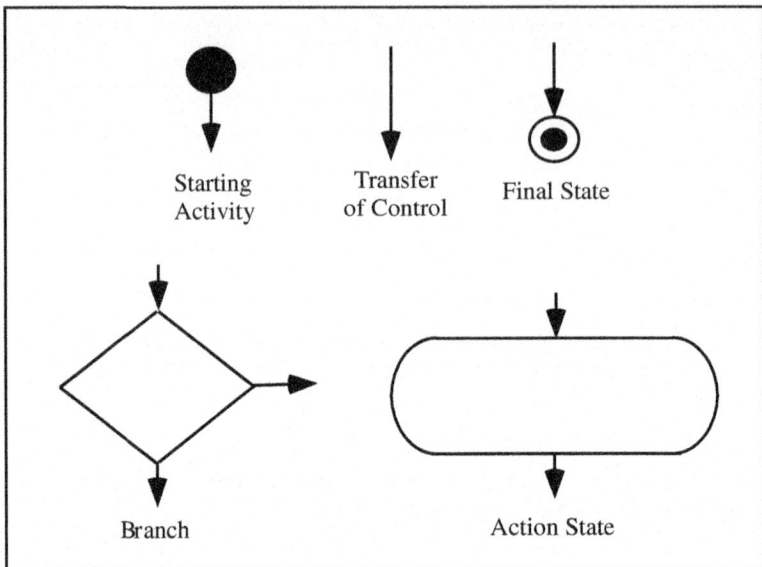

Fig. B.2 UML activity diagram symbols. Adapted from (Barrett and Pack)

In the design step, a detailed circuit diagram of the entire system is developed. It will serve as a roadmap to implement the system. It is also a good idea at this point to investigate available design information relative to the project. This would include hardware design examples, software code examples, and application notes available from manufacturers. At the completion of this step, the prototype design is ready for implementation and testing.

B.3.5 Implement Prototype

To successfully implement a prototype, an incremental approach should be followed. Again, the top–down design, bottom–up implementation provides a solid guide for system implementation. In an embedded system design involving both hardware and software, the hardware system including the microcontroller should be assembled first. This provides the software the required signals to interact with. As the hardware prototype is assembled on a prototype board, each component is tested for proper operation as it is brought online. This allows the designer to pinpoint malfunctions as they occur.

Once the hardware prototype is assembled, coding may commence. As before, software should be incrementally brought online. You may use a top down, bottom up, or hybrid approach depending on the nature of the software. The important point is to bring the software online incrementally such that issues can be identified and corrected early on.

It is highly recommended that low cost stand–in components be used when testing the software with the hardware components. For example, push buttons, potentiometers, and LEDs may be used as low cost stand–in component simulators for expensive input instrumentation devices and expensive output devices such as motors. This allows you to insure the software is properly operating before using it to control the actual components.

B.3.6 Preliminary Testing

To test the system, a detailed test plan must be developed. Tests should be developed to verify that the system meets all of its requirements and also intended system performance in an operational environment. The test plan should also include scenarios in which the system is used in an unintended manner. As before a top–down, bottom–up, or hybrid approach can be used to test the system.

Once the test plan is completed, actual testing may commence. The results of each test should be carefully documented. As you go through the test plan, you will probably uncover a number of run time errors in your algorithm. After you correct a run time error, the entire test plan must be performed again. This ensures that the new fix does not have an unintended effect on another part of the system. Also, as you process through the test plan, you will probably think of other tests that were not included in the original test document. These tests should be added to the test plan. As you go through testing, realize your final system is only as good as the test plan that supports it!

Once testing is complete, you might try another level of testing where you intentionally try to "jam up" the system. In another words, try to get your system to fail by trying combinations of inputs that were not part of the original design. A robust system should continue to operate correctly in this type of an abusive environment. It is imperative that you design robustness into your system. When testing on a low cost simulator is complete, the entire test plan should be performed again with the actual system hardware. Once this is completed you should have a system that meets its requirements!

B.3.7 Complete and Accurate Documentation

With testing complete, the system design should be thoroughly documented. Much of the documentation will have already been accomplished during system development. Documentation will include the system description, system requirements, the structure chart, the UML activity diagrams documenting program flow, the test plan, results of the test plan, system schematics, and properly documented code. To properly document code, you should carefully comment all functions describing their operation, inputs, and outputs. Also, comments should be included within the body of the function describing key portions of the code. Enough detail should be provided such that code operation is obvious. It is also extremely helpful to provide variables and functions within your code names that describe their intended use.

You might think that a comprehensive system documentation is not worth the time or effort to complete it. Complete documentation pays rich dividends when it is time to modify, repair, or update an existing system. Also, well–documented code may be often reused in other projects: a method for efficient and timely development of new systems.

B.4 Summary

In this appendix, we discussed the design process, related tools, and applied the process to a real world design. It is essential to follow a systematic, disciplined approach to embedded systems design to successfully develop a prototype that meets established requirements.

B.5 References

M. Anderson, Help Wanted: Embedded Engineers Why the United States is losing its edge in embedded systems, *IEEE–USA Today's Engineer*, Feb 2008.
Barrett S, Pack D (2006) Microcontrollers Fundamentals for Engineers and Scientists. Morgan and Claypool Publishers. https://doi.org/10.2200/S00025ED1V01Y200605DCS001.
M. Fowler with K. Scott "UML Distilled – A Brief Guide to the Standard Object Modeling Language," 2nd edition. Boston:Addison–Wesley, 2000.
N. Dale and S.C. Lilly "Pascal Plus Data Structures," 4th edition. Englewood Cliffs, NJ: Jones and Bartlett, 1995.

Appendix C
Getting Started

Objectives: After reading this chapter, the reader should be able to do the following:
- Successfully download and execute a simple program using the Arduino Development Environment; and
- Write programs for use on the Arduino UNO R3.

C.1 The Big Picture

We begin with the big picture of how to program different Arduino development boards as shown in Fig. C.1. This will help provide an overview of how chapter concepts fit together.

At its most fundamental level, the Arduino Development Environment is a user–friendly interface to allow one to quickly write, load, and execute code on a microcontroller. A barebones program need only consist of a setup() and loop() function. The Arduino Development Environment adds the other required pieces such as header files and the main program construct. The ADE is written in Java and has its origins in the Processor programming language and the Wiring Project [www.arduino.cc].

C.2 Arduino Quickstart

To get started using an Arduino–based platform, you will need the following hardware and software.

- an Arduino–based hardware processing platform,
- the appropriate interface cable from the host PC or laptop to the Arduino platform,
- an Arduino compatible power supply, and

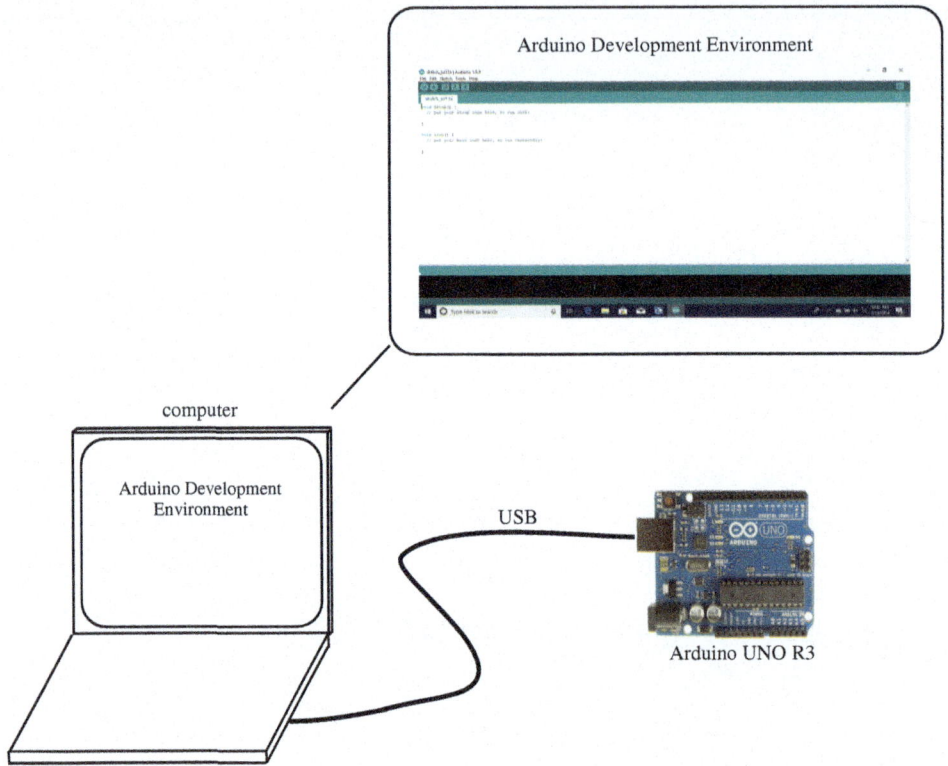

Fig. C.1 Programming the Arduino processor board. (Arduino illustrations used with permission of the Arduino Team (CC BY–NC–SA) www.arduino.cc)

- the Arduino software.

Interface cable. The UNO R3 connects to a host PC via a USB cable (type A male to type B female).

Power supply. The Arduino processing boards may be powered from the USB port during project development. However, it is highly recommended that an external power supply be employed. This will allow developing projects beyond the limited electrical current capability of the USB port. For the UNO R3 platform, Arduino www.arduino.cc recommends a power supply from 7–12 VDC with a 2.1 mm center positive plug. A power supply of this type is readily available from a number of electronic parts supply companies. For example, the Jameco #133891 power supply is a 9 VDC model rated at 300 mA and equipped with a 2.1 mm center positive plug. It is available for under US$10. The UNO has an onboard voltage regulator that maintains the incoming power supply voltage to a stable 5 VDC.

Appendix C: Getting Started

C.2.1 Quick Start Guide

The Arduino Development Environment may be downloaded from the Arduino website's front page at www.arduino.cc. Versions are available for Windows, Mac OS X, and Linux. Provided below is a quick start step–by–step approach to blink an onboard LED.

- Download the Arduino Development Environment from www.arduino.cc.
- Connect the Arduino UNO R3 processing board to the host computer via a USB cable (A male to B male).
- Start the Arduino Development Environment.
- Under the Tools tab select the evaluation **Board** you are using and the **Port** that it is connected to.
- Type the following program.

```
//*************************************************************

#define LED_PIN 13

void setup()
{
pinMode(LED_PIN, OUTPUT);
}

void loop()
{
digitalWrite(LED_PIN, HIGH);
delay(500);                       //delay specified in ms
digitalWrite(LED_PIN, LOW);
delay(500);
}

//*************************************************************
```

- Upload and execute the program by asserting the "Upload" (right arrow) button.
- The onboard LED should blink at one second intervals.

C.3 References

Arduino homepage, www.arduino.cc

Index

A
ABZ encoder, 156
ADC conversion, 34
ADC encoding, 35, 36
ADC in ladder logic, 263
ADC quantization, 35
ADC resolution, 37
ADC sampling, 35
ADC sampling rate, 35
Advanced Encryption Standard (AES), 61
American Standard Code for Information Interchange (ASCII), 74
Analog sensor, 159
Anemometer, 239, 244
Anti-aliasing filter, 35
Arduino IDE, 11
Arduino UNO R3, 326

B
Background research, 318
Battery capacity, 152
Baud rate, 73
Bell Laboratory, 35
BLE UUID, 86
Bluetooth Low Energy (BLE), 85
Bottom up approach, 321

C
CIDR addressing, 108
Cloud, 100
Code re-use, 324

Coils, 253
Comparator, 221
Contacts, 253
Controller Area Network, 126
Counters, 255
CRC checksum, 58
CRC polynomial, 58
Cybersecurity, 8

D
Data integrity, 58
DC motor, 195
DC motor control, 195
DC operation, 151
Design, 320
Design process, 318
DFRobot DFR0580 Solar Power Manager, 283
Digital sensor, 153
Digital Subscriber Line (DSL), 102
DIN rail, 19
Direct Sequence Spread Spectrum (FHSS), 86
Documentation, 323
Domain Name System (DNS), 103, 110
Dynamic Host Configuration Protocol (DHCP), 103

E
Electrical Safety Foundation International, 311
Electrical safety principles, 314
Electronic Control Units (ECUs), 126
Embedded system, 318

Encryption and decryption, 143
Environmental sensors, 185
Ethernet, 112

F
Flex sensor, 161
Fluid level sensor, 175, 286
Foreground and background processing, 54
Frequency hopping, 86
Full duplex, 74

G
Greenhouse, 86, 277
Greenhouse control, 283
Greenhouse Control System (GCS), 280
Greenhouse theory, 278

H
Harry Nyquist, 35
H-bridge, 200
Honey humidity sensor, 286
Honeywell humidity sensor, 236
Hysteresis, 263

I
I2C LCD, 77
IEC IEC61131–3, 250
Incremental encoder, 156
Industrial Internet of Things (IIoT), 2, 7
Inter–Integrated Circuit (I2C), 74
International Electrotechnical Commission (IEC), 10
International Society of Automation (ISA), 10
Internet, 100
Internet Corporation for Assigned Names and Numbers (ICANN), 111
Internet Service Provider (ISP), 102
Interrupts, 51
Interrupt Service Routine (ISR), 52
IoT architecture, 5
IPv4 addressing, 105
IPv6 addressing, 105
ISA/IEC 62443, 10
ISM frequency band, 85
ISO/OSI, 105
IT vs OT, 2

J
Joystick, 190

L
Laser light show, 207
LD editor, 257
LED biasing, 194
LightBlue, 91
Light Emitting Diode (LED), 194
Light sensors, 176
Linear actuator, 200
Linear Feedback Shift Register (LFSR), 59
LM34 temperature sensor, 236
Local Area Network (LAN), 103

M
MAC address, 110
Macroshock, 312
MAX232, 95
Metropolitan Area Exchange (MAE), 104
Microshock, 312
Misting system, 289
Motor speed control, 226
MQ sensors, 185

N
Network Interface Controller (NIC), 110
Network/ladder rung, 253
NRF Connect, 91
NRZ format, 74
Nyquist sampling rate, 35

O
Op amp configurations, 224
Op amp, ideal, 221, 222
Op amp, nonideal, 222
Op amp, saturation, 221
Operational amplifiers, 220
Operational Technology, 4
Optical encoder, 155
Optical fiber link, 180
Output device, 194

P
Parity, 74

Index

Passive heating, 278
Photodiode, 180, 219
Photo resistor, 177
Photo transistor, 180
PLC IDE, 250
PMC analog input, 33, 37
PMC analog output, 43
PMC digital I/O, 31
PMC digital input, 24
PMC digital output, 24
Portenta H7 processor, 23
Power supply, 326
Pre-design, 320
Preliminary testing, 323
Programmable Logic Controller (PLC), 4
Project description, 318
Prototyping, 322
Pulse width modulation, 17, 47, 197
Pushbutton switch, 154
PWM, multiple, 229

Q
Quadrature encoder, 156

R
Ragazzini, John, 220
Rain gauge, 243
Real Time Clock (RTC), 50, 244
Resistance Temperature Detectors (RTDs), 170
RGB color, 67
Rijndael algorithm, 61
Rotary switch, 155
RS-232, 95
RS-485, 131
RS-485 MAX485 transceiver, 142
RTC to I2C LCD, 79

S
Sampling rate, 35
Security, IoT and IIoT, 10
Sensor, level, 286
Sensors, 152
Sequencer, 267
Serial communication, 73
Series and parallel, 253
Servo driver, multiple, 230

Servo motor, 195, 203
Servo with feedback, 205
Skin resistance, 312
Slide switch, 154
Smart home, 7
Smoke detector, 188, 265
Soil moisture sensor, 288
Solar power, 283
Solenoids, 213
Sonalert, 194
Stepper motor, 195, 207, 268
Structured Query Language (SQL), 103
Switch debounce, 27
Switch debouncing, 155
Switches, 154

T
Temperature scales, 164
Temperature sensor, 164, 263
Test fixture, 271
Test plan, 323
Thermistors, 173
Thermocouple, 167
Tilt sensor, 183
Timers, 257
Top down approach, 321
Top-down design, bottom-up implementation, 321
Transducer interface, 218
Transmission Control Protocol/Intenet Protocol (TCP/IP), 105

U
Ultrasonic sensor, 161
UML activity diagram, 321
Unified Modeling Language (UML), 321
Uniform Resource Locator (URL), 111
URL hierarchy, 111
USB data logger, 143

V
Vent fan, 289

W
Water harvesting, 280
Water valves, 213

Weather station, 235
Weather vane, 239
Wi-Fi, 121

Widlar, Bob, 220
Wireless Locan Area Network (WLAN), 102

The manufacturer's authorised representative in the EU is Springer Nature Customer Service Centre GmbH, Europaplatz 3, 69115 Heidelberg, Germany. If you have any concerns regarding our products, please contact ProductSafety@springernature.com

Printed and bound by CPI Group (UK) Ltd, Croydon, CR0 4YY

26/03/2026

02078989-0003